Text Book Of

COMMUNICATION SKILLS - II

For

Semester - II

First Year Diploma Courses in Engineering & Technology

As Per SBTE's New Revised Syllabus for Jharkhand, July 2017

Prof. B. V. Pathak

Ex. Lecturer in English, Humanities, Industrial Management
and Engineering Economics at Govt. Polytechnic, Pune
and Visiting Lecturer in Government and
Other Engineering Colleges.

NIRALI PRAKASHAN
ADVANCEMENT OF KNOWLEDGE

N4414

Communication Skills - II **ISBN 978-93-87686-17-5**

Second Edition : January 2020

Published By :
NIRALI PRAKASHAN
Abhyudaya Pragati, 1312, Shivaji Nagar
Off J.M. Road, PUNE – 411005
Tel - (020) 25512336/37/39, Fax - (020) 25511379
Email : niralipune@pragationline.com

➢ DISTRIBUTION CENTRES

PUNE

Nirali Prakashan (For orders within Pune) : 119, Budhwar Peth, Jogeshwari Mandir Lane, Pune 411002, Maharashtra
Tel : (020) 2445 2044; Mobile : 9657703145
Email : niralilocal@pragationline.com

Nirali Prakashan (For orders outside Pune) : S. No. 28/27, Dhayari, Near Asian College Pune 411041
Tel : (020) 24690204; Mobile : 9657703143
Email : bookorder@pragationline.com

MUMBAI

Nirali Prakashan : 385, S.V.P. Road, Rasdhara Co-op. Hsg. Society Ltd.,
Girgaum, Mumbai 400004, Maharashtra; Mobile : 9320129587
Tel : (022) 2385 6339 / 2386 9976
Email : niralimumbai@pragationline.com

➢ DISTRIBUTION BRANCHES

JALGAON

Nirali Prakashan : 34, V. V. Golani Market, Navi Peth, Jalgaon 425001, Maharashtra,
Tel : (0257) 222 0395, Mob : 94234 91860; Email : niralijalgaon@pragationline.com

KOLHAPUR

Nirali Prakashan : New Mahadvar Road, Kedar Plaza, 1st Floor Opp. IDBI Bank, Kolhapur 416 012
Maharashtra. Mob : 9850046155; Email : niralikolhapur@pragationline.com

NAGPUR

Nirali Prakashan : Above Maratha Mandir, Shop No. 3, First Floor,
Rani Jhanshi Square, Sitabuldi, Nagpur 440012, Maharashtra
Tel : (0712) 254 7129; Email : niralinagpur@pragationline.com

DELHI

Nirali Prakashan : 4593/15, Basement, Agarwal Lane, Ansari Road, Daryaganj
Near Times of India Building, New Delhi 110002 Mob : 08505972553
Email : niralidelhi@pragationline.com

BENGALURU

Nirali Prakashan : Maitri Ground Floor, Jaya Apartments, No. 99, 6th Cross, 6th Main,
Malleswaram, Bengaluru 560003, Karnataka; Mob : 9449043034
Email: niralibangalore@pragationline.com

Other Branches : Hyderabad, Chennai

niralipune@pragationline.com | www.pragationline.com

Also find us on www.facebook.com/niralibooks

Preface ...

All students aspiring to be engineers, managers, or teachers require an ever-increasing range of skills to maintain relevance with the global environment of the new millennium. Communication skills are a vital component of this, recognised by academia and industry alike. Such skills are essential for anyone who wants to carry out his/her professional practice in the global arena.

Communication skills are therefore essential for the successful future career of a student. In today's competitive world, communication skills in business are the most sought after quality of an educated person. Reading, writing and listening carefully are the three most important communication skills for students.

It thus gives me immense pleasure in bringing out the first edition of this book as per the new syllabus effective from June 2017. Great care has been taken to cover the whole syllabus concerning different aspects of Communication Skills - II as prescribed in the new curricular syllabus for Jharkhand. It has been my endeavour to present the content in a simple, lucid and easy to understand manner and I hope all my readers enjoy the learning process.

My thanks are due to Shri. Dineshbhai Furia, Shri. Jignesh Furia, my publishers, and the entire team of Nirali Prakashan for their support and encouragement throughout the process of publishing of this book.

I have put in a lot of thought in the preparation of this book, but would like to point out here that any errors or omissions are unintentional and will be taken care of in the next edition. I will be delighted to receive suggestions for the improvement of the text of this book from teachers as well as students so as to consider them for inclusion in the next edition of this book.

I extend my good wishes to all the teachers and students with a genuine hope that this book will be warmly received by them.

Prof. B.V. Pathak

■■■

Syllabus ...

Chapter 1 : Introduction to Communication **[04 Hrs; 08 Marks]**

1.1 Definition, Communication Cycle

1.2 The Elements of Communication: Sender – Message – Channel – Receiver – Feedback

1.3 Concept of Communication Process

1.4 States in the Process: Defining the Context, Knowing the Audience, Designing the Message, Encoding, Selecting Proper Channels, Transmitting, Receiving, Decoding and Giving Feedback (Case Lets).

Unit 2 : Types of Communication **[06 Hrs; 10 Marks]**

2.1 Formal Communication

2.2 Formal: Types (a) Vertical Communication
(b) Horizontal Communication

2.3 Informal: Types (a) Diagonal Communication

2.4 Verbal Vs Non-Verbal Communication

2.5 Verbal: Types (a) Oral Communication
(b) Written Communication

2.6 Non-Verbal: Types (a) Body Language
(b) Graphic Language

Unit 3 : Principles of Effective Communication **[04 Hrs; 08 Marks]**

3.1 Principles of Effective Communication (One example each)

3.2 Communication Barriers and How to Overcome them

3.3 Developing Effective Message, Thinking about Purpose, Knowing the Audience, Structuring the Message, Selecting Proper Channels, Minimizing Barriers and Facilitating Feedback

(Examples: Writing articles for Newspapers, Magazines)

Unit 4 : Non-Verbal Graphic Communication **[06 Hrs; 12 Marks]**

4.1 Non-verbal Codes: (A) Kinesecs, (B) Proxemics, (C) Haptics, (D) Vocalics, (E) Physical Appearance, (F) Chronemics, (G) Artifacts (One example each)

4.2 Aspects of Body Language. Types of Body Language. (One example each).

4.3 Interpreting Visuals and Illustrating with Visuals like Tables, Charts and Graphs

Unit 5 : Formal Written Skills **[10 Hrs; 20 Marks]**

5.1 Office Drafting: Circular, Notice and Memo

5.2 Job Application with Resume

5.3 Business Correspondence: Enquiry, Order Letter, Complaint Letter and Adjustment Letter

5.4 Report Writing: Accident Report, Fall in Production, Progress Report, Investigation Report

5.5 Defining, Describing Objects and Giving Instructions

5.5.1 Defining Objects: Appearance, It's Use

5.5.2 Describing Objects: Purpose, Components, Functions, Applications

5.5.3 Giving Instructions: Precise, Directive, Imagistic Statements of a Futuristic stance.

Unit 6 : Oral Skills **[12 Hrs; 22 Marks]**

6.1 Phonetics and Phonology

- Introduction
- Phonetics Symbols
- Consonants/Vowels/Dipthongs
- Stress and Intonation

6.2 Discussion Skills

- Importance of Group Discussion
- Leadership Skills
- Team Management

6.3 Presentation Skills

- Importance of Presentation
- Planning of Presentation
- Handling Stage Fright

6.4 Mock Interview

- The Interview Process
- Pre-Interview Preparation
- Answering Strategies

■■■

Contents ...

■■■

Chapter **1** ...

Introduction to Communication

Contents

1.1 CONCEPT OF COMMUNICATION SKILLS

- Communication, as a subject, has been introduced for First Year Diploma students, with a view to suitably equip them for successful future career development.
- Knowledge of this subject will enable them to become effective communicators in general and better engineers and managers in particular. This will help students acquire and develop methods of communication, both written and oral and will help them achieve professional success. Nearly seventy per cent of our working time is consumed in communicating in one form or the other viz. reading, writing, speaking or listening. We cannot imagine the existence of a group of people without communication. We convey our ideas, thoughts or feelings through the process of communication. Take the example of an imaginary group in which there is a member who speaks only English. The others in the group do not know the English language. In this case, the thoughts and feelings of the person speaking only English will neither be properly understood nor grasped by other members of the group.
- Perfect communication is that through which thoughts or ideas are transmitted in such a way that a certain picture is created in the mind of the listener, and the said picture synchronises with the one in the mind of the conveyor or speaker. The understanding of what is

communicated is reflected either through words, facial expressions or body language of the listener. Understanding a message is the basic purpose of effective communication. Effective communication is an essential tool to influence others, to express feelings and emotions and to provide, receive and exchange information. The integrity of the formal structure of an organisation is reinforced by the channels of communication laid down by that organisation. It is the most vital process that keeps the wheels of an organisation turning smoothly. The study of communication skills helps students become aware of the importance of communication in day-to-day business or professions. It equips them with both oral and written communication skills and develops in them an intellectual interest in the field of business and engineering. It also provides them with practical knowledge through discussions about practical situations in day-to-day life. Business and society would be better served if proper attention is given to the improvement of communication and development of a cordial worker-management relationship.

1.2 DEFINITIONS OF COMMUNICATION

- The word communication is of Latin origin. In Latin, *'communis'* means common. It means sharing of information or intelligence. All the acts of listening, speaking, reading, writing, facial expressions, movements, actions, colours, diagrams, gestures, ideas, facts, opinions, information and understanding others help individuals communicate their thoughts to other people and amongst themselves. Communication is a transfer or transmission of certain information and understanding from one person to another. Any effort to produce meaning is a form of communication. Communication is, in fact, an unending process in the world of human beings, animals and plants. It is continuous in nature. It is an inseparable part of life itself. The need for communication is as basic as the hunger for food and thirst for water. Scholars have defined communication in a number of ways. A few of these definitions are given below:

1. *"Communication is the process by which information is transmitted between individuals and/or organisations so that an understanding response results".* **– Peter Little**
2. *"Communication is an exchange of facts, ideas, opinions, or emotions by two or more persons."* **– W.H. Newman**
3. *"Communication is the process of passing information and understanding from one person to another. It is essentially a bridge of meaning between people. By using this bridge of meaning, a person can safely cross the river of misunderstanding that separates all people."* **– Keith Davis**
4. *"Communication is the sum of all things that one person does when he wants to create an understanding in the mind of the other. It is a bridge of meaning. It involves a systematic and continuous process of telling, listening and understanding."* **– Louis A. Allen**
5. *"Communication is the process of meaningful interaction among human beings. More specially, it is the process by which meanings are perceived and understanding is reached among human beings."* **– Dr. Mc Farland**
6. *"Communication is the process involving the transmission and accurate replication of ideas reinforced by feedback purporting to stimulate actions to accomplish organisational goals."* **– Scott and Sprigel**
7. *"Communication is the transmission and reception of ideas, feelings and attitudes both verbally and non-verbally eliciting a response. It is a dynamic concept underlying all kinds of living systems."* **– Norman B. Sigband**
8. *"Communication is an exchange of facts, ideas, opinions or emotions by two or more persons."* **– William Newman and Charles Summer**
9. *"Communication is a continuous thinking process dealing with the transmission and interchange with understanding of ideas, facts, and courses of action."* **– George R. Terry**

10. *"Communication is a process of transmitting ideas or thoughts from one person to another person for the purpose of creating understanding in the thinking of the person receiving the communication."* **– Brown C.A.**
11. *"Communication is the form of interaction which takes place through symbols. The symbol may be gestural, pictural, physical, verbal or any other which operates as stimuli to behaviour which would not be evoked by the symbol itself in the absence of special conditions of the person who responds."* **– Lundberg**

- The first definition by Peter Little uses an important word - information. Communication comprises transmission of information. But this definition does not indicate the objects about which the information is to be transmitted. This is exactly what is being tackled in the second definition by W. H. Newman. Communication transmits information not only about tangible facts and ideas and opinions, but also about emotions and the frame of mind of the communicator. In some cases, facial expressions also play an important part in the process of communication. After going through these definitions, one can come to the conclusion that in any communication, the following four important factors are involved:
  1. The process of communication involves communication of ideas.
  2. Communication should reproduce exactly what the transmitter intends to convey into the mind of the receiver.
  3. The transmitter must be assured of the exact replication of his ideas through feedback. Feedback envisages the receiver's response, which is communicated by him back to the transmitter. Communication is thus a two-way process consisting of transmission and feedback.
  4. The purpose of all communication is to elicit action.
- In any communication, the ideas being transmitted should be sufficiently enlarged to include emotions. In brief, it can be stated that the process of communication involves exchange of facts, ideas, opinions or emotions between two or more persons where feedback is an essential factor. The main purpose of communication is to inform or to bring the receiver around to a certain point of view or to elicit action. The entire process of communication can hence be summarized as follows:

Situation for Communication

1. There is a person who desires to pass some kind of information
2. There is another person to whom the information is to be passed on
3. The receiver wholly understands the message passed on to him
4. The receiver responds through feedback

Need for Communication

1. To transmit ideas, facts and feelings.
2. To fulfil the objectives of an organisation.
3. To carry out the day-to-day activities of human life.
4. For the function or process of management.

Importance of Communication

1. Communication is the life blood of any business. No business can function without effective communication.
2. In case of large organisations, the word "up-to-date" is used commonly. As per Webster's Dictionary as well as Oxford Dictionary, there is no such word as '*upto*' in any dictionary. In India however, this error is generally overlooked. Thus an effective and efficient network of communication system is a must, especially if local business is to come at par with global business.
3. Effective communication promotes a spirit of understanding and co-operation between the management and employees, and works as an effective tool of supervision.

4. It is important in case of negotiations and keeping in touch with distributors, retailers and individual customers.
5. Possessing appropriate communication skills is one of the most important qualifications both at the time of appointment and/or during promotions.
6. It is useful for motivating and for morale boosting.

1.2.1 Nature of Communication

The analysis of the above definitions implies that communication has the following characteristics, which define the nature of communication.

(a) Meaning based: Communication is meaning based. It has been very succinctly said by a specialist, "the most immediate need for communication is to be able to refer to things in the real world, that is, to be able to name things, states, events and attributes, by using words". Also, we must be able to link words together so as to create meaningful sentences and language. Thus, there is a need to be clear-headed about what we want to say.

(b) It is a process: The term process is defined as an identifiable flow of interrelated events moving over time towards some goal or an end. In other words communication, is a process in which there are some identifiable interrelated events which are initiated by the sender. This is followed by encoding messages, through some channel till the receiver receives the massage and ends the process with a feedback.

(c) It is inevitable: Communication is essential a social, physical and psychological exercise. If an individual is provided all physical comforts, but is not allowed to read, write, speak and or listen, he will become mentally deficient.

(d) Communication is systematic: Every part of communication is affected by one or the other. The one who sends the message, the message itself as well as the receiver of the message; all are interrelated and affected by each other.

(e) Communication can be intentional and unintentional: While some communication is intended, or much of the communication may be unintentional, making us convey, what we don't wish to communicate. It is rightly said "we cannot, not communicate".

(f) A two—way traffic: A significant aspect of communication is involvement of two people, i.e., sender and receiver. In fact, one person cannot communicate to himself. A receiver is a must to complete the communication act. There is no communication, unless the message sent by the sender is not being received by the receiver. A Businessmen or a Director may send hundreds of memos and warning letters to his employees, but communication will not be complete, unless it is received and read by them. Since communication is an exchange of views, opinions, and directions it involves both parties. Communication, may be upward or downward. Messages, directives, and opinions, are sent to lower levels in management hierarchy. Also, grievances, complaints, opinions feelings and points of view are communicated upward along the line, i.e., from workers (lower level) to management (higher level).

(g) Communication involves interaction and transaction: The participants of a communication exercise exchange ideas and information besides influencing each other in the process. They share and exchange thoughts and meanings.

(h) Communication is a social process: Hiuman being is a social animal. He cannot live in isolation. Communication is a process which helps him to interact and socialize making it a social process.

(i) **A dynamic process:** Dynamic means "ever changing". Communication is not constant, or a one time event. It is a dynamic process, which is dynamic and ever changing all the time.

(j) **It is spiraling process:** Communication between the receiver and sender usually does not start at the same level nor does it grow at the same rate. This is due to the difference in the abilities of the sender as well as the receiver. Moreover, noise and time have an impact on it. As a result, communication takes a spiral shape before it gets completed and reaches the receiver at the same level and space.

(k) **Continuous process:** Communication is a continuous process. More often, it is repeated to achieve desired results. It is an on-going process, in which one interaction is followed by the other.

(l) **Needs proper understanding:** There are numerous media of communication but the main purpose of conveying the message is to create a proper understanding of the message in the mind of the other party. For this purpose, it should be clearly and concisely worded.

(m) **Dispels misunderstanding:** It does providing this by clear understanding between persons and thereby building a bridge of co-operation among people.

(n) **It is contextual:** Communication happens with reference to a context. The same words would mean different things if they are said in different contexts. Hence, meaning may differ in different situations.

(o) **Leads achievement of the organizational objective:** Effective communication does this by creating a sense of object orientation in the organization.

(p) **It has specific skills:** Communication has four specific skills, namely reading, writing, speaking and listening.

1.3 FEATURES OF COMMUNICATION

Features of Effective Communication

- Effective communication is a part and parcel of any successful organisation. Communication should be free from barriers so as to be effective. Communication is a two way process where the message sent by the sender should be interpreted in the same terms by the recipient. The features or characteristics of effective communication are as follows :
- Communication is inescapable
  1. Communication is irreversible
  2. Communication is complicated
  3. Communication emphasizes content and relationships
  4. Communication is governed by rules
  5. It is a 2-way process.
  6. Communication process happens between or among two or more parties. (Sender and Receiver)
  7. Communication involves exchange of ideas, feelings, information, thoughts, and knowledge.
  8. Communication involves mutuality of understanding between Sender and Receiver.
- **Specific objectives:** Communication is always made to satisfy a predetermined objective(s). That is in any type of communication there must be a specific purpose.
  1. **Understanding:** Communication must include both the transfer and understanding of messages. Only transfer of message can not complete the communication process. To be effective the transferred message must be understood by the receiver.

2. **Barriers:** Where there is communication there must be some kind of barriers. In real world we can not think of communication without barriers. Infact we are to make communication by overcoming the barriers.
3. **Dynamic Process:** Communication is not a static process, rather it is a dynamic process which has different steps.

Main Objectives of Communication

- There are five main objectives of communication in a business organisation:
  1. It assists in attainment of the operating objectives of the business.
  2. It helps to improve performance and job satisfaction of employers at all levels.
  3. It improves the image of the organisation with the employees and the general public.
  4. It improves understanding, approval and support of the organisational position on vital, economic, political and social issues.
  5. It keeps the management informed about attitudes, trends and reactions among employees and the general public.

Important Functions of Communication

- Communication performs the following four important functions with a group:
  1. Communication gives information.
  2. It motivates individuals and groups.
  3. It provides an outlet for the expression of emotions.
  4. It works as a motivator.

  The first function that communication performs is that it plays a role in decision-making. It increases motivation by clarifying to employees what is to be done and how well they can do it. Communication within the group is a fundamental mechanism for the members to express their feelings. It provides an expression to their feelings for the fulfillment of social needs.

The Significance of Communication

- Organisations are totally reliant on **communication**, which is defined as the exchange of ideas, messages, or information by speech, signals, or writing. Without communication, organisations would not function. If communication is diminished or hampered, the entire organisation suffers. When communication is thorough, accurate, and timely, the organisation tends to be vibrant and effective.
- Communication is central to the entire management process for four primary reasons:
  1. **Communication is a linking process of management:** Communication is the way managers conduct the managerial functions of planning, organising, staffing, directing, and controlling. Communication is the heart of all organisations.
  2. **Communication is the primary means by which people obtain and exchange information:** Decisions are often dependent upon the quality and quantity of the information received. If the information on which a decision is based is poor or incomplete, the decision will often be incorrect.
  3. **The most time-consuming activity a manager engages in is communication:** Managers spend between 70 to 90 percent of their time communicating with employees and other internal and external customers.
  4. **Information and communication represent power in organisations:** An employee cannot do anything constructive in a work unit unless he or she knows what is to be done, when the task is to be accomplished, and who else is involved. The staff members who have this information become centers of power.
- The ability to communicate well, both orally and in writing, is a critical managerial skill and a foundation of effective leadership. Through communication, people exchange and share

information with one another and influence one another's attitudes, behaviors, and understandings. Communication allows managers to establish and maintain interpersonal relationships, listen to others, and otherwise gain the information needed to create an inspirational workplace. No manager can handle conflict, negotiate successfully, and succeed at leadership without being a good communicator.

1.4 THE COMMUNICATION CYCLE

- Communication is necessary to life, not only of human beings but also of animals and plants. Communication is the name given to the countless ways of maintaining relationships with the concerned person or group or organisation, within day-to-day life.
- It means communication is natural to all living creatures. It can be established even between human and non-human beings in case they live together. Pet animals, for example, communicate with their human companions. Methods of communication may however, differ from individual to individual, group to group, or community to community. A civilized group possesses better communication ability. It is central to everything that we perform in day-to-day life. In order to achieve goals effectively, it is necessary to adopt the process of effective communication.
- An organisation can be held together only if proper communication exists among its members; communication is necessary to bring about such co-ordinated and co-operative efforts. Hence, without communication, no organisation can function properly. It is a tool which can be exercised to influence others. Communication can help establish and maintain relations with others. Thus, there can be no society without communication. For transferring thoughts and ideas, certain means and methods are required to be adapted. These tools of communication include:
  1. Symbols, 2. Signs, 3. Pictures, 4. Sounds, 5. Words
- The process of communication begins with the emergence of a thought or idea in one's mind. However, the mere emergence of a thought or idea does not start the process of communication. The concerned individual must intend to communicate it to somebody. This intention to convey ideas or thoughts starts the process of communication.
- In a broad sense, the communication process takes place in the following general form:

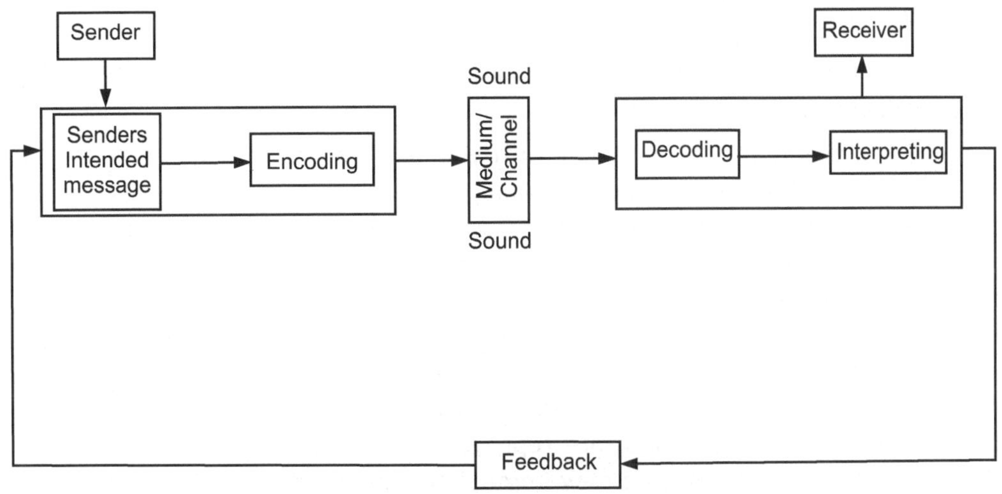

Fig. 1.1: General Communication Process

1.5 ELEMENTS OF COMMUNICATION: SENDER, MESSAGE, CHANNEL, RECEIVER, FEEDBACK

- **The Sender**

  A person who intends to convey or communicate his thoughts or ideas is called the sender. He has to translate his message into a language which is known or understood by the receiver. The effect of his message not only depends on the language he uses but also upon his status and reputation. Status relates to authority, knowledge and prestige while reputation relates to the kind of person he is known to be i.e. whether he is co-operative or non-co-operative (hostile). For example, a message from a popular boss who is liked will be appreciated or believed by the subordinates much more than the one received from a boss who is not held in very high esteem by the subordinates.

- **The Message**

  The need for communication exists because there is a requirement to transfer thoughts and ideas. The thoughts and ideas exist in the mind of the sender and can be transferred to the other person only through communication. These thoughts or ideas may comprise orders, instructions, feelings, guidance or any such things. These are called 'messages'.

  Although thoughts and ideas are abstract, they take a physical form like a letter, a picture or an order on being communicated. When this physical form is received by the receiver, he will interpret it based on his frame of mind and it will now become an abstract thing in his mind. This *conversion of a message from the abstract form to a physical form and then again to the abstract form is called encoding and decoding respectively*.

- **Encoding**

  The method which initiates the message is called encoding a thought. The sender puts his thoughts in a certain logical and coded manner. It may be either oral or written communication. It can be stated in a specific form, may be in letter, telegram, verbal talk, (gestures) or some other form of expression. As the sender intends to know the effect of his message, he considers various factors such as the receiver's knowledge about the subject matter, his working conditions, his job responsibilities and other background information. There are four aspects which are important for encoding the message:

  1. The sender's skill of encoding the message.
  2. His knowledge as well as the socio–cultural system of which he is a part.
  3. Success in sending the message. This depends upon the sender's skill of writing, speaking, reading, listening and reasoning.
  4. The attitude of the sender which substantially influences the receiver.

- **Decoding of message**

  The work of interpretation is concerned with the receiver of the message. He accepts the message and tries to find out its exact meaning by analysing the sender, his intent, his role, knowledge, experience and authority. The receiver of the message tries to translate the symbols and ideas into a form that can be understood by him. The process of translation and interpretation of the signals is called decoding. The process of decoding requires that the receiver must possess skill in reading or listening and must also be able to apply reason. The receiver's attitude and cultural background can distort the message. In case, the encoding is efficient and the receiver is conversant with those codes and symbols, his perception can be good and there can be a fairly accurate reproduction of the message. When the message is properly decoded by the receiver, he may immediately respond to it. This response exists in his mind in the form of ideas or emotions.

- **Dispatch and Reception**

  A message is the actual physical product at the encoding source. When one speaks, writes or paints a picture, it is considered as a message. Gestures, movements of arms, facial expressions etc. also convey messages. It is affected by the code or a group of symbols used to transfer meaning. The meaning of what is to be communicated itself and the decisions that the sender makes in choosing, arranging and selecting both contents and codes, constitutes a message.

- **Channel**

  For transmitting the message, some kind of medium or channel is essential. This process involves certain media. It may be a face-to-face message, a conversation or a written or typed matter or simply some sort of gestures. The channel bridges the gap between the source and the receiver. In case of face-to-face communication, a talk over the telephone or a radio broadcast or a telecast on television, reception almost synchronises with the dispatch of the message. In other forms of communication, it depends upon the efficiency of the medium used. There may be a time interval between the dispatch and actual reception of the message. A letter may take a few days to reach its destination while a telegram may reach within a few hours, and an e-mail in a few seconds.

- **The Receiver**

  Just as the sender has his own frame of references, so has the receiver. What the receiver decides to select to listen and give attention to, depends upon his own opinions, attitude and relationship with the sender. It also depends on the type of message sent, the channel employed, the manner of its reception and the situation in which it is received. The frame of mind of the receiver also carries equal importance in the process of interpretation or decoding of the message as that of the sender in encoding.

- **Feedback**

  Feedback is the response sent by the receiver to the sender of the message. It determines whether understanding has been achieved or not. When the message reaches the stage of decoding, the process of understanding starts. The sender expects the receiver to show some response or reaction so that he can act or reply or behave in a particular way as a result of the response he receives. The sender of the message needs to know whether the receiver responds in the desired way or not. He can find this out only through the feedback he gets from the receiver. Thus, feedback is the knowledge of the result of communication, which can be useful in undertaking or modifying future communication.

- **Storing**

  Sometimes there is neither possibility nor necessity to give instantaneous response to the message received. Under such circumstances the message received is properly stored and feedback effected later at an appropriate time.

- **Retrieving**

  When the feedback to a message is sent sometime later, the information stored is retrieved and feedback is given after studying and reflecting on the message.

- **Context**

  The message, encoding and decoding are all affected by the context of the message. The encoding and decoding of a message might change depending on the context in which it is generated. For example, the meaning of any word or sentence will change depending on the context. The word 'train' can imply a 'mode of conveyance', 'training' or even 'training one's guns', depending on the usage.

- **Medium/Channel**

  Another important element of communication is the medium or channel, which is nothing but vehicle or medium which facilities the sender to convey the message to the receiver. It could be oral, written or non —verbal. Prior to the composition of the message, the medium/channel should be decided. Each medium follows its own set of rules and regulations. For example, in oral communication, one can afford to be a little informal, but when using the written mode, all rules of communication need to be observed. It must be remembered that anything in writing is a document that would be filled up for records or circulated to all concerned

1.6 CONCEPT OF COMMUNICATION PROCESS

- The process of communication can be defined as *"Transferring of thoughts or ideas from one person (or group of persons) to another person so that it can be understood and acted upon."* It always has a certain purpose. In order to transfer an idea, which one gets in one's mind, and to make the communication effective, one has to use certain words, symbols, pictures or sounds that stand for those ideas. The symbols one uses must be such that the receiver can understand them.
- Communication is a dynamic process. Simply looking at something involves analysing it for description in terms of other matters or events related to it. Everything changes with the changes in the surroundings and circumstances. Communication, which must be considered as a whole, is a process of dynamic interaction both affecting and being affected by many variables. It is a basic social process necessary for the growth and development of individuals, groups and society.
- Communication links called channels bind the groups. It is not a single process but part of a set of processes. It uses various means like oral, written, verbal, non-verbal etc. The means of communication are determined by many factors such as personal, social, political, economic, cultural etc.
- Business organisations use mass communication media to establish and maintain contact with various people within and outside the organisation. It can be done with the help of various means.

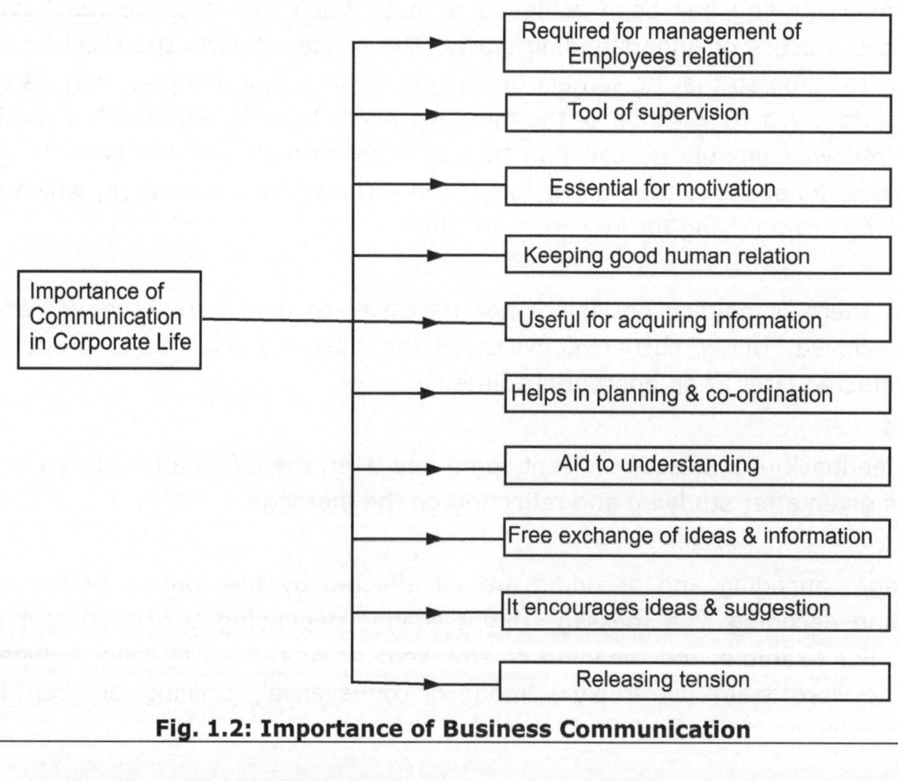

Fig. 1.2: Importance of Business Communication

- In the case of subordinates, communication is established through intercom, memo, orders, face-to-face communication etc. Business communication is concerned with the study of communication between individuals and groups created in the process of administration of an organisation. One has to be alert, careful and cautious about what employees or people say and how it affects all other employees or people. In certain cases, gestures and other non-verbal signals convey meaningful messages.
- The success of any communication, either personal or corporate, lies not in sending and receiving it, but in its appropriate understanding and reciprocation by the receiver.
- Business communication is thus a process of linking superiors and subordinates in various sections of the organisation, and all outsiders associated with the organisation, in one way or the other. Communication, in this sense, is the dynamic interactive process of relating people to each other, relying on feedback.

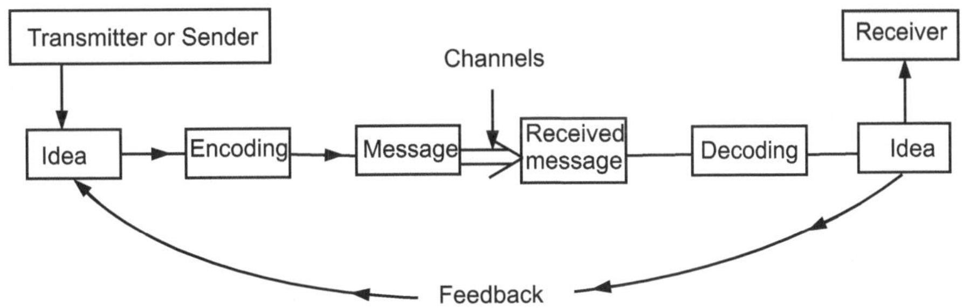

Fig. 1.3: Communication Process

- Messages can be transmitted:
  1. By verbal communication (in-person or by telephonic conversation)
  2. In writing (through letters, circulars, bulletins, signboards etc.)
  3. Non–verbal signs (patting, winking, shaking hands etc.)
  4. Through gestures (by pointing fingers, nodding etc.)
  5. Noise
  6. Feedback

  The sender of a message has to put his ideas in a proper language or symbols so that they can be transmitted through verbal or non-verbal medium.
  1. The process of packing ideas is called encoding.
  2. The receiver of the message has to interpret the message. This is called decoding the message.
  3. When the encoded message is the same as the decoded one, communication is considered to have taken place effectively.
  4. When the message is not effectively communicated i.e. mis-communicated, it can be referred to as 'noise'. There are four kinds of noise. Physiological noise is distraction caused by hunger, fatigue, headaches, medication, and other factors that affect how we feel and think. Physical noise is interference in our environments, such as noises made by others, overly dim or bright lights, spam and pop-up ads, extreme temperatures, and crowded conditions. Psychological noise refers to qualities in us that affect how we communicate and interpret others. For instance, if you are preoccupied with a problem, you may be inattentive at a team meeting. Likewise, prejudice and defensive feelings can interfere with communication. Finally, semantic noise exists when words themselves are not mutually understood. Authors sometimes create semantic noise by using jargon or unnecessarily technical language."

Working of the Process of Communication

It is One Way Process:

- The sender, according to his ideas, behaviour patterns and intention, selects a message.
- He then encodes the message.
- After encoding the message, he transmits or send it to the receiver through a medium — be it verbal or non-verbal.
- As the message reaches the receiver, he decodes it and gives an internal response to the perceived message. The response is not in relation to the actual content but rather to the "perceived content" of the original message. This first phase of the communication process is completed. The manner in which the sender and receiver perceive the same give rise to differences in encoding and decoding.

  The one way process of communication is completed.

One Way Communication Process

SENDER → ENCODING → MESSAGE → CHANNEL → DECODING RECEIVER

It is Two Way Process:

- The one way communication process is incompleted as the sender cannot tell whether his message has been understood by the receiver or not. The process completes only after the sender receives feedback from the receiver. In the second phase, the receiver formulates his message, encodes it and transmits it to the original sender to the receiver. This stage is reoffered to as providing feedback and is most important. If, the feedback is in tune with the original intent of the sender, communication can proceed without a hitch. However, there could be moments when receiver does not agree with the message of the sender. It does not mean that there is a breakdown in communication. In such instances; effective communication is held up for the time being and would resume after subsequent discussions.

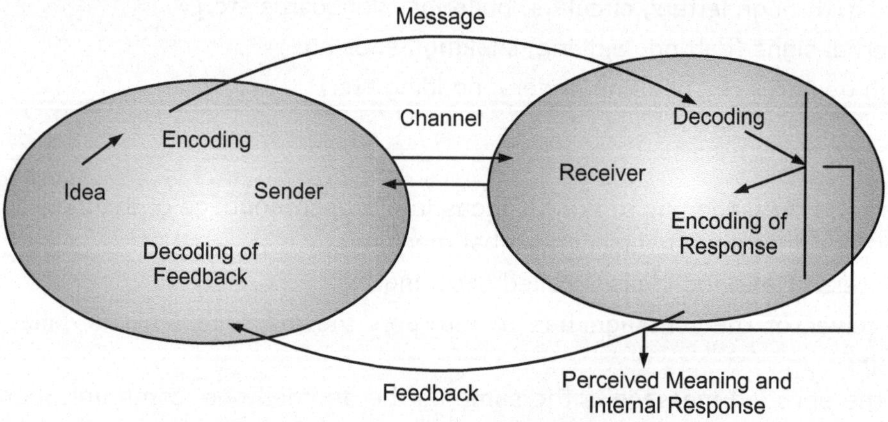

Fig. 1.4: Two Way Communication Process

- For communication to be effective, there should be a well defined goal in the mind of the sender. Harmony between the goals of the two communicators makes for good and easy progression of ideas and concepts. The sender necessarily needs to look to the following stages:
  1. Create awareness in the receiver on the said subject.
  2. Propose his point of view with clarity and preciseness, to eliminate confusion in the mind of the receiver.
  3. Enable smooth flow of discussion.
  4. Reinforce or correct ideas in the mind of the receiver.
  5. Achieve the goal of communication.

1.7 STAGES IN THE PROCESS: DEFINING THE CONTEXT, KNOWING THE AUDIENCE, DESIGNING THE MESSAGE, ENCODING, SELECTING PROPER CHANNELS, TRANSMITTING, RECEIVING, DECODING AND FEEDBACK

1.7.1 Stages in the Process of Communication

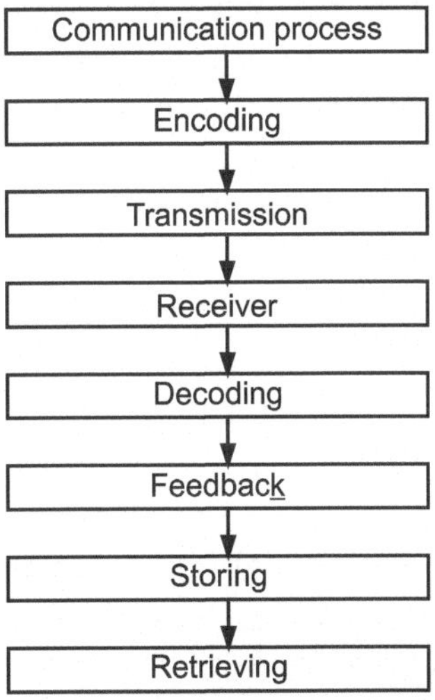

Fig. 1.5: Stages of Communication Process

- **Encoding:** The function of symbolising a message in words, pictures etc.
- **Transmission:** The action which transfers the message through a medium.
- **Receiver:** The person for whom the sender intends a message. The receiver is responsible for listening to, understanding, and responding to the message.
- **Decoding:** This is the main function of interpreting as well as understanding the message.
- **Feedback:** The action of sending a reply to the message.
- **Storing:** This is the function of recording and filing the message so that it can be used in future.
- **Retrieving:** The act of obtaining and recalling the stored message.

1.7.2 Defining the Purpose

- Conscious or intended communication has a certain definite purpose. People communicate their thoughts and ideas as well as feelings to make someone do something or take some action or think or feel in a certain way. The source has to be clear about the purpose or goal of the communication as well as the intended audience. The source must put the message in words so that it can be clearly understood by the receiver. The process of encoding should also be audience specific.
- An objective is something that we want to accomplish through our efforts. Thus when we speak or write to, we may not have a specific purpose, except to keep in touch; when we chat with a group of friends, we want to socialize and express ourselves. But in official or business situation, when we speak, listen or write to customers, our subordinate/superiors, we have a purpose of accomplishing something. Communication could have many objectives depending on the context and persons involved. Communication within a family, classroom, seminar, boardroom, theatre or organization has different objectives depending on the purpose that has to be achieved.
- The objectives of business communication would include the following:
  - **To inform:** This is the first objective of any communication. Information needs within and outside the organization and are usually met through communication.

- **To persuade:** It is important to persuade employees to work efficiently, to persuade customers to buy our product and so on. The objective of communication is also to persuade.
- **To educate:** To disseminate knowledge and develop skills and attitudes among people working in the organization is another objective of communication.
- **To train:** Communication is part of any training programme. Training is required to achieve proficiency in specific skills. Instruction, demonstration, practice and discussion during training require communication to be integral part.
- **To motivate:** Morale and motivation are must to ensure high levels of productivity and efficiency, on a sustainable basis. Communication helps to keep these motivation levels high.
- **To integrate:** Large organizations have different business units, departments and territorial divisions which pursue different targets. Communication helps for an integrated approach in pursuing organizational goals.
- **To relate**: Good business relations are a must for the continued success of any business organization. Communication provides the means for building and nurturing mutually beneficial relationships.
- **To entertain:** In every business, there is always a time entertainment. Communication facilitates social bonding and help in creating lighter moments and in relieving tension.

- Thus objectives of communication are dynamic and ever-changing. Some of the for objectives of official communication are to get or give information; ask for or give instructions advice or suggestions; to make requests and to persuade other people to agree us.

1.7.3 Knowing the Audience

- The goal of speaking to a group is to inform, persuade or entertain it. To achieve any of these objectives, one should know the group's interests, likes and dislikes. For this, one should learn as much as possible about the group and should get to know the group. This is not a one-time affair, but a continuous process and happens not only before the speech but even during the speech in the form of audience feedback. If the talk is boring or the audience is not interested in the talk, it will be evident.
- Before commencing the speech, in fact, even before the speech is prepared, one should find out more about the background, level of knowledge on the subject, interest, prejudices and biases, etc. of the audience.
- Depending on the feedback received during the speech, one should be ready to modify the speech if required, rather than delivering a boring speech to the audience. Knowing the audience well enables the speaker to use his speech to arouse their interest.

1.7.4 Designing the Message

- Language is the expression of thoughts in the context of its environment. The same language if used in different cultures, may imply a different meaning. For example, the English used in India and that in Britain carries different shades and meaning. In India, the English language has different connotations. Pronunciation also changes, as it comes under the influence of local languages.
- Each language independently shapes the reasoning of its speaker. Many a time, the speaker changes the language according to the person they are talking to. There are different situations where one is required to modify the language. For example, a businessman's language is the central element of his environment. The success of his business depends on the way and nature in which he uses: written and spoken communication.

- In sending messages, words and letters are the most common factors. A letter is written because it is one of the most common and important means of keeping in touch with those who are separated by distance. In the days before the emergence of telecommunication, letters were used for sending messages. Even though other means of sending messages are now available, a letter is still the most commonly used and important method of conveying messages. It helps maintain personal relations, serves as a record and as a means of keeping other parties informed. The speed of modern media has definitely changed and made its impact on our day-to-day life, especially on our daily communication. Media like telex, fax, e-mail have made a big difference.
- Modern forms of writing messages have become easy to understand and help cultivate and maintain personal relations. Progressive companies believe in getting favourable responses if they write in a personal and informal manner rather than by using old fashioned business jargon. A large amount of printed and non-printed matter that reaches the business people, executives, managers is simple, informal and sincere in style as it receives greater responses. Following are some useful and important hints for preparing the form and content of a message:
  1. The message (written or oral) must create goodwill.
  2. It must be as brief as possible. A brief message is easy to remember.
  3. Meaningless or outdated words and phrases should be avoided.
  4. It should be simple in style so as to be easy for understanding.
  5. Communication must be up-to-date and efficient.
  6. The beginning of the message must try to establish personal touch with the receiver.
  7. Unfriendly and pompous terms should be avoided.
  8. Use of technical terms without enough explanation should be avoided, especially when a message is addressed to a layman.
  9. The content of a message should make a proper impression and convey its purpose within the short period of attention it gets.
  10. The manner in which a message is written is more important than the matter it carries.
  11. It must attract a favourable response from the receiver.
  13. The message must be clear, precise and courteous.

Message (Form)

Oral	**Written**
(Using words) In-person orders, telephone talk etc.	1. Letters 2. Reports 3. Memos 4. Notes 5. Drafts 6. Notices 7. Resolutions 8. Minutes

1.7.5 Encoding

- Refer to Section 1.5 of this chapter.

1.7.6 Selecting Proper Channels

- All communication takes some way, or passage through some well-defined, easily identifiable stages. Everyone is familiar with the phrase "through proper channel," that is used in almost all business letters/applications/representations. Here the sender of the written communication to passes it on to the addressee through somebody occupying an important position in the hierarchical system. The word 'channel' therefore, becomes synonymous with the position or point through which the communication passes. These positions or points exist at different levels in the pipeline.

- Channels of communication may be classified into two categories:
  1. Formal
  2. Informal

1.7.7 Transmitting

- Refer to Section 1.5 of this chapter.

1.7.8 Receiving

- Refer to Section 1.5 of this chapter.

1.7.9 Decoding

- Refer to Section 1.5 of this chapter.

1.7.10 Feedback

- Refer to Section 1.5 of this chapter.
- Feedback is the response from the receiver to the sender of the message. It is critical for effective communication. It indicates that the receiver has received the original message, understood it in his/her own way and is sending the response. It is a very important phase in the communication process. The sender of the message wants to know if the receiver of the message has understood what is conveyed, as that would help in completing the communication process. He would also be interested in knowing his reactions or impressions about the manner in which it is expressed, and whether it was clear, so that he can improve his communicating skills. *The main idea behind feedback is to motivate behaviour or purpose*.
- Feedback can be in verbal or non-verbal form, including written communication or even by way of action.

  Example 1: When a superior instructs his subordinate (original message sent), he expects an answer. The response could be as follows:
  1. Silence (indicating that the receiver [the subordinate] has chosen to ignore),
  2. Action as expected in the original message,
  3. Request for further clarifications,
  4. Excuses for not following the instruction (based on genuine reasons like lack of adequate time, manpower, supplies, power supply etc.)

  Example 2: During the warranty period, a customer sends a complaint letter to the company regarding problems encountered while using their product. The customer (original message) asks the company to either replace the piece or get it repaired. If the company acknowledges the message and sends a service person, this feedback makes the customer comfortable and if the replacement or repair is completed, he is satisfied. On the other hand, if the company ignores the complaint or refuses to accept that the product is defective and argues about it, it is a negative feedback and an unsatisfied customer, which can destroy the goodwill for the company in the market.

  Thus the feedback can be **positive** or **negative**. While giving the feedback, the receiver in turn becomes a sender. In the first case above, the superior receives the feedback (message from the subordinate) and in turn, sends his feedback. The superior (receiver of the feedback) can show frustration when his instructions are not acted on. If the feedback he has got is action, he will acknowledge it and support the action. If the feedback falls in the categories 4 or 5, he will initiate suitable action.

In the second case above, the customer has sent his message (complaint). The company's feedback is positive when it acknowledges and acts on it. It is negative when it disputes the complaint. The frustrated customer can spread the message by word of mouth or even in the social network, damaging the reputation of the company, whereas the satisfied customer will help build goodwill for the company.

Thus, for every message, there is a feedback, which completes the process and can start a new communication cycle. *Lack of feedback can lead to frustrated customers, unmotivated employees and a negative corporate culture*.

Feedback is a two way street. A person should know when and how to give feedback and also how to receive it constructively. It is also important to proactively seek feedback from your superiors, colleagues, and customers. *Handled properly, feedback has the potential to make your workplace a much more productive and harmonious place*.

- **Customers** need to be provided with an avenue for expressing their needs, concerns and opinions. When a customer has an issue with a product/service, he needs a proper outlet to provide feedback to the company. That is why many companies use Information technology and services to run help lines. At these help lines, they can even have access to an elevated level if the reply at the initial level is not satisfactory. Many help lines provide for live chat with executives of the company. The only way to know if the customers are happy and satisfied is to gather accurate and honest feedback from them.
- **Employees** who are given the opportunity to provide feedback feel appreciated, important and understood. They will feel empowered. They will not be fearful of sharing their opinions. Generally as they are the hands-on persons and are aware of the ground realities, they can come up with doable solutions to problems. At the same time, they also get feedback about their performance or behaviour from their superiors or peers. Opportunities for upward, downward and horizontal feedback create a committed workforce. People who feel appreciated and respected are more motivated than those who think their efforts go unnoticed. They take more interest in the work they do, and they're more committed to their teams and organisations, because they know that they are making a real difference. And the people involved will improve and become more satisfied because they are being coached to do better. This is especially important if the department is working on tight budgets and so raises and bonuses cannot be offered.
- **Corporate Culture:** At work, it is important to build a culture that values feedback. It is necessary to reinforce the feeling that everyone deserves feedback and that providing feedback is a primary function of a supervisor. With a lack of opportunities for feedback, the corporate culture can become depressing and unmotivated. Every business has a particular culture which can be the result of the communication systems set in place. Feedback allows both customers and employees to voice their opinion, creating a healthy corporate environment which leads to greater productivity and motivation. By making a conscious choice to give and receive feedback, the company can use it as a powerful tool for employee development and customer relationship. Feedback creates a supportive environment built on trust.
- **Profitability:** When the employees are motivated and the customers are satisfied, it gets reflected in the profitability of the company. Customer feedback helps the company to identify problem areas and fix them. It will also help in identifying the strong points on which it can build its future strategies. Hence, in the long run, feedback will create greater productivity and profits.

Tips for Effective Feedback

> *"The leader of the past was a person who knew how to tell. The leader of the future will be the person who knows how to ask."*
>
> ***- Peter Drucker, author and consultant.***

- **Try to make it a Positive Process and Experience:** It is necessary to remember the basic purpose of feedback, viz., to improve the situation or performance. Nothing can be accomplished by being harsh, critical, or offensive. *A lot more can be achieved when your approach is positive and focused on improvement.* It is good practice to start with a positive statement.

  For instance, while giving a complaint about a product, the customer can start with the statement that it is a great product, but has developed some problems. It is a great product, otherwise you would not have bought it in the first place. Such a positive beginning takes the edge out of the complaint, but gets better attention.

  While commenting on an employee's behaviour, a positive approach in the beginning and at the end of the meeting will put the other person at ease and make him want to improve and succeed. Very often people do not give enough praise in the workplace. However, it is natural for people, whether they are team members, colleagues, customers, suppliers or even the boss, love to get sincere recognition for a job well done. However care should be taken to make sure that the sweeteners are not too many. Otherwise, it will take away the expected sting in the main message in the feedback.
- **Negativity:** There is a role for negative feedback and even anger if someone is not paying sufficient attention to what is being said. However this should be used sparingly.
- **Be Timely:** The feedback should be as close to the event as possible. For instance, if an employee is not performing satisfactorily, it would be easier to handle if he or the agency he represents gets the notice immediately and not after observing his performance for a whole year. Similarly, the customer should contact the supplier as soon as the defect is noticed. The exception is if the persons involved are emotionally disturbed, in which case, it is better to wait till the situation calms down. There is no point in making a remark for which can cause unnecessary irritations.
- **Make it Regular:** Feedback is a process that requires constant attention. When something needs to be said, it has to be said, so that people know where they stand and there are few surprises and the situation is under control. In an organisation, except in the case of formal feedback, feedback should not be treated as a scheduled, a once-a-year or a once-every-three-month type of event. Depending on the situation informal, simple feedback should be given much more often than this. With frequent informal feedback, there will be no surprises during formal feedback sessions.
- **Prepare Your Comments:** It helps to organise your response, so that there is clarity and smooth flow of points. It may not be written down and read, but a little bit of preparation helps you stay on track and stick to the issues.
- **Be Specific:** The feedback should be based on firsthand experience, without any exaggerations and be aimed at bringing about specific improvements. Tell the person exactly where the improvement is necessary, so that specific facts can be presented and there is total clarity. Vague expectations should not be made.
- **Focus should be Limited:** Only the issues on hand should be discussed. It is advisable to limit feedback to one or two issues, so that the receiver can concentrate on these issues and bring about improvements. If too many issues are taken up, the receiver may feel that he is being attacked and insulted.

- **Discuss Issues that can be changed:** Only behaviour issues that can be changed should form part of the feedback. Otherwise, it becomes a senseless criticism.
- **Do not get Personal:** While giving feedback, it is necessary to be factual and objective and not subjective or personal. It should not be used as an opportunity to settle personal scores.
- **Avoid Absolute Statements:** Usage of words like 'never', 'always', 'all' will put the other party on the defensive. For instance, if the customer says "I'll never buy from your company", he has closed the doors permanently.
- **Use "I" Statements and be direct:** The feedback has to be from the person's perspective, and not include general statements. In an employee review, the suggestions for improvement should be specific, with achievable goals and milestones, giving the feeling that the superior is interested in the development of the subordinate.
- **Criticise in Private:** Public scrutiny or criticism is never appreciated, but public appreciation is always valued. The principle to be followed is based on the saying: "Tell me if you are not satisfied, but tell the world if you are happy with my product or service." Hence, it goes without saying that a negative feedback should be given in a place where no one would interfere or be able to overhear. Especially in these days of social networking, the slightest criticism of a product or the company you are working for can create gigantic ripples with irreparable damages.
- **Listen:** The feedback may bring about some response from the other person, which may be very valid.
- **Document and Monitor:** It may be a good practice to document the entire feedback session so that it can be monitored.

1.8 FUNCTIONS OF COMMUNICATION

- Communication plays a very important role in every walk of life. According to Luthans it covers about three —fourths of an active human life, and even higher percentage of a typical manager's time. The important functions of communication are:

  (a) **Assists in decision making:** The most important function of every manager is to take critical decisions. To make accurate and appropriate decisions, one needs to obtain information that is available in various channels of communication. The way how, decision is communicated will have an impact upon the outcomes in terms of co-operation and support of the people to achieve organizational goals.

  (b) **Information sharing:** The important purpose of communication is to transmit information to individuals or groups. These could be in the policies and rules; changes and development in the organization. There need for fast diffusion of some information in the organization, like rewards and awards given, settlements with the union, and other changes in the organization.

  (c) **Problem solving:** In many cases communication helps in solving problems. Communication between the management and unions on issues like to bring about negotiation helps, find solutions and consensus.

  (d) **Facilitating change:** Change-introduced in an organization depends to a large extent on the clarity and spontaneity of the communication. Communication between managers and employees helps in recognizing difficulties in the planned change, and in taking corrective action.

  (e) **Group building:** Communication helps in building relationships. Communication provides the necessary lubrication for the proper functioning of a group.

  (f) **Gate keeping:** Communication helps to build linkages of the organization with the outside world. The organization can use its environment to increase its effectiveness.

(g) **Feedback:** To provide feedback to employees on their achievements, to the department on their performance, and to the higher management on fulfillment of goals. It also motivates people in developing challenging and realistic plans.

(h) **Influence:** Information is power and a significant purpose of communication is to influence people. The manager communicates to create a conducive environment, right attitudes, and congenial working relationship.

(i) **Helps in co-ordination of effort:** Communication is an effective tool for co-ordinating activities of different persons engaged in running a business. In organizations, there exist many differences, which are often formally created by an organization through their departments, divisions, delegation, decentralization, authority and power. Various effective communication mechanisms like letters, circulars, meetings conferences, telephone, cellular phones come into play. These differences are minimized and activities are properly co-ordinated to achieve organizational goals.

(j) **Good industrial relations:** Communication develops good and harmonious industrial relations as it conveys feelings, opinions, ideas and viewpoints of different parties. The two parties namely the management and the subordinates come closer through communication. They understand each other and dispel any misunderstandings in order to promote better co-operation and industrial relations.

(k) **Conveying the right message:** The objective of communication is to convey the right message to the right person. The message conveyed should be understood and accepted by the receiver. It should carry the same meaning which has been conveyed, so that it may be translated into action effectively.

(l) **Ensuring effectiveness of policies:** An organization makes policies and programmes to guide the work force. It should be conveyed properly to those who are really responsible for the execution of work to achieve the organization's larger objectives. Only effective communication can translate policies into action. Effectiveness of policies are judged from the success which depends upon an effective communication system.

(m) **Motivating people:** The people working in organizations, are not regularly informed about their management's expectations, plans and policies with respect to their future career and growth, promotion and welfare measures, they feel frustrated and demotivated. Through various communication devices, managers declare rewards and incentives to motivate employees.

(n) **Job instruction:** Managers need to communicate to their subordinates job instructions from time to time. Failure on the part of managers in communicating these instructions may lead to confusion, wastage and inefficiency, affecting productivity and bottom lines.

(o) **Controlling people:** Each and every organization has its own rules, regulations and procedures to perform various activities to regulate the behaviour of its people. An organization issues notices, circulars, letters etc., to communicate the existing or changed rules, regulations and procedures. The management information system is well-known as a control mechanism. Information is transmitted to ensure that plans are being carried out according to the original design. Communication helps in ensuring such control.

(p) **Performance feedback:** People working in an organization need to know how well they are performing and what needs to be done to achieve and exceed standards set by the management? Through measures like letters of appreciation or suggestion, subordinates are given feedback about their performance status.

(q) **Development of managerial skills:** Communication helps to understand human behaviour at work. The facts, ideas, opinions, information and feelings add value to the knowledge of managers about happenings, within the organization and the behaviour of people. This is done sharing and communicating. Communication is also a process of learning.

Important Points

- **Introduction (Concept of Communication Skills)**
  1. Communication skills are important for one's career.
  2. Most of a person's time is used for communication.
  3. It is important to understand the different meanings and aspects of words and sentences.
  4. To make communication effective, it must be understood.
  5. Communication must influence others.
  6. Communication is essential for keeping the wheels of an organisation turning.
  7. Communication is the process of information transmission.
  8. It is useful for transferring ideas, facts, thoughts and feelings.
  9. It is essential for fulfilling the objectives of an organisation.
  10. It is necessary for management.
  11. It is the life blood of any business.
  12. Effective communication promotes co-operation.
  13. Communication is useful for motivation, supervision, negotiation etc.
  14. Minimum of two persons are involved in the process of communication.
  15. It improves understanding.
  16. It enhances the image of the organisation.
- **Functions of Communication:**
  1. To provide information.
  2. To express emotions.
  3. Induce motivation.
- **Communication Cycle:**
  Important factors in the Communication Cycle – symbols, signs, pictures, sounds and words.
- **Elements of Communication :**
  1. Important elements in the Communication cycle - message, sender, encoding, medium, receiver, decoding, feedback, storing, retrieving.
  2. Symbols play an important part.
  3. Natural to all living creatures.
  4. Holds the organisation together.
  5. Useful to transfer ideas.
  6. A dynamic process.
  7. Communication may be oral, written, verbal, non-verbal etc.
  8. Business organisations use mass communication.
  9. Understanding between the sender and the receiver decides the success of any communication.
- **Communication Process**
  Messages can be transmitted through:
  1. Verbal communication, 2. Written media, 3. Non-verbal signs, 4. Noise, 5. Feedback
  Messages are transferred through language and symbols, etc.
- **Stages in Communication:**
  1. Sending, 2. Encoding, 3. Transmission, 4. Receiving, 5. Decoding, 6. Feedback, 7. Storing, 8. Retrieving.
- **Knowing the Audience:**
  1. Natural to all living creatures.
  2. Civilised groups possess better communication abilities.
  3. Effective communication motivates.
  4. Symbols, signs, pictures, sounds, words are important in communication.
- **Designing Messages:**
  A properly designed message either written or oral:
  1. Creates goodwill.
  2. Is easy to remember.
  3. Is easy to understand.
  4. Establishes personal relations.
  5. Properly conveys a purpose.

- **Precautions to be taken for designing messages:**
  1. Must be brief.
  2. Simple language must be used.
  3. Outdated and meaningless words and phrases must be avoided.
  4. Should be simple in style.
  5. Must be up-to-date.
  6. Must be efficient.
  7. Unfriendly and pompous terms should be avoided.
  8. Avoid technical jargons.
  9. Must be easy to understand.
  10. The style used should be appealing.
  11. Must be easily readable.
  12. Must get a positive response.
  13. Must have a courteous approach.
  14. Must be clear and precise.
- **Importance of Communication:**
  1. Communication occupies an important place in an organisation.
  2. Communication is required to get desired results.
  3. Communication is the binding agent of social systems.
  4. It shows the efficiency of an organisation.
  5. Flow of communication measures the efficiency of an organisation.

Practice Questions

1. State the importance of communication for the development of an engineering career.
2. Define communication and state the importance of communication for the development of a career in engineering.
3. What is communication? What are the different meanings that can be ascribed to the term?
4. What are the major objectives of communication? Explain them in brief.
5. State in brief the different features of effective communication in fulfilling the objectives of an organisation.
6. Write short notes on:
   (a) Need for effective communication
   (b) Importance of communication
   (c) Features of communication
   (d) Functions of communication
   (e) Basic types of communication
7. Explain in brief, the communication cycle and define communication.
8. Explain in brief the communication process.
9. Write short notes on:
   (a) Definition of communication
   (b) Need for communication
   (c) Importance of communication
   (e) Functions of communication
   (f) Communication cycle
   (g) Elements of communication
   (h) Process of communication
   (j) Designing a message
10. Explain communication on the basis of direction of flow.
11. Explain communication on the basis of relationship, form and direction.

■■■

Chapter **2...**

Types of Communication

Contents

INTRODUCTION

- Communication assumes a very important place in any organisational set up. At every stage, communication is essential for achieving the desired objectives. The efficiency of any organisation, whether government, private or commercial, largely depends on effective communication. It can be said that communication is the binding agent of all social systems. As a result, the success of management processes in any organisation depends upon the effectiveness of the communication flow within that organisation, and how the management makes its plans, instructions, directions and devices known to its subordinates.
- Although man communicates with other individuals since infancy, the process of transmitting information from an individual (or group) to another is a very complex one, with many sources of potential errors. People in organisations spend most of their time in interpersonal situations, and poor communication becomes the root of most organisational problems. Therefore, effective communication is an essential part of organisational success, whether it is at the interpersonal, intergroup, intragroup, organisational or inter-organisational level.

2.1 FORMAL COMMUNICATION

- Effective communication is essential for any organisation to achieve its objectives, and it should be remembered that in a person's life, communication failures are more frequent than communication success. Communication is an integral process of any management function, and it is interwoven with planning, organising, directing and controlling – the basic elements of the management process.
- The route, path or method through which a message flows from a sender to a receiver is called the *channel of communication*. Messages flowing through these channels of communication

should be brief, so that the possibility of their distortion is minimised. It is therefore, essential to organise the flow of the message carefully. The message should pass through a well organised network of channels of communication to ensure a smooth freeflow. It is thus important to establish a proper and understandable relationship between one end of the message, viz. the sender, and the other end of the message, viz. the receiver.

- Organisational communication can be either external or internal. *External communication* is concerned with correspondence with those outside the organisation. *Internal communication*, as the name indicates, is concerned with communication within the organisation and can be classified into different groups.
- On the basis of relationship, internal organisational communication can be:
  (a) Formal
  (b) Informal
- On the basis of form, it can be:
  (a) Written
  (b) Oral
- On the basis of direction of flow, it can be:
  (a) Upward
  (b) Downward
  (c) Horizontal/Vertical
- These different groups or organisational communication can be graphically represented as follows:

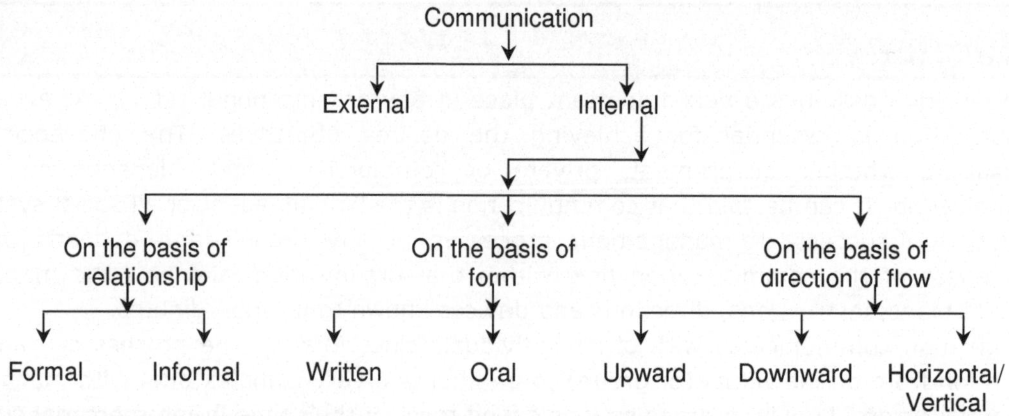

External Communication

- When an organisation deals with people outside the organisation, the communication involved is called *External Communication*. It comprises of communication that is directed from the organisation to the people outside, viz. Outward Communication, and the communication which is received by the organisation from people outside, viz. Inward Communication.

Outward Communication:

- The communication that takes place between the organisation and those other than within the organisation is termed as *outward communications*. All messages that are meant for suppliers, customers, insurance agents and/or companies, banks, transporters, other companies, government offices, the media and the general public fall under this category. The style, content and approach of the outward communication reflects the status and image of the organisation in the eyes of the public at large. Much of this communication takes the following forms:

1. face-to-face discussion
2. telephone calls
3. meetings
4. formal written reports
5. formal written proposals
6. letters
7. press releases
8. advertisement

Inward Communication

- As against outward communication, an organisation constantly receives communication from people other than within the organisation such as customers, clients, dealers etc., and from other organisations, in the form of letters, fax messages, e-mails, presentations, reports, magazines, journals, circulars, government letters, orders, telephone calls, sms etc. These fall under the category of *Inward Communication*.

Internal Communication

- The flow of information or messages within the organisation is rather complicated. The direction and flow of the messages are generally decided on the basis of levels of authority or the hierarchy of the organisation, and requirements or the aims to be achieved by the organisation.

Horizontal (or Lateral) Communication:

- Horizontal communication, also called lateral communication, refers to the flow of communication among the people at the same level of authority (peers). Often, people with the same level of authority feel it necessary to exchange thoughts or views amongst themselves. This channel is very common in big organisations and is essential for smooth functioning. This communication takes the form of inter-personal dialogues, management meetings and conferences, inter-departmental correspondence, telephone or cell-phone conversations, etc. Face-to-face conversations on phone or in person, help solve common problems, create an informal atmosphere, allow freedom of expression, clear doubts and become helpful in teamwork. This type of communication also helps keep every department informed of the needs, activities, developments taking place in other departments and builds a feeling of co-operation and an atmosphere of coordination among all those responsible for achieving the targets of the organisation at large.
- In any organisation there are set procedures for the exchange of thoughts between people at different levels. In a hierarchical organisation, the messages would typically flow from the higher level to the lower level and vice-versa. For such communication, there is always a set procedure, termed as 'proper channel'.
- The communication which flows in this hierarchical system through the proper channel is called formal communication. Such communication primarily comprises of orders/instructions and feedback/reports.
- Higher authorities do not generally convey messages to the workers at the lowest level. Messages from the higher level management are first passed to the middle-management level, then to the supervisory level and in the end to the workers. Such channels are designed or created so that they are received and understood by all the people in the hierarchy. This helps to avoid any misunderstanding or missing links in the communication. Such channels are essential for connecting with and coordinating people at different levels. Each and every person

who is part of the channel must necessarily know 'what is going on' in the organisation. This helps to achieve smooth, interruption-free working in the organisation. If this is not done, the whole working process may lead to chaos and the organisation may fail to achieve its goals.

- Formal communication has its advantages as well as disadvantages. These are as follows:

Advantages/Merits of Formal Communication:

1. Formal communication allows the flow of information in an orderly and authentic manner since it takes place according to officially prescribed rules.
2. A formal channel covers all the sub-systems of an organisation. It is useful for easily reaching out to the branches.
3. In formal communication, the information is filtered and delivered appropriately.
4. Formal communication satisfies the people in managerial positions and helps in controlling subordinates.

Limitations/Demerits of Formal Communication:

1. Formal communication inhibits free flow of information as messages can be transmitted only through the channels laid down.
2. It is slow and rigid and therefore time-consuming.
3. It is expensive as it is in the written form.
4. Distortion of information may occur because it has to pass through several levels of authority.

2.2 FORMAL TYPES

2.2.1 Vertical Communication

- Vertical communication is communication that flows both up and down the organisation, usually along formal reporting lines-that is, it is the communication that takes place between managers and their superiors and subordinates. Vertical communication may involve only two persons, or it may flow through several different organisational levels.

Downward communication

- Traditionally organisational communication meant only downward communication, as the belief was that the management commanded and the employees worked. Today, organisations have realised that they have to facilitate and encourage their personnel to willingly contribute and to learn, to be participative, responsible and accountable in their activities. For the success of an organisation all the people in it should change the way they think and operate.
- Downward communication occurs when information flows down the hierarchy from superiors to subordinates. Examples include circulars, emails, memos, etc.

Objectives of Downward Communication

- Communication from a superior to a subordinate may be about orders or instructions, guidance, advice or solving problems. The following are the objectives of downward communication:
    - To give specific job instructions and directives.
    - To inform about organisational rules and regulations.
    - To apprise subordinates about their performance.
    - To highlight the link between the job, the employee's performance and the organisation's success.
    - To be supportive in their organisational and personal problems.
- Downward communication may be conveyed by writing or speaking. Writing could be in the form of letters, notice, memos, circulars, bulletins, posters and even annual reports. Oral downward communication could be by issuing orders and instructions, informing, educating or

training through meetings. Technology is frequently used for downward communication today. It is easy to telephone a subordinate and easier still to e-mail information to several sub-ordinates at the same time.

Merits of Downward Communication

1. **Valuable for the organisation:** Downward communication is the most essential component of an organisation. The organisation uses downward communication to communicate essential official information to employees.
2. **Valuable for Employees:** Downward communication is valuable for employees as they are well informed about their work from time to time. It provides them motivation and enhances the morale of employees.
3. **Disciplined organisation:** Through downward communication, division of responsibility and accountability brings about discipline, satisfaction, harmony and co-operation among the employees.
4. **Involvement in Organisation:** In downward communication, the employees carry a feeling of participation in the management of the organisation. They regard themselves as an important part of the organisation.

- Though downward communication flows from a higher authority to a lower one and is the earliest channel of organisational communication it has its drawbacks and limitations.

Limitations of Downward Communication :

1. **One-way communication:** In several cases a superior still behaves in an authoritative manner and does not encourage feedback (upward communication). He may instruct but will not entertain clarifications. To avoid face-to-face interaction most of his communication may be in writing or through e-mails. The reasons could be that the superior is not sure of the information he is passing on or that he is diffident of facing subordinates. He may not want to be physically present because the information he is conveying is unpleasant or that he would like to be only heard not seen. Whatever the reasons, a superior has to be responsible and is accountable for his actions. So he has to ensure that he maintains adequate contact with his subordinates and encourages two-way communication.
2. **Difference in values and perceptions:** It is common for the superior to be committed to the total organisation, while the subordinate relates to his department or sub-group alone. The superiors visualise their performance in terms of long-term goals while subordinates see theirs in terms of immediate outcomes. Superiors typically view their contributions in terms of achievement while subordinates are more likely to see themselves as only contributing long hours and hard work. Such dissimilar viewpoints can be barriers to downward communication because subordinates may filter out parts of the downward messages conveyed to them.
3. **Mistrust:** The feeling of mistrust appears when there is lack of frequent superior – subordinate contact. The subordinate knows that the superior controls his rewards and each downward message is viewed with mistrust and some concealed motive. For e.g. when an employee receives a transfer order, he views it as a punishment. He feels that the superior needs to promote somebody else so he is being moved out to pave way for 'somebody'.
4. **Inner conflicts of Leadership:** The pressures of their 'position' can cause inner conflicts or status anxiety in a superior. He is torn between the responsibilities of his 'status' and the desire to be popular among subordinates. In his attempt to be a 'responsible' superior, he may end up being under communicative. Decisions made without consulting departmental heads, information passed down withholding relevant background details and actions ordered before discussions with employees concerned could cause confusion in subordinate's mind. On the other hand, the desire to be liked by his subordinates could create the over communicative superior.

5. **Resistance to authority:** An employee generally believes that all communication from a higher authority will be anti-employee and if at all there is a positive one, there would be an unpleasant motive for it. So any downward communication is viewed with hostility. They accept or acknowledge only parts of the message that they are comfortable with, ignoring the other parts.

Upward communication

Upward communication is the interpersonal process, which is a response to the downward flow of communication. Upward communication consists of messages from subordinates to superiors. Although it is the channel for communication from the subordinate to the superior for several other matters as well, it is predominantly considered to be only a feedback channel. One study has shown that employees felt that communication with superiors to be the most important and satisfying type of communication, yet the same study showed this communication to be the most difficult to perform. Upward communication can be effective only with the encouragement of the higher authority.

Objectives of Upward Communication

Just as downward communication has certain objectives to fulfill so does the upward communication. Subordinates need to inform superiors on various aspects of organisational development besides responding to downward communication. The objectives of upward communication are:

- To respond to downward communication like orders, directives and instructions.
- To inform about commencement and progress of departmental activities.
- To suggest and submit ideas for organisational development.
- To encourage participation of subordinates in the decision making process.
- To improve inter personal relationships.

Upward communication will depend on downward communication. The response to orders and instructions will be positive only if they have been conveyed clearly. Subordinates can improve their performance if they are guided and helped to overcome their weaknesses periodically. Inculcating in them a sense of belonging to the organsiation makes them more responsible. They must be encouraged to give suggestions especially on matters related to their work. It is the attitude of the superior towards his subordinate that will decide on the quality of upward communication in an organisation.

Merits of Upwards Communication

1. **Feedback:** Through upward communication the management receives requisite feedback. The management ensures whether the instructions given to the employees have been properly understood and followed. Upward communication helps the employees to express what they think of the organisation and its policies.
2. **Positive recommendations:** A positive suggestion given by the employees helps the organisation to grow. In turn, it promotes the welfare of the employees as well.
3. **Conducive and Healthy organisational environment**: Upward communication creates harmony and improves employer and employee relations in the benefit of the organisation.
4. **Channel for unexpressed emotions:** Upward communication provides an opportunity to the employees to express their grievances. The management has to consider the employees problem for smooth functioning of the organisation. The authentic and critical grievances are redressed because of upward communication

It has been found that the flow of upward communication has several barriers. It is possible to overcome these limitations if the management provides employees the right environment for communication.

Limitations of Upward Communication:

1. **Attitude of the Subordinate:** There is always a tendency on the part of the subordinates to transmit those messages upwards, that they feel will improve their standing with their bosses and to withhold information that will damage their image. Thus, there is deliberate distortion of information. Stronger the interest in advancement or promotion, greater is the chance for distortion. Subordinates are also reluctant to be bearers of unfavourable information, as they fear it will reflect negatively on them and their performance.
2. **Attitude of the Superior**: Superiors are also responsible for this distortion of upward communication. They maintain threatening, mistrustful and distant relationships with their subordinates. They do not encourage subordinates who are willing to face them with challenging or unpleasant information. They perceive such subordinates as 'troublemakers'.
3. **Organisational structure:** Some organisations follow rigid structures and procedures. They could block upward communication. The physical distance between superiors and subordinates also restrict upward communication. In some larger organisations the senior executives could be functioning at the corporate/ divisional headquarters while workers are in field offices or retail outlets. On the other hand, within the same office, superiors remain inaccessible to subordinates. There may be some subordinates who decide what information should reach these superiors. They ensure that only pleasant messages reach the superiors.
4. **Skipping ranks:** Subordinates find it convenient to relate to superiors who are friendly. They ignore their superior and approach the more senior person who may be friendly but is not their boss. For e.g. a junior executive who is not friendly with the officer above him approaches the senior manager with a request. If the senior manager accedes to the request there are bound to be breaks in the communication channels – between the officer and the junior executive and between the officer and the senior manager.

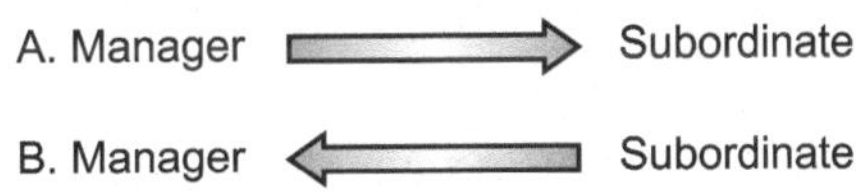

Fig. 2.1: Vertical Communication

2.2.2 Horizontal Communication

- Communication with persons of relatively equal status in an organisation is termed as horizontal or lateral communication. Horizontal communication ensures coordinated cross-functional efforts in achieving organisational goals. Communication with peers provides the much required social support for a person. They tend to turn towards their peers for support than to those above and below them. This may work well or adversely for an organisation.

Objectives of Horizontal Communication

- Studies have shown that in organisations, horizontal communication takes up more time than vertical communication. Personnel who occupy almost equal status and handle similar responsibilities will have much to share and learn. Some of the objectives of horizontal communication are:
  1. To coordinate functions throughout the organisation.
  2. To share information.
  3. To solve problems.
  4. To resolve conflicts.
- For coordination of activities, information has to be shared. Data collected by one department could be useful for the functioning of another department. Problems affecting the departments could be solved by employing brainstorming sessions. Such interactive sessions could bring out valuable ideas. Conflicts between individuals or departments could be resolved if the two

parties concerned or their departmental representatives meet, discuss and find amicable solutions. Sometimes horizontal communication serves as a substitute for vertical communication when it encourages subordinates to resolve conflicts and to take decisions without the aid of their superiors.

Merits/Importance/Advantages of Horizontal Communication:

- Advantages of horizontal communication are as follows:
    1. **Better understanding:** Horizontal communication ensures better understanding, accordingly employees of different divisions are able to perform their job more efficiently.
    2. **Coordination:** Direct communication by breaking the vertical channel strengthens the bondage between employees of different departments which helps to boost up co-ordination.
    3. **Increase productivity:** Horizontal communication minimizes the problems of understanding among various departments as a result, productivity increased.
    4. **Better implementation of decision:** Decisions are made by top management but they are executed by the lower level employees. When lower level employees allowed communicating each other through horizontal communication, it helps to realize the policy and ensures better implementation of decision.
    5. **Motivation:** Horizontal communication not only increases the level of coordination but also facilitates the task of motivation.
    6. **Cordiality:** Due to the free flow of information, cordiality among the employees increases considerably.
    7. **Employee morale:** Better understanding and cordial environment also boost-up the morale of the employees.
    8. **Teamwork:** In any organisation for some specific purposes teamwork is essential. Horizontal communication is a prerequisite for successful teamwork.
    9. **Job satisfaction:** Another major advantage of horizontal communication is, it increases job satisfaction of the employees. By ensuring cordial atmosphere and lifting the morale of employees horizontal communication plays a vital role in increasing the job satisfaction.

Disadvantages/Demerits of Horizontal Communication:

- Disadvantages of horizontal communication are as follows:
    1. **Lack of control:** Sometimes too much horizontal communication creates problem for the management and controlling may become difficult as the horizontal communication increases.
    2. **Conflict:** By passing the vertical channel horizontal communication allows the employees to communicate freely but sometimes it leads to personal conflict.
    3. **Time consuming:** Communication in addition to vertical instruction may delay the implementation of decision.
    4. **Lack of discipline:** As strict rules are not followed, organisational discipline also can be hampered in horizontal communication.

2.3 INFORMAL COMMUNICATION OR GRAPEVINE COMMUNICATION

- Informal Communication is also termed as the Grapevine. The need for Informal Communication arises from the desire to transmit information while remaining outside the formal structure. It takes place with or without the approval of the manager and can have both destructive as well as constructive effects on the enterprise. Since it cannot be eliminated when required, it should not be ignored. Instead it should be mobilised to transmit and receive information supplementary to the formal communication.

- Man is an entity who is constantly talking or writing or listening or reading. The volume of communication increases in geometrical progression as the organisation expands. A manager's whole day is spent largely in reading, writing, speaking and listening. It means that the maximum time of the manager is spent in sending or receiving messages.
- Man is a social animal and moves in groups. When two or more persons come together, they start talking on various topics and subjects. Man also creates a network of information in various industries and organisations. This network represents an informal or unofficial channel of communication. These channels are created and controlled by people who are working in the organisation rather than the management of the organisation. Grapevine Communication becomes a routine when there is a:
  1. Feeling of uncertainty among the working group.
  2. Lack of confidence related to work.
  3. When managers favour a certain individual and/or group more than others, giving rise to a feeling of insecurity among other employees.
- Sometimes, messages are required to flow formally from the superior to the subordinate. However, this channel is not always sufficient or reliable for handling all types of messages quickly and with complete understanding. Therefore, at times the manager contacts the subordinates informally, either personally or through others in the organisation. *As long as no orders are given or implied through such a form of communication, this practice is highly useful*. The work requires voluntary application by all members of the organisation. This requires motivation and it is necessary to provide training to the subordinates in the communication process. In an efficient organisation, there is no place for poor communication.

Types of Grapevine Chains:

- Following are the four types of grapevine chains:
  1. **Single Strand:** In this type, the information moves through a single chain/strand of persons to the final receiver. As shown in this Fig. 2.2 below, the final receiver is 'D', the originator is 'X' and others are intervening persons.

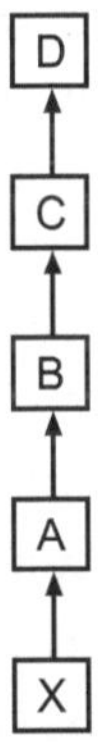

Fig. 2.2: Single Strand Grapevine Chain

  2. **Gossip Wheel:** In this case, information is transmitted all around the person. This network is just like a wheel, as shown in Fig. 2.3.

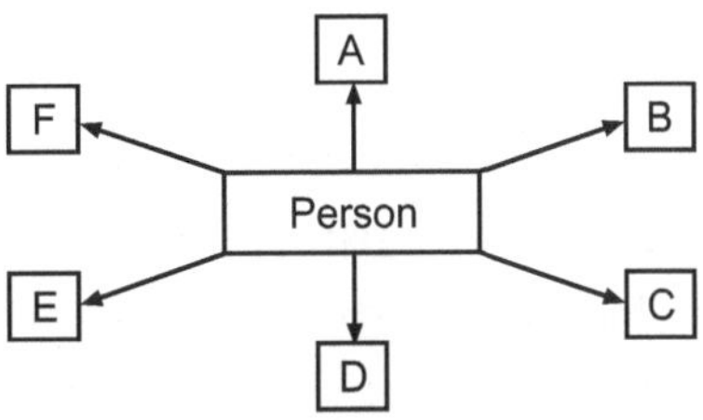

Fig. 2.3: Gossip Wheel Grapevine Communication

3. **Probability Chain:** In this case, one person creates a certain message, passes it on randomly after which it gets spread in the same way and is transmitted further among others. This is shown in the Fig. 2.4 below. Here the person is the originator and all the others are the recipients.

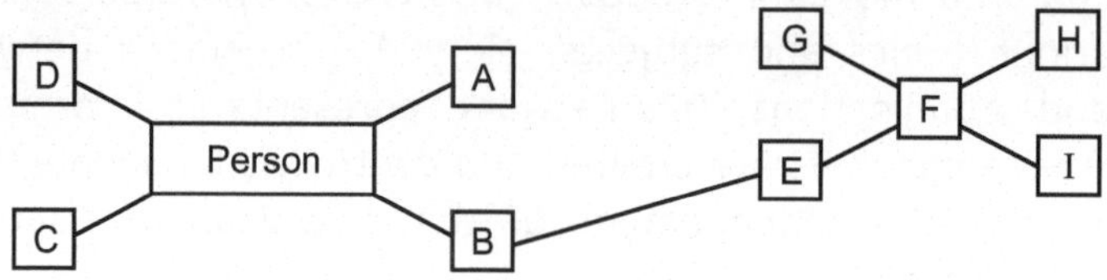

Fig. 2.4: Probability Chain Grapevine Communication

4. **Cluster Chain:** This is selective communication in a group. A person transmits the information to those whom he trusts or finds trustworthy. This is a typical pattern of grapevine which moves in the group and shown in Fig. 2.5.

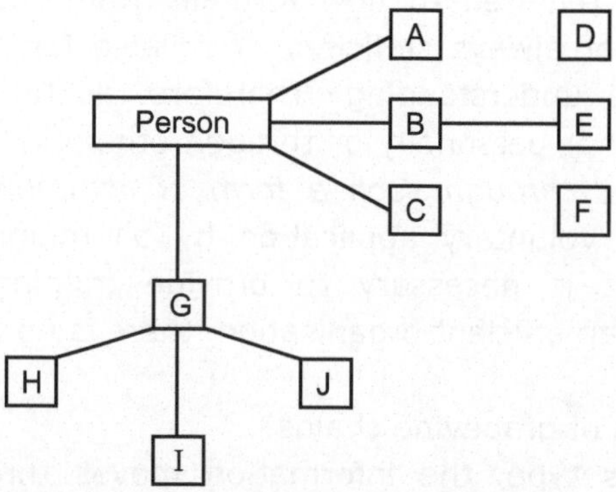

Fig. 2.5: Cluster Chain Grapevine Communication

Features of Grapevine Communication:

1. Grapevine communication can be used in an organisation as a tool which could be useful in creating a co-operative atmosphere in the organisation.
2. Management can implement an open door policy.
3. It can identify the group leaders in the organisation who would be helpful in the management's decision-making policy.
4. It gives opportunity to the members to freely express their ideas.
5. It can work as a feedback system of communication.
6. Workers may feel free to talk with their superiors.

- The merits and demerits of Informal or Grapevine Communication are as follows :

1. Advantages/Merits of Grapevine (Informal) Communication:

(a) It moves very fast because there is no definite route. In English language there is a saying '*rumours spread like fire*' is indeed true!

(b) Management can get feedback through grapevine communication without wasting much time.

(c) It brings certain workers immediately together and creates a sense of group identity.

(d) It can work as an additional channel of communication.

2. Disadvantages/Demerits of Grapevine (Informal) Communication:

(a) It may create misunderstandings within the organisation as it lacks complete information.

(b) It has no credibility among the members.

3. **Gossip and rumour** are an essential part of grapevine communication and fail in providing a complete picture of the organisation. Sometimes these may even give a wrong impression of the organisation and discredit it.

Rumour:

1. Rumours are an undesirable feature of grapevine communication.
2. Rumours are not authentic.
3. They may constitute untrue part of a grapevine communication.
4. Rumours are generally incorrect.
5. They are generated because of several reasons.
6. They arise because of uncertainty of future.
7. Rumours may bring undue pressure on management.
8. Rumours may be detrimental to the progress of an organisation.
9. It is difficult for the management to erase the bad effects of a rumour from the minds of subordinates.
10. Management must act quickly against rumours in an organisation.

So far, the book has dealt with communication classified on the basis of organisational structures i.e. Formal Communication and Informal Communication i.e. Grapevine. On the basis of the mode of expression, communication is further classified into verbal (oral) and non-verbal (written).

Effective Use of Grapevine Communication:

Keeping in view, its merits and undeniable presence in every organization, there is a need to use the grapevine effectively for we have to learn to live with it. A tactful manager has to take positive measures to get the best out of this informal channel of communication. The following few important points in this regard are:

(i) A tactful manager will keep the employees well-informed about organizational policy matters, plans and prospects. This will check the tendency of speculation that can sometimes assume dangerous situation because of fear and anxiety, on the part oaf employees.

(ii) Fruitful group activities enhances self-worth and update knowledge should be given as frequently as possible. This will help to boost the morale and self-confidence of the workers but also check their inclination to indulge in small talk.

(III) The managers should, have an open-door policy without indulging in an impression of cheap popularity or favouritism.

(iv) He should create a healthy environment where there is room for personal talk. It must be made clear that work is of paramount importance. Nothing should be allowed to interface with the progress of the organization. For this purpose regular timings should be fixed up for meetings with employees.

(v) The manager should tactfully identify leaders and win their confidence so as to feel the pulse of their followers.

(vi) The employees, through their leaders, should be associated with decision making. This will frustrate any negative aspect of the grapevine.

(vii) The manager must keep trying to give clues about his style of functioning through regular interaction with employees in as tactful, and diplomatic a manner as is possible.

(viii) Rumour spreading are aimed at character assassination or maligning somebody in the organization should not be encouraged. Showing distaste for such talk will earn praise for the manager's leadership qualities.

(ix) A manager must learn to be a good listener. It is worthwhile to enumerate three types of listening:

(a) Discriminative listening.

(b) Appreciative listening.

(c) Empathic listening.

Empathic listening is the least practised by managers. They are too busy to listen empathetically. This is a 'relief-seeking' conversion, and important to build meaningful human relations. It does not require any regular training as a psychologist. Today's manager can always benefit from the insight of modern psycho-analytical research and learn to practice empathic listening.

2.4 VERBAL COMMUNICATION

2.4.1 Verbal or Oral Communication

- Verbal or oral communication is one of the most important types of communication. The examples of verbal communication are numerous. Oral orders, face-to-face discussions, telephonic talk, lectures, interviews, counselling, public speeches, audio–visual aids etc. fall in this group. Quite often, more than one medium is required to be employed to make the communication effective. Face-to-Face communication combines the oral form with the visual. Graphs and posters often combine the oral form with the visual. It encompasses gestures and facial expressions. Oral communication is carried out face to face, over the microphone/ loud speaker, telephone, radio, internet and so on.

Oral Communication

- Skill in oral communication is an important asset. In face-to-face interaction, a person does not communicate through words alone. The person's whole personality, his general bearing, facial expressions, gestures etc. are involved in the process. A properly dressed man with pleasant expressions will be able to establish contact with others with greater ease. Once a person secures the attention of his audience, more than half of the battle is won.

  For example, when someone is asked to address a gathering, he must walk straight and stand at ease, keep his hands on the sides and try to appear relaxed. When a microphone is to be used, one must keep oneself eight to ten inches away from it to ensure the proper transmission of voice. Once the speaker realises he has complete attention of the audience, after a few minutes, he is likely to feel relaxed and develop confidence. His speech will be easier, more natural, and therefore, effective.

Features of Oral Communication

- **Eye Contact:** It is one of the most immediate and effective means of establishing rapport with the audience. A good speaker first looks at the audience and takes a pause before beginning his speech. It helps to make a good impression on the audience. When a speaker gives a break and puts his ideas into smaller units, rather than presenting them in a lengthy narration, it may allow listeners some intervening time to grasp them. After the pause, the speaker can proceed to explain the next point or idea. While speaking, it is also necessary to maintain eye contact with the audience. If the listeners feel that the speaker is ignoring them, reciprocally, they are also likely to ignore the speaker and what he says. Through eye contact, the speaker gets an idea whether the channel of communication is open or not. This sensing of signals itself works as feedback.
- **Voice Modulation:** The speaker must know how to effectively use his own voice. A good voice is a must for an effective speaker. A good voice is no doubt a natural gift; but one can very well

make efforts to improve the quality of one's own voice with skilful training and practice. Proper attention to this aspect will help in acquiring confidence and power.

1. Variation in the pitch and tempo of the voice is essential to transmit a message effectively.
2. Fast delivery of speech indicates lack of confidence, therefore, betrays the objective of the speech and the image of the speaker.
3. The speed of delivery must be such that the audience can keep pace and understand the matter properly. The speaker should not go beyond a speed of 125 to 150 words per minute. This skill can be practiced by speaking within this range through proper practice.
4. Pronunciation of the words used must be proper with due stress wherever required.
5. Everyone, including the last man in the audience, should be able to hear the speech clearly.
6. Words like "You see", "I mean," "Do you understand," "is it clear" etc. must not be used repeatedly. Such utterances hamper the smooth flow of speech.

- **Audience Awareness:** Before choosing a speech topic, speakers should perform an audience analysis or be aware of the audience to determine what topics might be appropriate and of interest to this specific group. Sometimes this involves gathering new information about an audience; sometimes it simply requires taking into account information they already have as they plan and develop a presentation.
- Common aspects of audience analysis include typical age, gender, group affiliations, education, type of employment, knowledge of the topic, attitudes about the speaker and the topic, and personal or professional interests that might be relevant to the topic.
- Occasionally, a competitor will give a presentation intended for a general audience. This is usually a problem. A general audience is an anonymous audience; it is not known who the audience members are, what they care about, or how they relate to the topic. The speaker has little information to decide what to cover and how to approach the audience; the competition judge has little information to assess the likely effectiveness of the effort. Speakers with specific audiences in mind generally present a clearer message.
- Before starting the speech, the speaker should:
  1. Size up the listeners.
  2. Consider the age, sex, background and interest of the listeners.
  3. Ascertain whether the audience is friendly or hostile etc.
  4. Make the listener feel that the speaker is talking to each member of the audience individually.
  5. Make the audience feel the speaker's sincere concern about their interests.
  6. While narrating a joke, narrate it in such a way that it should appear to be a part of the speech and that the audience should receive it almost unprepared.
  7. Do not get disturbed if a listener smiles or whispers to a neighbour.
  8. Concentrate on ideas.

1. **Presentation Plan:** The art of speaking is the fruit of constant efforts. There is no definite formula to be adopted to become an effective speaker. Enough planning and preparation is essential for successful presentation. Preparation is required for overcoming nervousness.
   1. As far as possible, the speaker should not read out a written speech, word to word.
   2. A written speech seldom sounds fresh.
   3. It is not necessary to memorise the speech. It never exhibits spontaneity.
   4. Memorizing the speech can hamper flexibility. Consequently, communication will suffer.
   5. Face-to-face conversation or interaction requires thinking and speaking and not mere repetition.

6. Even while reading a speech, the speaker has to lift his/her head occasionally and observe the audience.
7. The speaker must make notes of all the important points and arrange them in a proper order before beginning the speech. The notes should be written down, and should bear proper headings and sub-headings.
8. The speaker must devote enough time for the introduction and for the main body of the talk.
9. A positive attitude should be exhibited while speaking.

2.4.1.1 Characteristics/Features of Oral Communication

1. It is the cheapest and fastest form of communication.
2. It is instant and immediate in its conveying and receiving of information.
3. Use of non-verbal cues adds another important dimension to the communication experience.
4. Talking to people face-to-face makes it easy for those engaged in the conversation to observe and study the nonverbal communication cues. Individuals actually communicate more through nonverbal cues, which is why it is important to be able to see people while speaking. Nonverbal communication involves the use of body language, postures and eye contact to convey a message or an emotion. They tell people more than the spoken words. The results of a 1968 landmark study conducted by Dr. Albert Mehrabian and published in "Psychology Today" showed that people pay more attention to body language and the way words are said than to what is actually being said.
5. The audience can immediately judge and communicate not only the content of the speech, but also how it is put across. They can assess whether the speaker is effective and persuasive or not. They can provide instant feedback through body language, allowing speaker to gauge the level of understanding and attention given by the audience. This will help the speaker to fine-tune his way of speaking and make his speech effective.
6. It creates transparency, adding to the chances of the audience understanding the meaning the speaker is trying to communicate.
7. Builds trust and confidence between the audience and the speaker.
8. Builds interpersonal and intrapersonal relationships by allowing the speaker to interact with the audience directly.
9. Discussion is a type of verbal communication that allows people in a group to express their thoughts and views, as well as to consider the opinions offered by others. Individuals can work together to examine the possible solutions to a problem and then decide on the course of action.
10. Dialogue among members of a group or team allows them to exchange ideas and collaborate with one another. It helps in understanding different points of view. Dialogue involves a free flow of ideas among individuals within the groups, and is particularly effective in business.
11. However, direct person-to-person communication requires the ability of the participants to think on their feet, as in interactive communication, spur-of-the-moment responses may be called for.
12. However, with all the new technologies and different means of communication, including social media now available, people tend to forget the value of face-to-face communication.

2.4.2 Verbal Communication : Written Communication

- English writer Francis Bacon had said, "Reading maketh a full man, writing an exact man, conference a ready man." To make himself 'exact' i.e., to say effectively what he wants to say, the writer has to make considerable effort. Speech comes to us naturally and spontaneously, writing comes after practice and careful organization of thought. The word 'write' has been formed by the old English word 'writan', that means to scratch, draw or inscribe. Man learnt writing through a long process of drawing, scoring and incising symbols rock faces, dried skins, tree barks and clay tablets.

- Every language has evolved its own rules of grammar, though many languages grouped tighter have more or less similar rules. In writing these rules you have to be rather strictly followed. Speech, on the other hand, is more flexible. It also does not have the permanence of writing. Unless there is a typescript or types of simultaneously taken down notes, speech is heard and sooner or later forgotten.

Written Communication in an Organisation

- Written communication has acquired a great significance in the lives of individuals as well as business organizations. It reaches across vast geographical areas and targets readers around the world. Many businesses rely on records and written documents, rather than on verbal contracts and oral commitments.
- It is impossible to think of business or an organization without written communication. There are various reasons for this, namely:
    1. In an organization, there are too many people to have face-to-face communication. They are spread over wide geographical distances, and are sometimes not even connected by telephone. Though the situation is changing fast. But, even then, exchange of letters remains important component of communication.
    2. People have to function with defined limits of authority and responsibility. In the absence of written communication, it is difficult to fix responsibility. Therefore it is an essential part of any manager's responsibility to communicate on paper.

Methods of Written Communication

- Written communication is an essential part of organizational life. Telephone, telex and fax machines have not in any way affected the importance of letters. They have only changed the mode of transmission and made the exchange of letters or memos faster. Methods of written communication in an organisation include:
    - letters
    - memoranda
    - notices
    - minutes
    - circulars
    - manuals
    - handbooks
    - reports
    - orders
    - agenda
    - enquiries
    - complaints
    - contracts
    - forms

Characteristics of Written Communication

1. There are not many rules of grammar for written communication. Written communication is conventional in the sense that it has to follow definite pattern as per rules laid down by the language.

2. In an organization, documentation of records and decisions made from time to time are very important for which written communication comes handy.
3. It is used for circulation of information in theorganization. Written communication makes it possible to circulate information without distortions and misrepresentations.
4. The important feature of written communication is the presence of just the sender or the receiver is sufficient at a given point in time, to continue the process of communication.
5. Usually most of the informal, casual conversation or friendly conversation is done orally. Whenever there is need for formal communication, it is the written mode that is preferred.
6. Another salient feature of written communication is the time factor it involves. While in a face-to-face communication, situation, the sender's encoded messages are instantaneously decided by the receiver in a written communications some delay necessarily takes place. There is no fixed time limit to this delay. The sender or encoder takes his own time in formulating the message which in turn takes quite sometime to reach the receiver or decoder. The receiver will take his own time in filtering it through his mind and responding to it.
7. Written communication is essentially a creative activity which requires conscious and creative effort. The creativity of this effort comes from the stimuli produced by the mind. The stimuli of oral communication are picked up from outside by sensory receptors. In other words, written communication is more specifically, more carefully thought out than oral communication that is based on spontaneous reaction to signs picked up from outside.

   As an example, let us take up the writing and production of a report that we want to present. For this purpose, we gather all the necessary information and data. Then, we process it through our logical thought processes and encode our communication. This is not a face-to-face communication situation. There is no interchange of message or external stimuli rather, an entirely creative activity of the mind.

Advantages of Written Communication

(a) Legal defence: Maintenance of proper records, letters, reports and memos builds up legal defences of the organization. Organizations usually have their legal advisors who cannot be of any help unless proper records are made available to them.

(b) Ready reference: Written communication has the advantage of providing records, references, etc. In the absence of ready reference, great confusion may be created and the working of the organization will virtually comes to a halt.

(c) Mass access: It gives access to a large audience through mass mailing. It is a common practice on the part of a well known organization to reach out to people at large and win customers through wisely drafted 'mail shots' or unsolicited circulars.

(d) Promotes uniformity: Written communication promotes uniformity in policy and procedure. It is the only means of laying down clear guidelines for the working of the organization.

(e) Image building: Written communication builds up the organization's image. It is not at all surprising, therefore, that the outgoing letters/messages of certain well known companies are cited as examples to be emulated.

(f) Suitable for distance communication: Written communication is suitable when the receivers are spread over vast geographical distances and oral communication is either not possible or will cause the message to lose its effectiveness.

(g) Permanent in nature: Written communication is permanent. The growth of an organization is promoted, to a large extent, by reference to its old, well-maintained records and minutes of meetings.

(h) **Accurate and unambiguous:** Written communication must be accurate and unambiguous. Care has to be taken in drafting any letter, memo or report so that the message is effectively conveyed. Oral communication may often give rise to confusion because every speaker has his own way of putting himself across. While speech is very personal written communication rises above the person, especially in a business organization where precision is the norm.

(i) **Permits substitution and revision:** While spoken words cannot be taken back or erased, since they have to be instantaneously and thoughtfully planned out, the sender of a written message gets time to conceive the idea and formulate his message after due substitution and revision of his text.

(j) **Facilitates to assign responsibility:** It facilitates proper assignation of responsibilities. Moreover, the lower staff behaves with more responsibility, and also feels secure, when communication is sent in writing. There is greater pursuance and seriousness attached to something that is "in writing".

Disadvantages of Written Communication

- Some of the disadvantages of written communication are as follows:

  (a) **Time consuming:** Both encoding and transmission of the message take time, resulting in delays. It is, therefore, a time-consuming process. Postal delays cause a lot of time to be elapsed between despatch of written message and its receipt by the target.

  (b) **Limited to literate world:** Written communication is limited to literate people, who have the capability to read and write. Though, inspite of literacy levels rising across the world, there still remains a great number of people who are still devoid of this skill.

  (c) **Needs expertise in expression:** Written communication runs the risk of becoming ineffective in the hands of people otherwise good in their job, but poor in expression. It is a serious concern of a modern organization to recruit people who are very good in expression, especially in letter and report-writing ability.

  (d) **Lot of paper work:** Written communication creates mountains of paper gathered around the premises of an organization. It is a common sight in offices. Very often valuable papers get lost. Therefore managers have to be extra careful to keep sensitive or confidential material in their own custody.

  (e) **Costly:** Written communication is a costly process in terms of stationery and the people involved in typing and sending out letters.

  (f) **Lack of immediate feedback:** Written communication is rather handicapped by its inability to get immediate feedback. The receiver of the message takes his time to understand and respond to the message.

  (g) **No immediate clarification:** If the receiver of a written message at a distance seeks some clarification, he cannot have it, as quickly as he would like to. He will also have to write back and wait for the reply to his query. Immediate clarification is also not possible in exchange of written communication.

  (h) **More man hours needed:** It is costly in terms of man-hours lost in taking dictation, typing, entering in the diary, dispatching etc., The same job can be done efficiently and expeditiously by harnessing modern technology like computers and e-mails.

2.5 VERBAL COMMUNICATION VS. NON-VERBAL COMMUNICATION

Difference Between Verbal and Nonverbal Communication

- Communication is a natural phenomenon, it is an act of interacting with people and sharing information with them. Do you know, whether you speak or not, it communicates the message to the other party. There are two types of communication Verbal and Non-verbal. **Verbal communication** is a form of communication in which you use words to interchange the information with other people either in the form of speech or writing.

- On the contrary, **Nonverbal communication** does not use words for communicating anything, but some other modes are used, i.e. where communication takes place by way of unspoken or unwritten messages such as body language, facial expressions, sign language and so forth. In this article excerpt, given all the important differences between verbal and non-verbal communication in detail.
- Comparison Chart

Basis for Comparison	Verbal Communication	Non-verbal Communication
1. Meaning	The communication in which the sender uses words to transmit the message to the receiver is known as verbal communication.	The communication that takes place between sender and receiver with the use of signs is known as non-verbal communication.
2. Types	Formal and Informal	Chronemics, Vocalics, Haptics, Kinesics, Proxemics, Artifacts.
3. Time Consuming	No	Yes
4. Chances of transmission of wrong message	Rarely happens.	Happens most of the time.
5. Documentary Evidence	Yes, in case of written communication.	No
6. Advantage	The Message can be clearly understood and immediate feedback is possible.	Helpful in understanding emotions, status, lifestyle and feelings of the sender.
7. Presence	The message can be transmitted through letters, phone calls, etc. so the personal presence of the parties, doesn't make any change.	The personal presence of both the parties to communication is a must.
8. Use of word	Verbal communication uses oral or written words	Non-verbal communication not uses any oral or written words.
9. Types	Verbal communications two types: Oral and written.	Non-verbal communications may include various types: visual, audio, audio-visual, silent etc.
10. Understand	Easy to understand.	Difficult to understand.
11. Structured	Verbal communications is highly structured.	Non-verbal communication lacks informal structure.
12. Distortion of information	Less possibility of distortion of information.	High possibility of distortion of information.
13. Continuity	Verbal communication begins and ends with words.	Non-verbal communication continues until the purpose achieved.
14. Feedback	Verbal communication gives a less and delayed feedback.	Non-verbal communication gives a lot of feedback.

2.6 NON-VERBAL TYPES

- Non-verbal communication is usually defined as the process of sending and receiving wordless messages. Such communication, without the use of words, is usually simple but limited. These wordless messages can be communicated through gestures, body language or postures, facial expressions or eye contact, object communication such as clothing, hairstyles, architecture, symbols, emotions, voice quality, etc.
- Non-verbal communication can occur through any sensory channel - sight, sound, smell, touch or taste. It comes in many forms at the same time. For example, a person's dress, tone of voice, attitude, movement, etc.
- The first scientific study of non-verbal communication was found in Charles Darwin's book, *'The Expression of Emotions in Man and Animals (1872)'*.

Non–Verbal Communication

Visual Auditory

- Visual methods are those which can be seen; for example signs, pictures, colours etc. Auditory are those which can be heard; for example bells, horns, whistles etc.
  1. Non–verbal methods of communication have an instant effect because they can be quickly perceived by the receiver.
  2. Non–verbal methods of communication are less time-consuming to understand, like seeing a picture, colour, or listening to a bell, horn than verbal methods of communication.
  3. Non-verbal methods can present data in a compact form.
  4. Several pages of written material can be instantaneously conveyed through graphical, pictorial, tabular or diagrammatic methods. For the purpose of comparison, such non-verbal methods are more convenient.
  5. Pictorial symbols can be quickly understood and easily remembered; hence they are more useful in semi-illiterate or illiterate population.
  6. Pictorial symbols can be more suitable.
  7. Graphs and charts of different kinds are nothing but a pictorial presentation of statistical information, where special skill and technique is applied. Through them the overall situation can be understood at a glancc.
  8. Non–verbal communication is more effective in representing figures and tables. It needs less space than what is required for a description.
  9. Maps are also a form of non-verbal communication. They are used for numerous purposes like showing transport routes, climatic conditions, population distribution, crops, vegetation etc.
  10. Many a time, visual signs and signals are used for communication if the sender and receiver are not within hearing distance of each other. Such signs are made with hands, lights, colour or any other means which is large enough and is easily visible.
  11. One part of non-verbal communication falls in the area of auditory symbols. Auditory symbols are those which are heard by the ears i.e. sound. Tribals use drum-beats to convey the messages to distant places in the forests. Its modern form is the siren which is used in times of emergency situations like fire, accidents, war etc.

Realia (Real objects, people, events etc.)

- A number of non-verbal clues play an important role in oral communication. Upon meeting a stranger and interacting with him, a person observes the stranger's physical characteristics, body movements, facial expressions etc. Non–verbal communication is no substitute for verbal communication. However, it supplements and reinforces what is being said by the communicator. There are occasions when non-verbal clues may contradict verbal communication. For example, a tiny, shabbily dressed person may create a poor impression but he may later prove to be an interesting person/good speaker.

(A) Body Language

- There is a language beyond verbal, linguistic and visual ones which can be expressive and loaded with denotative and connotative meanings. It has come to be known as "body language". Non-verbal language can be classified as:

Non-verbal Language →
- → Body language or Kinesics (gestures, body movements, facial expression, eye movements, postures etc.)
- → Paralanguage (laughing, yawning).
- → Proxemics (perception of physical space).
- → Olfaction or body odour.
- → Skin sensitivity (touch and temperature).
- → Dress (Cosmetics and grooming).

- Body language is a comprehensive term used for ability of living beings to communicate without verbal communication, through different parts of the body; it includes gestures, facial expressions, different body postures, miming, touch, dependence upon the physical space between the communicators, dressing style/attire, as well as all sorts of body movements. In certain cases, silence also functions as body language. Silence plays an important role in certain exchanges. Body language, by and large, is unintentional and more truthful than verbal messages.
- **Gestures:** In certain cases, gestures are used extensively when a person does not know the local language. Deaf and dumb people mainly depend on gestures for communication. In certain discussions, gestures of a person convey much more than what he speaks, for e.g. enthusiasm, emotions, etc. A gesture like shaking hands displays friendship and warmth. Other such examples are holding of hands, patting on the shoulder, waving, etc.
- **Facial expressions:** A smile expresses friendliness and affection. Raising of eyebrows conveys surprise and furrowed forehead expresses worry and anxiety. Various parts of the human face are significant in communicating information, especially the eyes. Eyes are the most significant organs which communicate subtle messages. Eyes by themselves do not show emotion; but when associated with other parts of the face, they are the windows to the soul. Squinting, winking, staring and gazing express a world of meaning. Looking at the eyes, face and other verbal signals during conversation perform certain important functions:
  1. Looking helps the synchronising of utterances.
  2. It provides definite feedback.
  3. It provides additional non-verbal information which elaborates verbal messages.
- **Postures:** Each movement or posture of the body has expressive and defensive functions. The way in which one sits or stands, walks in or walks out, speaks a lot about the person's emotions and attitude. But no body movement itself has any exact meaning. Just as gestures add to the meaning of spoken words, postures also make sense when considered together with other verbal and non-verbal clues. For example, lowering of the head indicates the end of a statement or raising of the head indicates the end of a question. A large number of postural changes indicate end point in interactions; for example, the end of a thought or the end of a statement. For example, when talking to friends/peers, our body posture is relaxed and at ease. However, while talking to our superiors, our body posture is more erect and formal.

Standing with the arms folded across the chest is a common posture that indicates hostility, disagreement or being unresponsive to suggested ideas/thoughts. A shift in posture indicates dynamism. This aspect of body language is known as "Proxemics."

- **Dress and Appearance:** The way one dresses and the way in which they groom themselves shows the person's status and attitude. Most people are influenced by how people look and the clothes they wear. Physical attractiveness plays an important role in one's assessment of people. Attractive people are generally considered to be more successful, more intelligent, and more competent than those who appear less attractive. A person's physical appearance and dress convey a great deal of information about him, for example his/her status, designation, qualification, etc. For many if not all the manner in which a person is judged or evaluated depends greatly on his/her physical appearance. A person's attires hence exercises a definite impact on the communication process.
- **Silence:** *"Speech is silver, but silence is golden".* It means silence speaks louder than words do; it is non–verbal language *par excellence*. It describes the relationship between communicators and their attitudes towards one other. Silence – defined here as an absence of speech or noise – has been generally ignored as a form of communication because it represents inaction or non-behaviour. But it's not necessarily inaction. Nor is silence, as many believe, a failure to communicate. It can, in fact, be powerful form of communication. It can mean someone is thinking or contemplating a response to a question. It can mean a person is anxious and fearful of speaking. It can signal agreement, dissent, frustration or anger. In case of public speaking, when the speaker punctuates his remarks with periods of silence, it may mean that they are used to stress the importance of his message, or it may indicate a momentary pause given for the audience to grasp the meaning of what the speaker just said. In some cases, for example, a worker maintains a generalised silence when he fails to complete his assigned job and the supervisor makes enquiries about it.

  Failing to pay close attention to the silent portion of a conversation can result in missing a vital part of the message. Astute communicators watch for gaps, pauses, and hesitations. They hear and interpret silence. They treat pauses, for instance, as analogous to a flashing yellow light at an intersection – they pay attention to what comes next.

(B) Visual Communication

- Before the primitive man learnt how to use verbal communication, he might have used non-verbal and visual signs for making others understand. He might have used pictures, images, paintings etc. Pre–literate man might have expressed himself visually. The invention of the letters of the alphabet led to a form of writing, and abstract symbols replaced visual representation of objects.
- Any attempt to reach us through our eyes, whether it is a roadside billboard or a flier posted for a meeting, is considered visual communication. In face-to-face communication, visual communication can take the form of gestures and body language. A gesture such as a "thumbs up" or the body language of a head nod indicates approval, while a "thumbs down" or a head shake indicates disapproval. We rely heavily on visual communication, and for good reason: research indicates that it is very effective. People can recall information presented to them with visual images better than if they are just told the information.
- In communication that is not done face to face, visual communication usually takes the form of printed, video, or computer-generated projected messages. In printed form, some examples include signs (traffic signs, billboards, restroom signs), fliers, posters, exhibits, and logos. Projected visuals are those used with computers, using such programs as Microsoft PowerPoint.

Visual Aids

- Visual aids are used for the following reasons:
  1. To seize the viewers' attention and help them to focus on major points. Sometimes it takes an eye-catching visual to grab the viewer's attention in order for the viewer to pay attention to your message.
  2. To translate words into meaning. A graphic, photograph, or chart may be easier to understand than words on a page or someone's vocal description of a topic.
  3. To get your point across. Visual aids help to educate, inform, and persuade.
- In addition to the more traditional forms of visual communication, such as videos and photographs, here are some other examples of visual aids:
  1. Your body can be a visual aid. Body language and gestures communicate messages more effectively than voice alone.
  2. Objects and models can be used in a variety of situations to communicate. In a chemistry class, a model of a water molecule can help to show how hydrogen and oxygen bond to form water. If you're showing how to hold a football properly, and hold up a real football to show where to put your fingers on the laces, you're using an object to communicate a message.
  3. Computer-generated slides are extremely popular for formal presentations.
  4. Posters, fliers, and displays work well when they feature eye-catching images to draw attention to a text message and, better yet, add a layer of meaning to it.
  5. Writing boards (chalkboard, dry erase boards, smart boards) let you illustrate your points off the cuff to support educational presentations.
- The important visual forms of communication today:

  In visual communication, elementary and simple ideas, orders, and warnings are used. It can be effectively used only in combination with other media. The diagrammatic representation combines the written with the visual. Pie charts and pictograms also combine the two, because there is always a written indication of what is shown in them, commonly referred to as 'key'. Such forms are very useful in the presentation of statistical data.

Visual Communication Design Principles

- Any visual communication form must be designed properly for your audience like use of large, eye-catching text or images, or something unusual. Good visual communication designs should attract the audience. Remember a few key design principles for visual communication: simplicity, unity, emphasis, readability, organisation, balance, accuracy, and clarity.

  2. **Simplicity:** A poster, display, or computer-generated slide should not be crammed full of text or images. Keep it simple. Usually—especially with computer slides—the fewer the elements (text, images, illustrations) on the page or screen, the easier it is for your audience to understand and recall your message.
  3. **Unity:** Work to enhance harmony in all of your visual communication. If you are going to design several different posters for an event, use the same font style or the same image on all of the posters to show unity. Keep the background and text the same and use similar graphic images on all the slides for a computer-generated presentation.
  4. **Emphasis:** Similar and harmonious materials create unity in the overall presentation, which makes it easy to differentiate and emphasize different and important information. The interplay between similarity and occasional eye-catching difference clarifies information for audiences. Your careful and sparing use of a few strategies for emphasis will help the audience catch on more quickly when information is unfamiliar and when concepts are

particularly important. Emphasis is most effective when used sparingly: if everything is emphasized, then nothing stands out. (And if you use too many strategies for emphasis, you risk confusing the audience.) Focus the audience's attention to a specific place on your poster, exhibit, or slide by using a few, carefully chosen elements:

- **Animation:** On some slides, animation or transition "reveals" might draw attention to text or images on the screen.
- **Underline, italics, and boldface:** Used consistently these are excellent attention-getters. Pick one or assign them to different jobs. (For instance, you could use boldface for headings, italics for unfamiliar terminology, and underline for emphasis.) If you use all three indiscriminately, the audience may become confused.
- **Bullet points:** For some posters, exhibits, and computer slides, bullet-pointed text helps readers digest information quickly. People have become accustomed to bullet points and automatically direct their attention to bullet points because they believe that important information will be presented there.
- **Colour:** Placing some text in a different colour than the rest of the text will immediately draw attention to the page or screen. Colour images also draw attention. Certain colours also are interpreted in specific ways by most people. Warm colours are red, yellow and blends of those two colours (orange). These colours are related to heat, fire, and the sun. Cool colours like blue and green call to mind the sky, sea, and wilderness. Depending on what message you are trying to get across to your audience, you may want to use one of the following colours:
  1. Red is the most dramatic colour. It excites and stimulates people. Red is often associated with aggressive behaviour, passion, success, and impulse. Use red sparingly, because it is such a "hot" colour.
  2. Bright yellow often is associated with health and well being. Yellow is also associated with caution. For example, all traffic signs that pertain to caution—such as yield signs are in yellow.
  3. Blues are associated with tradition, orderliness, and stability. Light blues are associated with cleanliness. Dark blue colours seem to have a calming effect. Purple is seen as daring, royal, and elegant.
  4. Greens provide a sense of nature and regeneration. Green is used by many environmental organisations.
  5. White is a symbol of purity and innocence in Western countries.
  6. Black usually signifies finality, ending, and death in Western countries.

Advantages/Merits of Visual Communication

1. To overcome language difficulty.
2. Charts and graphs are easy to grasp and understand.
3. Retention period of pictures and graphs is longer.
4. Comparison can be easily made.
5. To make publicity work more effective, for example in case of advertising and media, charts and graphs are introduced. These are very effective in changing attitudes.

Disadvantages/Demerits of Visual communication

1. Charts, graphs etc. cannot be used for conveying all kinds of information. They are also not suitable for discussions or for giving instructions etc.
2. They are not suitable for explaining ideas and their details.

Importance of Communication in Corporate Life

Importance of Communication in Corporate Life →

- Required for management of employee relations
- Tool of supervision
- Essential for motivation
- Maintains good human relations
- Useful for acquiring information
- Helps in planning & co-ordination
- Aid to understanding
- Free exchange of ideas & information
- It encourages ideas & suggestions
- Good for releasing tension

- Communication is a process of dynamic interaction both affecting and being affected by many variables. It is a basic social process necessary for the growth and development of individuals, groups, society and people.
- Communication links called "channels" bind the groups. It is not a single process but a set of processes. It uses various means like oral and written, verbal and non–verbal etc. The means of communication are determined by many factors. They are personal, social, political and economic. Business organisations use mass communication media to establish and maintain contacts with various people within and outside the organisation. It can be done with the help of letters, telephone and similar means.
- Business communication is concerned with the study of communication between individuals and groups created in the process of administration of an organisation.
- Process of communication is dynamic rather than static.

  Besides the basis of form and relationship, communication is also classified on the basis of the direction of its flow. It is categorised as Downward, Upward, Horizontal (or Lateral) and Diagonal (or Crosswise) communication.

2.7 SILENT COMMUNICATION – NON-VERBAL COMMUNICATION, ADVANTAGES AND DISADVANTAGES

Words alone are, in many cases, not adequate to express our feelings and reactions.

'Non-verbal Communication' refers to the process of 'Communication by means of elements and behaviours that are not coded into words'. It is the process of sending and receiving wordless messages through gestures and touch, by body language or posture, by facial expression and eye contact as well as through objects or artifacts. Oral communication also has non-verbal elements known as paralanguage, which includes voice quality, rate, pitch, volume, style, emphasis etc. These are perceived by the receiver. Even written communication has some aspects which add to the meaning of the words. However, the study of silent communication is generally with respect to face-to-face interaction.

Silent communication is a medium of communication, which adds another dimension to the spoken word and so it is also known as 'Communication by Implication'. Non-verbal messages express true feelings in a more accurate manner than the spoken or written language. Non-verbal Communication is a very powerful means of communicating and hence, it is important to study this as a separate topic. Most people believe that the manner in which you say something is more important than what you say.

The significance of non-verbal communication can be seen from the fact that the content of a message is composed of:

- 55% from the visual component
- 38% from the auditory component
- 7% from language

When people communicate, listeners gain information about the speaker's attitudes towards them from visual, tonal, and verbal cues. Similarly, the listeners can send their feedback using these cues. Much depends upon a number of other factors, such as context of the communication, and how well the communicators know each other. It also depends on the cultural background of the communicators.

Scientific research on nonverbal communication and behavior began with the 1872 publication of Charles Darwin's *The Expression of the Emotions in Man and Animals*. Since then, there has been an abundance of research on the types, expressions and effects of unspoken communication behaviour.

Edward T. Hall's (1959) influential book *The Silent Language* placed a heavy emphasis on cultural nuances of nonverbal communication. Hall is generally acknowledged to be the founder of this field of communication studies. The book studies the previously hidden dimensions of human communication, particularly proxemics (how space affects communication) and chronemics (how time affects communication).

In the mid 1960s, Paul Ekman studied emotions and discovered six facial expressions that almost everyone recognises world-wide: happiness, sadness, anger, fear, disgust, and surprise. Although they were controversial at first, they are now widely accepted. Paul Ekman (born February 15, 1934) one of the 100 most eminent psychologists of the twentieth century is a pioneer in the study of emotions and their relation to facial expressions.

Significance of Silent Communication

Stated briefly, the manner in which a message is expressed may carry more significance and weight than what is said, i.e., the words themselves. Depending on the situation, the speaker can reinforce his message with a smile or a frown, raise his/her voice, or keep it pleasant, gentle, and easy. In short, depending on the context, the communicating parties can frame the message in such a way as to convey the entire meaning. These signals are often so subtle and involuntary that the communicator is not consciously aware of them.

- In recent management literature, nonverbal communication has received much attention especially in the areas of business meetings, negotiations, business presentations, sales and marketing, and also in the area of social skills. It is extremely important in general communication as well. It is estimated that less than ten percent of interpersonal communication involves words. The remaining 90 percent is made up of message carriers, such as voice tone, sounds and a variety of devices such as kinetics (movement), haptics (touch), oculesics (eye-contact), proxemics (space) and chronomics (time) as well as posture, and sound symbols. Even silence can communicate some feeling. In many cases, the communicator may use groups of behaviors to mean the same emotion or feeling, such as combining a frown with crossed arms and unblinking eye gaze to indicate disapproval.

- For managers, working in this flat world, where businesses operate from various locations and interfacing with different cultures, the need to understand non-verbal cues gains further importance because of the major differences in the usage and interpretation of these forms based on cultural nuances. Body language, expression, personal space and other nonverbal tools are an integral part of the communication process, irrespective of whether the person giving such messages is the speaker or the receiver.
- Nonverbal communication occurs not only between people, but also ***internally***. People grimace, stand in certain postures, and show other behavioural stances in order *to reinforce to themselves certain positions, attitudes, and implicit beliefs*. Unconsciously, they suggest to themselves the role they choose to play, such as being submissive or dominant, trusting or wary, controlled or spontaneous.

Research has identified several different types of nonverbal communication. The following is a relatively simple classification of non-verbal cues in communication:

Kinesics:	Body motions, such as shrugs, eye movement, foot-tapping, drumming fingers etc.
Proxemics:	Spatial separation (in relation both the social and physical environment)
Haptics:	Touch
Oculesics:	Eye contact
Chronemics:	Use of time, punctuality, waiting, pausing
Olfactics:	Smell
Vocalics:	Tone of voice, timbre, volume, speed
Sound Symbols:	Mumbling and making sounds such as: grunting, mmm, er, ah, uh-huh,
Silence:	Absence of sound (muteness, stillness, secrecy)
Adornment:	Clothing, jewelry, hairstyle
Posture:	Position of the body (characteristic or assumed)
Locomotion:	Walking, running, staggering, limping, sudden and jerky movements
Expression:	frowns, grimaces, smirks, smiles, pouting, tears in eyes

Functions of Non-verbal Communication

To be able to understand how people use non-verbal communication, it is necessary for managers to realise their functions and the significant role they play in conveying the meaning. Only then can they understand messages in complete and within the overall context. Nonverbal communication in general performs three basic functions, viz.,

- Managing identity,
- Defining relationships, and
- Conveying attitudes and feelings (but not ideas).

There is a close relationship between non-verbal cues and the words they accompany.

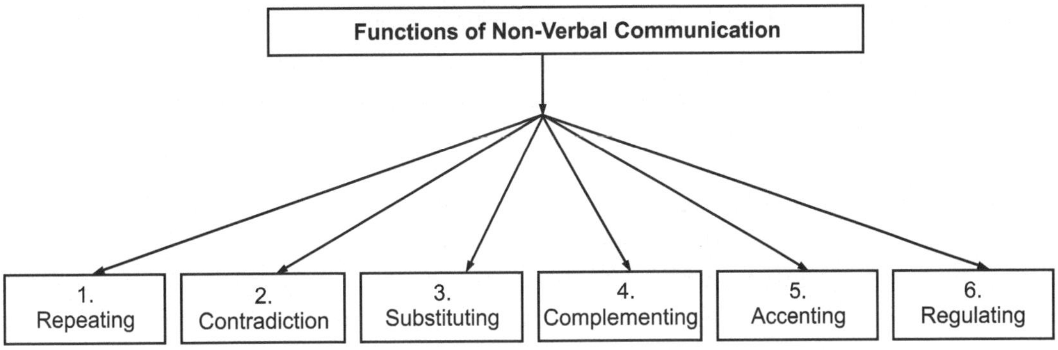

Fig. 2.6: Function of Non-verbal Communication

1. **Repeating:** Nonverbal cues are used to reinstate the meaning conveyed in words.
2. **Contradiction:** When the sender contradicts his words by using nonverbal means.
3. **Substituting:** Nonverbal cues serve as substitutes for spoken words or messages, such as signs to denote peace, victory, hostility (clenched fist), submissiveness (stooped position), depression (bowed head), etc.
4. **Complementing:** Non-verbal cues help complete a message. *Baird* states that the phrase *"I love you"*, spoken sincerely is usually accompanied by physical cues that demonstrate the feeling behind the message. When something is said in anger, the feeling is shown not only in the spoken message, but also by way of clenched fists, flashing eyes and strained voice.
5. **Accenting:** To accentuate the verbal message or give it more force, like increasing the volume or pitch of the voice while asking someone to leave the place.
6. **Regulating:** Controlling the flow of communication is another important function. A forward lean, a nod, a vocal inflection, pointing to the next speaker or a change in eye behaviour can indicate to the other person that his/her turn is over.

To Sum up: *Ray Birdwhistell* estimates that in face-to-face interaction the words spoken account for less than 35 percent of the total meaning produced while the remaining 65 per cent is obtained from non-verbal cues. The sender or speaker transmits messages through appearance, gestures, postures, facial expression, vocal characteristics and words expressed. However, it should be remembered that the interpretation of these symbols can differ even within the same cultural setting.

Form	Main Function (in some cultures)
Nod (Yes)	Repeating
Shrug (I don't know or I don't care)	Substituting
Scratch head, perplexed look	Substituting or Complementing
Tone of voice	Emphasis, accentuation
Hand raised	Regulating, warning
Head shake	Contradicting
Eye movements	Deceiving or loss of interest
Staring	Dominating
Looking down or away	Submitting
Raised fist	Aggression
Hand-shake	Socialising
Touching, kissing	Friendship, Arousal
Over-adornment	Boasting

Need to Understand Non-verbal Means of Communication

People react to the unspoken, as much (if not more) to how something is said and try to find out the explicit meaning of the words. Failure to identify and understand them can lead to misunderstandings. Hence, to be effective communicators, it is very necessary for managers to learn about how the human body and actions can send signals, either voluntary or intentional, from reading and through training programmes, and also about the effects of different methods of nonverbal communication. The following figure shows the elements of body language.

> The same gestures can be misinterpreted due to cultural differences. Look at these examples:
>
> - Making a circle of index finger and thumb generally means 'everything is fine', 'perfect'. However, in France, it means 'worthless'; in Japan, it means 'money'; in Germany it signifies rudeness; and in Malta and Brazil it is an obscene sign.
> - 'Thumbs up' generally means 'all OK'; but in Australia, it is rude; and in Japan, it indicates number five.
> - Palm facing, all fingers open in general means 'stop'; in Turkey, it says 'you get nothing from me', and in W. Africa, it indicates 'you have five fathers!

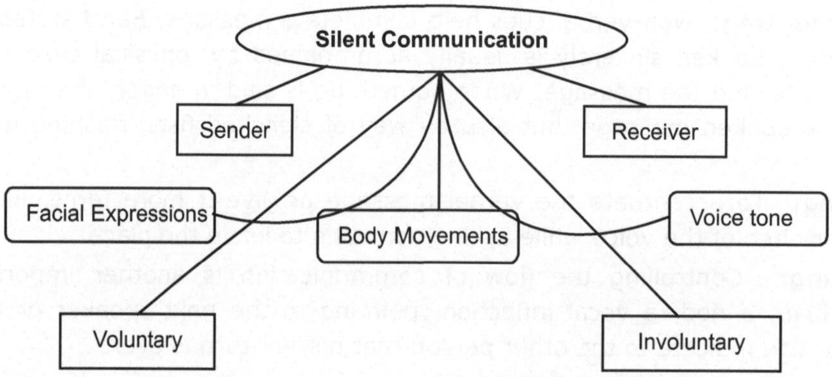

Note: It can be in agreement with or contradict spoken words

Fig. 2.7: Elements of Body Language

Advantages of Silent Communication

- It would be possible to communicate with someone who is hard of hearing.
- In places where it is necessary to maintain silence such as in a meeting or in a hospital, silent communication helps.
- It would be possible to communicate without being heard by others.
- It is possible to communicate with someone who is out of hearing range, but can see the sender.
- It is helpful in communicating with someone who does not understand the sender's language.
- It can be understood instantly and instant response can also be obtained. It makes conversation short and brief, as actions speak louder than words.
- It helps in clarifying verbal communication.

Disadvantages of Silent Communication

- Conversation with only silent conversation becomes very brief and it is not to communicate details.
- It may communicate small ideas. Detailed discussions are not possible.
- The receiver has to observe the changes of expression, body language, adornments, space etc., carefully or the meaning is lost.
- The receiver may not understand, requiring repetition of the message.

- It cannot be used as a public tool for communication. It cannot be used everywhere.
- It may fail to create an impression on the listeners.
- It cannot be preserved and recorded for further use.
- Involuntary body language due to force of habit may be misunderstood by the others. A person may have natural stooping shoulders, which may be understood by the others as weakness. Involuntary tapping on the table may be mistaken for nervousness.
- It may overemphasise the spoken word and create misunderstandings. Hence it has to be used very carefully.
- Non-verbal communication may be multi-channel. For instance, while observing the facial expression, one may disregard body movements. (*see box below*)
- The signs and symbols are not universal in all cultures. They may be misunderstood. Care has to be taken to understand the nuances of cultural background of the receiver. For instance, eye contact is expected in some cultures while it is regarded as impudence or arrogance in others. Cultural expectations should be understood and respected.

> Generally, poker players do not show their expressions on their face and so 'poker face' has become a generic term indicating a blank face. Recent research shows that these players indicate their level of either confidence or hopelessness or anxiety in many other ways like how they pick up the tokens and place them as stakes, how they change their stance etc. Such insights can help players to assess their own potential better. Similarly, managers have to observe the behaviour pattern of people with whom they are interacting in totality.

Body Language

Body language is a term used to describe a whole set of nonverbal means of communication, such as body posture, gestures, facial expressions and eye movements. Human beings possess this unique mental and physical capability to send as well receive such signals and also interpret and respond to them almost entirely subconsciously. They can 'read' others' body language and even mirror them to put the other person at ease. In the 1990s Paul Ekman expanded his list of basic emotions like sorrow, happiness and anger and included a range of positive and negative emotions such as:

1. Amusement
2. Contempt
3. Contentment
4. Embarrassment
5. Excitement
6. Guilt
7. Pride in achievement
8. Relief
9. Satisfaction
10. Sensory pleasure
11. Shame

1. Facial Expressions

Of all the types of nonverbal communication, this may be one of the most noticeable. Sometimes, fleeting, involuntary expressions appear on the face which a person is not intending to communicate or even aware of. *People examine each others' faces as they talk, trying to get some information to confirm that the meaning is received in the manner in which it is delivered*. The human face is more highly developed as an organ of expression than any other animal.

For instance, *Smiling* is one facial expression that is likely to put other people at ease and make them feel accepted and comfortable. A smile can add happiness and comfort to the conversation.

It is a powerful cue that transmits happiness, friendliness, warmth, and liking. The person who maintains a smiling face while talking or delivering a speech will be received warmly by the audience. Such a countenance is indicative of a likable person, pleasant, friendly, warm and approachable. People will react to that person more favourably. It is said that a smile is a universal language. But let us not forget, a person can have a sarcastic smile while talking to someone. A smile with a smirk or grimace can indicate quite the opposite of a pleasant conversation. Scowling, lip chewing, and raising eyebrows can all signal different meanings. A person can express a number of negative feelings such as unhappiness, frustration, anger etc., through facial expressions. So it is important to be aware of how the face looks during a conversation. This is true of both the sender and the receiver.

Here are just a few types of feelings that a human face can express:

(a) Pensive
(b) Warning
(c) Startled
(d) Amused
(e) Pouting
(f) Confused
(g) Sad
(h) Anxious
(i) Sleepy
(j) Barely tolerant
(k) Attracted
(l) Intoxicated

2. Eye Contact

It is an essential tool in nonverbal communication as it **makes the other person feel heard, respected, and important**. It helps to open and regulate the flow of communication. It signals interest in the other person/persons and increases the speaker's interest, concern, warmth and credibility. Making good, consistent eye contact can elevate a person's status as a great conversationalist. However, it can make the other person feel uncomfortable if it goes on for too long. Modern American business culture values a fair degree of eye contact in interpersonal relations, and looking away is considered as avoidance or even deviousness. In some cultures, it is not appreciated if children look at their elders' face, as it can signify rudeness. For some it may feel too intimate, for others it may feel challenging. When cultures interact, this inhibition of gaze may be misinterpreted as "passive aggressive" or worse.

3. Gestures

Gestures are another type of nonverbal communication. They can add warmth and personality to a conversation. Hand gesture, nodding appropriately etc., can be an easy way to show that the listener is understanding, and connecting with the speaker. Failing to use gestures while speaking may make the person appear boring and stiff and this perception can hinder the communication process. A lively and animated speaking style captures the listener's attention, makes the conversation more interesting, and facilitates understanding.

There are many kinds of gestures:

(a) Clenching fist
(b) Biting fingernails
(c) Rubbing chin
(d) Raising eyebrows

(e) Scratching head
(f) Hands behind head
(g) Sticking out tongue
(h) Shaking a finger
(i) Tugging at hair
(j) Smoothing hair
(k) Pursing lips
(l) Looking away
(m) Rubbing nose
(n) Tugging earlobe
(o) Pointing
(p) Squirming
(q) Folding arms
(r) Narrowing eyes
(s) Hands on hips
(t) Rocking
(u) Waving

These, too, have many different meanings in different cultures, and what may be friendly in one country or region can be an insult in another (Morris et al, 1979, Maginnis, 1958).

Body language could be involuntary or 'accidental' body language covering all the different types of movements, postures and facial expressions. It could be intentional warm body language, such as empathic movements and touch, postures and facial expressions or intentional abusive body language, such as violent/abusive movements and touch, postures and facial expressions.

4. Posture and Body Orientation

The way a person walks, talks, stands, and sits sends numerous messages to others. Standing erect and leaning forward communicates to the listeners that the person is approachable, receptive and friendly. Interpersonal closeness results when the sender and the listener face each other. Not facing the listener while or looking at the floor or ceiling should be avoided as it communicates disinterest, or even dishonesty. Posture is always easy to understand.

A comfortable and confident person will stand or sit in a relaxed manner, and will have fluid smooth movements. Such a person will face the conversation partner to indicate confidence and engaging conversation skills.

5. Proximity

Personal space and interpersonal distance, i.e., physical distance between the communicating partners signals their level of intimacy and comfort. **If an unknown person stands too close or touches too often, it will make the other person feel uncomfortable**. Sitting on the table in the office, standing too close to the superiors etc., may not be appreciated. There are unspoken and invisible rules governing personal distance. People stand too near when they are more intimate. When they are not so close, they should be at a distance.

Edward T. Hall is most associated with proxemics, the study of the human use of space within the context of culture. In *The Hidden Dimension* (1966), he argued that human sensory apparatus is the source of all perceptions of space, and these perceptions operate at the subconscious level and are moulded and patterned by culture. These perceptions can impede communication and understanding in cross-cultural settings. He analysed both the personal spaces that people create around their bodies as well as the macro-level sensibilities that shape cultural expectations about the organisation of streets, neighbourhoods and cities.

Edward Hall's theory of proxemics suggests that people will maintain differing degrees of personal distance depending on the social setting and their cultural backgrounds.

Hall's most famous innovation has to do with the definition of the informal or personal spaces that surround individuals:

- **Intimate space -** the closest "bubble" of space surrounding a person. Only the closest persons can reach this level.
- **Social and consultative spaces -** the spaces in which people feel comfortable conducting routine social interactions with acquaintances as well as strangers.
- **Public space -** the area of space beyond which people will perceive interactions as impersonal and relatively anonymous.

Hall describes the handling of space during conversations: "A U.S. male...stands 18 to 20 inches away when talking face to face to a man he does not know very well; talking to a woman under similar circumstances, he increases the distance about four inches. A distance of only 8 to 13 inches between males is considered...very aggressive. Yet in many parts of Latin America and the Middle East, distances which are almost sexual in connotation are the only ones at which people can talk comfortably." Hall (1955) concluded: "If you are a Latin American, talking to a North American at the distance he insists on maintaining is like trying to talk across a room."

Hall argues that cultural expectations determine these spaces. Americans traveling to Europe would tend to back away instinctively, as the distance maintained there almost half of what Americans do.

Every culture has similar internalised expectations about how fixed and semi-fixed features, such as furniture, buildings and cities should be organised. United States cities, for instance, are customarily set out along a network, a preference inherited from the British, but in France and Spain a star pattern is preferred.

In general, distances differ according to culture and the closeness or otherwise of a relationship, such as friends, co-workers, sales people etc. and very often, this is expressed intuitively. Again it depends on the place of interaction, such as public places like parks, restaurants, or on the street and in private homes. It also depends on the gender of the communicating parties. It is necessary to study the cultural norms and look for signals of discomfort caused by invading the other person's space. It is necessary to be aware of the other persons' body language as well as their attitude and respond accordingly. It is always correct to maintain a dignified distance from the receivers.

6. Touch

The manner in which a person touches another communicates a great deal of information. Touch could take a variety of forms, indicating different meanings, such as a gentle grip or a firm one, holding the other person on the back of the upper arm, on the shoulder, or in the middle of the back. It could be a push or a tug. It could be a pat, a rub, or a grab. Each of these indicates different areas of personal intimacy, and this again can be gender- and age-specific. Many adolescents resent any touching that indicates a patronising stance or undue familiarity. Even the angle of one's holding another's hand might suggest a hurrying or coercive implicit attitude, or a respectful, gentle, permission-giving approach (Smith, Clance & Imes, 1998, Jones, 1994).

7. Locomotion

Physical expressions like waving, pointing, touching and slouching are all forms of nonverbal communication. This branch of study which covers body movement and expression is known as **kinesics**. These movements make communication much easier as they can be used to emphasise a point, or repeat one.

Examples:

- The classic example is *crossing of hands across the chest* to communicate that the person is angry or displeased, is through, and does not want to hear any more, thus putting an unconscious barrier between the communicating parties. If the person is also leaning away from the speaker, it could communicate opposition and if it is accompanied by frown or a harsh expression, it could be outright hostility.
- Eye contact can be interpreted in many ways depending on how it is done. Consistent eye contact may mean that the person has warm feelings and is thinking positively of what the speaker is saying. Interestingly, if the other person does not trust the speaker, his eyes will not wander and stay focused on him/her. On the contrary, lack of eye contact may indicate negative feelings for the other person. If the receiver has arms crossed across the chest while making eye contact, he may be signaling impatience and may want to talk about something that is bothering him/her. The act of fiddling while making eye contact may indicate that the person is distracted. Looking at the other person by moving from one eye to another eye and then at the forehead may signify an authoritarian posture, while if the focus shifts to the nose, the conversation is between equals. A simple explanation for lack of eye contact could be that the person has anxiety disorder and has problems focusing on the other person.
- Touching the ear or scratching the chin can indicate disbelief. It can also be seen if the person has averted the gaze.
- A tilted head may simply mean neck problem, but as part of nonverbal communication, it means loss of interest in what is being said. The person may engage the speaker in eye contact, but that does not mean anything.

The style of physical movement in space also communicates a great deal, as well as affecting the feelings of the person doing the moving (Morris, 1977). There could be a number of movements, such as:(a) Slither, (b) Stroll, (c) Jog, (d) Jump, (e) Swing, (f) Crawl, (g) Shuffle, (h) Spring, (i) hop, (j) Acrobatics, (k) Totter, (l) hurry, (m) Tiptoe, (n) skip, (o) Swim, (p) Walk, (q) run, (r) March, (s) Climb, (t) Slink

The pace or the manner in which an action is done can also indicate the mood of the person.

(a) Jerky, (b) Graceful, (c) Shaky, (d) Pressured, (e) Fatigued, (f) Deliberate, (g) Nervous, (h) Tense, (i) Furtive, (j) Gradual, (k) Easy, (l) Clumsy

A related variable to the pace aspect is the time it takes to react to a stimulus, called "latency of response." People can be slower or faster "on the uptake" than others while reacting to questions, or while interacting in conversations. Slow response may indicate that the person is organising his/her thought and is planning a response. It may also mean a superiority attitude almost saying. "How could you put this question to me?"

8. Objective Language, Adornment and Artifacts

Objective language is non-verbal message communicated through appearance. As a medium of non-verbal communication, it includes arrangement and display of material things. This method may include intentional or unintentional communication through material things like clothing, make-up, ornaments and other accessories, books, buildings, room furniture, interior decorations etc. These offer signals relating to the context, such as formal and informal settings. Dress and decoration communicate a great deal about the speaker's rank, status, personality, feelings, emotions, attitudes, opinions etc. What they reveal is something special about the person. For instance, dress of army men differs from civilians. The Service-regiment-seniority-specific uniform instantaneously communicates the rank of the person in the Defence Services. Religious leaders, political leaders, lawyers, judges, doctors, nurses, sports persons, porters, security persons and other workers have their own way of dressing, which identifies them distinctly. Dress by people of different countries varies. What is formal office wear by women in India may be considered informal and dressy in the western countries.

The way in which people carry cigarettes, pipes, canes or glasses also suggests different semiotic meanings. (Semiotics is the science of the emotional or psychological impact of signs, appearances and of how things look. This understanding is gaining a lot of importance, especially in audio-visual communication including advertisements, where a lot has to be communicated very effectively in a very short time.)

9. Context or Environment – Spatial Language

It is relating to the place or environment in which the actual process of communication takes place. It may be physical or psychological. The necessary requirements of environment are lighting, colour, ventilation, temperature, seating arrangement, display boards, public address system, audio-visual equipment etc., which would help a lot to attract and make listeners more attentive. While this category is not actually a mode of non-verbal communication, the setting up of a room or how one places oneself in that room is a powerfully suggestive action. Where one sits in the group in relation to the others is often useful in understanding that person's status or attitude. Group leaders as well as the participants need to study the way the room is organised, the seating and other arrangements and the manner in which it can affect communication. The variables to be considered in a group setting like a meeting, workshop, consultation or conference are:

- Size of the room
- Colors of the walls, floor, furniture, background etc.
- The source, amount, and direction of light,
- Color of the lighting,
- Obvious props such as the podium, display blackboard, cameras etc.
- The number of people present and seating arrangements,
- The numbers and ratios of high-status and low status people,
- The positioning of the various people in the space,
- Environmental sounds, smells, and temperature

10. Psychological Responses

Psychological responses are involuntary, and cannot be practiced voluntarily. Yet it would be useful for communicating partners to be aware of these subtle signs of emotion, so that efforts could be made to suppress them and make the communication process effective. Understanding these signs also helps in reviewing the message in all its aspects like timing, delivery etc. Some of the physiological responses include: (a) Shaking, (b) Sweating, (c) Blushing, (d) Blinking, (e) Flaring of nostrils, (f) Becoming pale, (g) Moisture in eyes, (h) Swallowing, (i) Trembling chin, (j) Cold clammy skin, (k) Flushing, (l) Breathing heavily

11. Silence

The question often arises whether silence is a mode of a communication or not. The answer is that one can communicate silently. Silence sometimes speaks louder than words. For instance, a speaker can stop his speech and remain silent if the audience is getting distracted. This works better than if the speaker appeals to the audience to pay attention. (Of course, silence on the part of the Speaker in a legislative body in India cannot have the same effect! The Speaker has to stand up to signal an unruly set of legislators to calm down and sit.) Silence, combined with an angry look can indicate admonition. Silence could also be in response to sudden shock, excitement, extreme grief etc. In a feedback situation, a written message may be ignored (silence) to indicate that the message had no value to the receiver. The practice of silence is usually taken as assent or approval in a number of personal, business and social transactions, depending on the issue, the participants and the culture they belong to. However, in the law of contract, silence, i.e., failure to respond to an offer is not proof that the offer has been accepted.

12. Paralinguistics

Sounds are the basis for paralanguage. Paralanguages include tone of voice, power or emphasis, pitch, rhythm, volume, pause or break in sentence, speed or delivery, loudness or softness etc. Speakers can use such nonverbal signals for effective communication. A manager has to learn to vary his/her voice according to the context. For instance, a person who speaks in a monotone voice does not command respect, as the listeners may perceive his/her as dull. A person who has a lively voice can catch the attention of the listeners and ensure that he/she is heard. On the other hand, a person may alter his/her voice quality to indicate something different than the words used in the communication. Then, the paralinguistics become most noticeable. The power of the voice tone over the meaning communicated can never be underestimated.

Paralanguage can be divided into four parts:

1. **Voice Qualities, which are in-born:** Including such factors as pitch, resonance, volume, rate, and rhythm.
2. **Vocal Characterisers:** Embracing laughter, coughing, throat clearing and sighing.
3. **Vocal Qualifiers:** Referring to variations in pitch, modulation and volume.
4. **Vocal Segregates:** Including the silent sound such as 'ahs' and 'ers' and pauses. These clues do much to influence meaning".

The lesson to be drawn by managers and students of communication studies is clear: Though they have some voice qualities which are in-born, they have to exert some control over the tone and pitch while communicating in the office. Using too many segregates can give the impression that the speaker is not confident about the content and also about how to deliver it. Secondly, they should try to keep the words spoken and the paralinguistics in sync or in agreement with each other. This will help keep the messages clear, understandable, and less likely to be misinterpreted.

Dialect

Related to paralinguistics is a semi-linguistic element, viz., dialect, with all its tonal variations along with difference in grammar and usage. This has deep cultural undertones, suggesting class, age, sophistication, etc., and the resultant prejudices. It affects understanding people from different backgrounds, generations and cultures. For instance, managers seeking to work in rural areas in India should not start with belittling them for their appearance, habits, dialect and style of talking. What matters most is the content of the message and how effectively it can be communicated. Managers should understand, accept and respect these nuances and not develop biases based on language and diction.

Sign Language

In the place of spoken words, numbers or punctuations marks when marks or symbols such as gestures are used, it is known as sign language. For instance, deaf, mute people communicate by making signs with their fingers. These signs are accepted and recognised and standardized. They are trained for this, as also their family members, teachers and care givers. Apart from this set of differently abled persons, under certain circumstances, sign language is used even by those who can speak, such as the sign used by hitch hikers, Morse code, traffic signals, outdoor advertisement hoardings etc.

Action Language

It is a language of movements. Action in a particular situation and context can be interpreted to give a meaning. By action, one may knowingly or unknowingly be communicating with others. For instance, anger is generally demonstrated by throwing things around, happiness with clapping etc.

Important Points

Formal Communication

1. Exchange of ideas at different levels through proper channels.
2. Messages flow from higher to lower levels and vice-versa.
3. Formal communication keeps everyone in the organisation well-informed about what is going on in the organisation.

Advantages and Limitations of Formal Communication:

1. Allows free flow of information.
2. Tries to filter the communication.
3. Maintains control on the subordinates.
4. Stops free flow.
5. Expensive when in written form.
6. Distortion of message possible.

Informal Communication (Grapevine Communication)

1. Also termed as grapevine communication.
2. May have good as well as bad effects on organisation.
3. Messages flow from subordinates to superiors or the other way.
4. No place for poor communication in an efficient organisation.
5. Channels of communication are created and controlled by the people.
6. Chains in the grapevine communication:
   (a) Single Strand Chain (b) Gossiping Chain (c) Probability Chain (d) Cluster Chain
7. Grapevine communication moves fast.
8. No fixed route for this type of communication.
9. Management must at times control grapevine communication.
10. Can become an additional channel.
11. May create misunderstandings.
12. Does not have credibility.

Gossip

1. May create a co-operative atmosphere in some cases.
2. Though good, it can be equally harmful.
3. May create leaders.
4. It is a part of the feedback system.

Rumours

1. Not desirable.
2. Not authentic.
3. Uncertainity is the main character.
4. They are dangerous.
5. Management must timely act against rumours.

Verbal and Oral Communication

1. It is an important type of communication.
2. Verbal communications are numerous in form.
3. Skill in oral communication is important.
4. In oral communication other things like body, face etc. play an important part.

Eye Contact

1. Effective means of communication.
2. It creates an impression (may be good or bad).
3. It gives a feedback signal.

Voice Modulation:

1. Voice is important.
2. Effective while speaking.
3. Good voice is a must.
4. A good voice gives self-confidence, tempo, etc. to the speaker.
5. Good pronunciation, stress etc. gives additional advantage to a speaker.

Audience Awareness

A speaker must always:

1. Size up the listeners.
2. Age group.
3. Sincerity of speech, sharing of views.

Presentation Plan

1. Art of speaking can be acquired only through constant efforts.
2. Preparation is very important and is the key to a good presentation.
3. Memorisation hampers natural speech.
4. Generation of a positive attitude in the audience is important.

Non-verbal Human Interaction:

1. Non-verbal communication means no use of words.
2. Maps, charts, graphs also convey information.
3. Non-verbal communication: (a) Visual (b) Auditory
4. Visual – non-verbal communication may have instant effect.
5. It takes less time to see pictures or listen to a bell etc.
6. Used to draw immediate attention.
7. Attention of human beings can be quickly drawn through pictures, colours and sounds which are more effective to draw human attention.
8. Television and Cinema are the most powerful means of mass communication.
9. Signals are more effective where instantaneous decisions are needed.
10. Non-verbal clues like body language, gestures, facial expressions, postures, appearance, silence play an important part in non-verbal communication.
11. In case of visual communication, charts, graphs, pictograms, tables, slides, film strips play an important part.
12. They are more effective in the education system.

Realia (Real objects, people, events etc.)

1. In case of strangers, the physical character, body movements etc. are observed first. These help in understanding non-verbal communication.
2. The language beyond verbal, linguistic and visual is also expressive.

Body Language:

Gestures, facial expressions, posture, dress, appearance, silence have definite effects on communication.

Communication Today:

1. In visual communication simple ideas like charts, graphs, etc. are effectively used.
2. Charts and graphs are easily understood and remembered. They help effective publicity.
3. Charts and graphs are not always useful as they are substitutes for discussions.
4. Sometimes charts and graphs may mislead in giving information.

Process of Communication:

1. It is a dynamic interaction.
2. Social process is necessary for the growth of group, society etc.
3. Communication links like "channels" bind the group.
4. Business organisations use mass communication.
5. Communication is established through memos, orders etc.
6. Business communication links superiors and subordinates.
7. In the communication process, messages can be transmitted by verbal, non-verbal, gestures, noise etc.
8. Process of packing ideas is called encoding.
9. Interpreting messages is called decoding.

Downward Communication

1. Flows from higher authorities to subordinates.
2. Helpful in achieving targets, updating knowledge, providing incentive etc.
3. Means like notices, manuals, bulletins, telephone etc. are used for downward communication.
4. Establishes mutual support between superiors and subordinates.
5. Useful for transmitting information.
6. Increases efficiency.
7. Effective if proper channels are followed.
8. Helps to establish decentralisation of authority.

Upward Communication

1. It is the flow of information from lower levels to upper managerial levels.
2. Encourages good suggestions, expectations, working policies etc.
3. Helps to understand the feelings of subordinates.
4. In spite of advantages, it never flows easily because of hesitation or fear among subordinates.
5. Useful for the progress of an organisation.

Horizontal Communication

1. Also known as lateral communication.
2. Takes place among the people at the same level of hierarchy.
3. Necessary for exchanging views and ideas.
4. Very common in large organisations.
5. Meeting, conferences etc. are important in horizontal communication.
6. It allows freedom of expression.
7. It increases co-operation.

Diagonal (Crosswise)

This type of communication takes place between people working at lower levels in one department and others working at higher levels in the other department.

1. It speeds up the work of the organisation.
2. May violate limit of command.
3. If properly managed, it is useful for increasing co-operation in an organisation.

Practice Questions

1. Describe, in brief, the process of formal written communication.
2. Explain the importance of effective communication.
3. Explain communication on the basis of direction of flow.
4. Explain communication on the basis of relationship, form and direction.
5. Explain Formal Communication and state its merits and demerits.
6. What is meant by Grapevine communication? Explain its merits and demerits.
7. Explain the difference between Formal and Informal Communication.
8. Explain the importance of verbal communication in day-to-day life.
9. Explain the importance of following in communication:
   (i) eye-contact (ii) voice modulation (iii) audience awareness (iv) presentation plan
10. Explain Non-Verbal Human Interaction and state the difference between Visual and Auditory methods of communication.
11. State the importance of the term 'Realia' in communication.
12. Explain the various aspects of 'body language'.
13. Explain the terms Kinesics and Proxemics.
14. Explain the importance of visual communication in daily life.
15. State the importance of communication in corporate life.
16. Explain the importance of communication process and the significance of Upward, Downward as well as Horizontal communication.
17. Write short notes on:
    - (i) Effective communication
    - (ii) External communication
    - (iii) Internal communication
    - (iv) Formal communication
    - (v) Informal communication
    - (vi) Grapevine chain
    - (vii) Grapevine communication merits and demerits
    - (viii) Formal and Informal communication
    - (ix) Rumours
    - (x) Verbal and Non-verbal communication
    - (xi) Eye contact
    - (xii) Voice modulation
    - (xiii) Audience awareness
    - (xiv) Non-verbal Human Interaction
    - (xv) Realia (Real objects, People, events etc.)
    - (xvi) Body language
    - (xvii) Gestures
    - (xviii) Postures

■■■

Chapter **3...**

Principles of Effective Communication

Contents

3.1 PRINCIPLES OF EFFECTIVE COMMUNICATION

- The main objective of any communication is to make others know what one thinks or feels, irrespective of its subject matter. The purpose of communication differs from person to person, person to group, group to group, person to community, group to community and vice-versa. The purpose of a friend interacting with his friend may either be to keep in touch with him or just to know more about him in his own personal capacity; this sort of communication has no bearing on his professional and public life. However, this sort of communication differs from business or organisational communication. In corporate life, the bearing or the nature of communication has a certain specific purpose. It cannot be vague or random.
- In business communication, speaking, hearing or writing does not take place without a particular objective. This objective has to be achieved through communication. The nature of communication should be such that it serves to attain prosperity, progress and proficiency for the organisation or the business concerned. If these desired objectives are achieved, the communication through which they are attained is called 'effective communication'. In short, communication has a specific objective in an official or business situation.
- The most common objective of any official or business communication is to convey information or to get information. It may be in the form of instructions, advice, suggestions or may contain requests to induce customers. Sometimes, the communication may be with an intention to complain, issue warnings etc. However, the latter part may involve displeasing one of the parties involved. It may not happen often but in some cases it may be essential.

Seven C's of Communication OR (Principles of Effective Communication)

- Effective communication is the most vital process that keeps an organisation operating smoothly and successfully. It underlies the two basic principles of: (i) giving information and (ii) persuading others. Effective communication can be briefly defined as the life-blood of any business that promotes the spirit of understanding and cooperation among all the elements concerned with business i.e. producers, distributors, retailers and above all the customers.
- Communication has four important attributes:
  (i) Intentional or Unintentional
  (ii) Dynamic
  (iii) Systematic
  (iv) Interactive and Transactive.

- It must be kept in mind that even if these attributes are fulfilled, the communication cannot achieve its desired effect unless it is properly formulated. It cannot be a random entity and requires proper thinking and planning. In order to become effective and useful, communication must fulfil certain conditions, known as Principles of Communication. There are seven C's of communication that must be borne in mind: Clarity, Consistency, Completeness, Conciseness, Consideration, Correctness and Courtesy. These, along with other principles for making communication effective are briefly discussed as under:

1. **Principle of Clarity:** Whenever an idea has to be communicated, attention should be given to its clarity and precision. The idea should not be confusing or vague. It should be easily understood by people at all levels of an organisation. No communication is said to be complete and effective unless it is understood by the receiver in the same sense with which it is conveyed by the sender.
2. **Principle of Consistency:** Whenever a person wishes to communicate a certain message, he must be consistent in his expression. It should be consistent with the overall objectives, policies, programmes and procedures of the organisation. There should not be any difference between a sender's statement and his action; otherwise the receiver may not be faithful in his acceptance of the message and its execution, and the sender will lose his credibility. The communicator must always try to take his subordinates into confidence and encourage them to follow his plan of action.
3. **Principle of Completeness and Adequacy:** Whatever is to be communicated should be adequate and complete in all respects. Insufficient or inadequate statements of communication may create misunderstanding in the mind of the receiver, resulting in delays. Consequently, original plans may not be successfully executed. The adequacy of the information also depends upon the power of understanding of the receiver.
4. **Principle of Conciseness:** It is necessary to put one's ideas, thoughts or feelings in a concise as well as precise manner so that the sender's and receiver's time is saved. It would also provide grace and beauty to one's speech and will help reinforce what one wishes to explain, whether orally or in writing. Unnecessary words reduce clarity of expression. Superfluous, irrelevant and inappropriate words can make the message unclear and incomprehensible. Repetition of words, sentences and thoughts should be avoid to minimize redundancy. It is always better to use simple sentences and words to make one's speech crisp and attractive.
5. **Principle of Consideration:** When a message is prepared, one has to consider the state of mind of the receiver – his background, his emotions or sentiments, his general approach when communicated to, etc. These things need to be paid special attention to especially while sending feedback or evaluation.
6. **Principle of Correctness:** Accuracy and authenticity of the message is a significant factor in transmitting messages. An incorrect message may create confusion in the mind of the listener and may ultimately spoil relations in the group or organisation. It is, therefore, necessary to take care of correctness of the message. There may be loss of time, money, energy etc. due to incorrect, untimely or delayed messages.
7. **Principle of Courtesy:** The idea of courtesy in transmitting a message includes polite manners, thankfulness etc. in conversation. In case mistakes occur, an apology should be made. As far as possible, harsh expressions should be avoided while preparing the message as well as while acknowledging the message received.

   [So far, we have looked at the *Seven Cs of Communication*. Apart from these, there are certain other principles of communication that must be followed. These are given below.]

8. **Principle of Information:** To start with, the sender of the message must have perfect clarity in his own mind about what is to be communicated. With proper understanding, comes effective communication. Communication is a systematic and continuous process of telling, listening and understanding. The communicator as well as the receiver should sincerely participate in the process of communication. There should be a perfect understanding between both the parties, i.e. receiver as well as the sender of the communication.
9. **Principle of Channel or Media Adaptation:** Appropriate channels of communication should be available and made judicious use of. The choice of a communication channel depends upon the content or matter of the message to be conveyed. For example, whenever the messages are lengthy, the written channel is suitable as it may not be possible to have face-to-face lengthy talks. On the other hand, simple messages can be conveyed through oral communication.
10. **Principle of Appropriate Time Element:** The sender of the message must consider the time element while conveying a message. Unsuitable communication time may result in failure of the communication. The system of communication should be continuous. Honest efforts on the part of the receiver and the sender are essential to make the communication successful.
11. **Principle of Integration:** In any organisation, it is necessary to prepare an integrated system of communication so that the enterprise can achieve its goals. Members of the management must give due importance to keep the communication frame, upward or downward, intact. Communication, in an organisation, functions like its blood vessels. As it is a means to an end, it should be free from any personal prejudices. It should not be offensive in its content or context, but should convey a sense of adopting the co-operative principle of "give and take".
12. **Principle of Flexibility:** Any organisation, no matter how big or small should have a perfect system of communication. It should be flexible so as to suit the changing requirements of the organisation. It should also be able to quickly and aptly adjust to new working techniques as well as new communication systems or methods without much resistance and difficulties.
13. **Principle of Informality:** The management of any organisation should ensure that formal communication must always be supported with informal contact. Informal communication helps to create a conducive and friendly atmosphere in the organisation.
14. **Principle of Feedback:** Communication is not a substitute for good management, but it requires good management to operate communication effectively and efficiently. Communication is a two way process which must provide enough opportunity for constructive criticism of the message from the receiver, leading to the emergence of creative ideas. The communicator must create an open atmosphere in the organisation. The organisation must create confidence in the minds of its employees so that they feel free to contribute to the working of the organisation. A proper feedback system ensures that there is a free and frank atmosphere in the company.
15. **Principle of Attentiveness:** While sending a message, all relevant details must be considered and included, in a correct and clear manner. No point should be left unattended. Communication will never be effective unless the receiver takes the message seriously.

Listening to the message and thinking about it, in other words 'active listening', is a must in the process of communication. In fact, communication is the joint reciprocal responsibility of both the sender and the receiver of the message.

16. **Careful Use of Body Language:** Body language plays a vital role, particularly in oral communication. Facial expressions and eye contact contribute in transmitting messages. They also help to create healthy atmosphere and helps gain confidence of the listener.

The principles of effective communication are given below in table form:

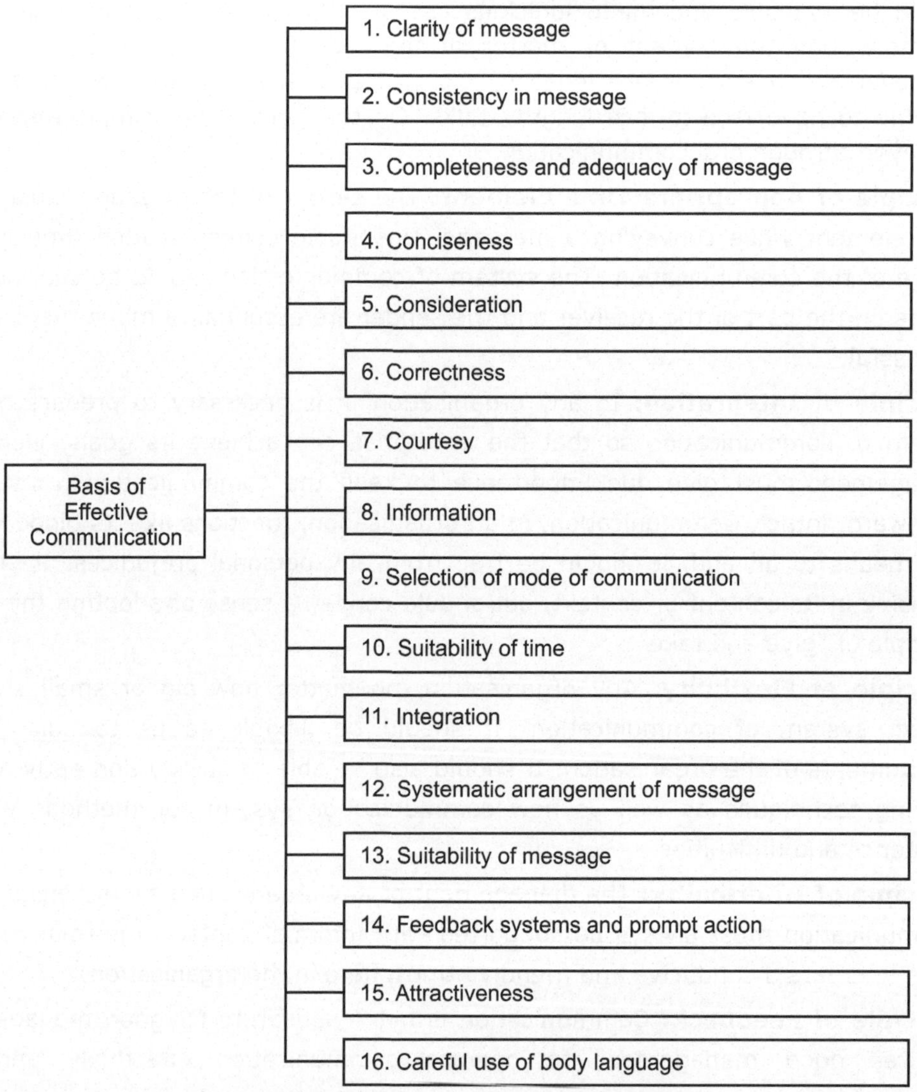

Fig. 3.1: Basis of Effective Communication

3.2 Communication Barriers and How to Overcome Them

- Communication is a complex, interactive process, involving assumptions and an unspoken agreement between persons. There are frequent errors and misunderstandings in the process of communication. Due to faulty communication, a large number of managerial problems arise. Poorly transmitted messages may lead to misunderstandings and frictions. This may have an adverse effect on the morale of subordinates. There are various obstacles or difficulties that come in the way of communication. They are termed as Barriers of Communication. They may be physical, mechanical, psychological, cultural or linguistic in nature.

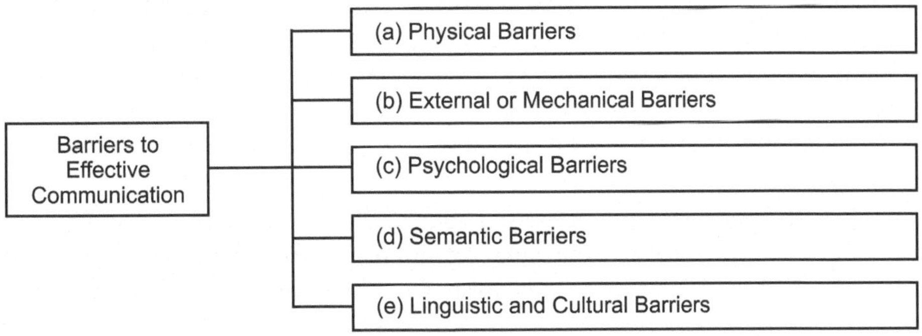

Fig. 3.2: Barriers of Effective Communication

Communication Barriers

(a) Physical Barriers:

In case of physical barriers, there are four main distractions:

1. **The competing stimulus:** In this case, another stimulus competes for attention of the receiver of the message. Examples of such stimuli are: another conversation going on within hearing distance, loud music, traffic noise etc.
2. **Environmental stress:** In this case factors like humidity, poor ventilation, strong glare etc. may disturb smooth communication, as the mind of the receiver is distracted.
3. **Subjective stress:** Sleeplessness, ill health, mood variations etc. disturb listening and may affect interpreting.
4. **Ignorance about the medium:** The use of a medium with which the receiver is not familiar, may become a barrier in understanding. For example, inability to read maps, graphs, charts etc.

Physical barriers can be on part of the sender of the message, as well as on part of the receiver. These are explained below.

Message Sender's Limitations:

1. The sender of the message may have limitations in use of language.
2. He may be providing unnecessary details without any logical order.
3. The voice of the sender may grate on the listener's ears.
4. The sender may be lacking in the art of composition of sentences. He may, therefore confuse the receiver.
5. His choice of words and their arrangement may not be appropriate and easy to understand.
6. The writing may be misinterpreted.
7. There may be too much of information in a single visual.
8. The design of the matter may not be simple and easy to understand.
9. The written material might not be properly arranged.

Message Receiver's Limitations:

1. The receiver may be inattentive.
2. The level of intelligence of the receiver to understand and memorise the message may be inadequate.
3. Each receiver enters into the communication process with ideas and feelings that will undoubtedly influence their understanding of a message, and their response.
4. The interest or the mood of the receiver can also be a barrier.

Physical or Physiological Barriers

- Physical defects in one's body are known to disturb communication. While communicating, an individual uses his/her:
  1. Vocal (speech) organs to produce sound/speech.
  2. Hand and fingers to write
  3. Ears to take in the spoken words eyes to absorb the written words

- Proper functioning of these body organs is essential for effective communication to take place. In case of any deficiency in any of these organs, the successful completion of communication will be difficult to carry out (speaking can be negatively affected by stammering, fumbling, utterance of improper sounds due to defective vocal organ/s, etc.) listening can be ineffective as a result of defective hearing. Deafness - total or partial- obstructs the absorption of sound signals. This leads to information loss.
- Writing can fail to express the message successfully because of illegible or unreadable handwriting due to hand injury, excessive trembling of the hand or fingers, numbness, etc. Reading can be affected due to poor eyesight. Defective vision impedes the reading ability of the receiver. Words may appear unclear, broken, overlapping, etc. to the receiver's eye.
- Therefore, Physical barriers also include Physiological barriers which can be caused because of bodily dysfunction. As mentioned above this can include hearing impairment, vision impairment and speech disorders. For an individual who is not able to hear, see or speak well or at all, communication becomes much more difficult.
- Communicating to someone who has a physiological barrier requires clear, direct messages, as someone with vision loss cannot perceive nonverbal cues and a person with hearing loss cannot hear the rate and tone of anyone's voice in expressing emotion.
- These barriers are related to personal, physical and semantic barriers. The challenges faced by those with physiological barriers and those communicating to persons with physiological barriers can be personal (e.g., bias against people with a disability), physical (e.g., being attentive about facing a person with hearing loss so one's lips can be read) and semantic (e.g., realising that those with physiological barriers may have diminished use of spoken terms, thus altering speech accordingly).

Physiological barriers thus include:

1. **Speaking:** Stammering, fumbling, utterance of improper sounds due to defect in speech, etc.
2. **Listening:** Hearing impairment, deafness, etc.
3. **Writing:** Illegible handwriting due to uncontrolled hand movements like trembling, numbness, etc.
4. **Reading:** Difficulty in reading due to poor eyesight, perceiving the written words as hazy, unclear and overlapping, etc.

(b) External or Mechanical Barriers (Machine Oriented):

There are certain defects in the devices used for communication. They are taken as external barriers, not within the control of either the receiver and/or the sender of the message. They are as follows:

1. Increase in difficulty in reception.
2. Defects in taking certain elements of the message to their destination.
3. The absence of communication facilities.
4. Disturbances interfering with the fidelity of transmission.
5. A defective telephone.
6. Cross connections often overheard (over) on intercom link.
7. Disturbances in the transmission channel:
   (a) In case of 'mass communication', failure of loudspeakers, disturbances etc. which are mechanical in nature.
   (b) Disturbances on the radio, ink spread on the newspaper, rolling of picture on TV etc. can be taken as barriers of communication.
   (c) Environment sounds like traffic noise, other conversations, door and window banging etc.

(c) Psychological Barriers:

The frame of the human mind through which one looks at others differ from individual to individual. Feelings like fear, desires, hopes, likes and dislikes, views, opinions are developed according to the social environment, education, family background, training and personal experience of the sender as well as the receiver. A certain individual acts in a particular way. So it is difficult to predict an individual's reaction because he/she might be influenced by different circumstances.

Self-image (self concept): The way a person looks at himself or at the picture he has of himself is called 'self-image'. It is this image which makes a person defend his point of view. The mind only retains information that is pleasant, or information which pleases the self-image, conveniently forgetting the information which is unpleasant, or which belittles the self-image. Following are the important points in respect to psychological barriers of communication:

1. **Wrong Assumptions:** There are certain assumptions which become barriers in communication. The speaker feels that he need not elaborate or explain certain points. This becomes a barrier in communication. For example, an engineer takes it for granted that workers understand various technical terms he uses. He may be wrong in his assumption.
2. **Wrong Presumptions:** Many a time, phrases like "It is well known" or "It goes without saying" are used. These and other terms like these may support wrongly presumed statements.
3. **Tendency of Generalisation:** Some people have the tendency to generalise. For example, someone may make a statement that 'Businessmen lack manners'. It is nothing but prejudice against a particular group.
4. **Wrong Judgements:** Sometimes, messages are interpreted in terms of the receiver's own frame of references. In that case, the receiver' judgment acts as a barrier in accepting facts.
5. **Snap Reactions:** A listener or reader may have the tendency to quickly or prematurely respond to the message. Such reactions are called as 'snap reactions.' They may prove to be barriers in communication.
6. **Tendency to Evaluate:** The main barrier in case of interpersonal communication is the natural tendency of humans to judge, evaluate and disapprove the statements of others.
7. **Group Identification:** Opinions are formed due to the influence of various factors such as a group, neighbours, colleagues, locality, city, religion, economic background etc. to which one belongs. For example, a person participating in a strike feels that to be in a group is more important for him, therefore she may not be willing to accept any communication contrary to the interests of his group.
8. **Self-Image:** When an image is created over years, it is difficult to accept any idea which goes against it. For example, consider a person who has an image of being efficient in performing his job. This person subsequently becomes an officer. As an officer, he would be required to control his subordinates and get the work done rather than doing it himself. However, if he fails to accept his shortfall in delegating work, it would affect his image adversely. In such a case, the image itself may become a barrier in communication between him and his subordinates.
9. **Closed Mind:** Intellectual background, narrow interest, failure in under-standing human nature, may become barriers to communication. Counselling can help overcome this difficulty.
10. **Poor Communication Skills:** Lack of writing and speaking skills prevents the speaker from encoding his ideas and attracting listeners. This skill can be improved upon by suitable training. Poor reading habits and faulty listening are psychological shortcomings. In case of oral communication, unfamiliar pronunciation, nervousness in facing an audience may become a major barrier in communication.

11. **State of Health:** State of health can also become a barrier in communi-cation. A person with ill-health has reduced efficiency of communicating; although temporary, this may prove a significant communication barrier. Reciprocally, a receiver may not properly respond for similar reasons.

- A barrier as has been mentioned before is anything that prevents something from getting through to its destination or receiver as intended. In communication, a barrier will prevent a message from travelling in a manner that gives it proper meaning. Although invisible, psychological barriers can be just as impenetrable as more obvious barriers, such as language differences, unless you become aware of them.

1. **Perception:** In a communication scenario, the communicators have to deal with two aspects of reality-the one as they see it and the other as they perceive it. The brain filters the message, that is, the words/symbols/signs, and assigns meaning to them, according to individual perception. Each individual has his/her own distinctive filter, formed by his/her experiences, emotional make-up, knowledge, and mindset that she /he has attained over a period of time. Because of this difference in perceptions, different individuals respond to the same word/symbol/sign based on their own understanding of the situation and assign meaning to it on the basis of their unique filter.

   Many a times, this variation in perception causes a communication gap, or an alteration, in the message. In face-to-face communication, this gap can be easily eliminated as there is immediate feedback. But in written communication, the gap between the intended meaning and the interpreted meaning remains unidentified, as the feedback is delayed or sometimes there is no feedback at all.

2. **Prejudices:** People with deeply fixed prejudices are very difficult to communicate with. They are not open to discussion or to new ideas, information, viewpoints and opinions. They have a closed mind and tend to react in a hostile way, thus ruling out all possibilities of communication. An unwilling to listen mind can, hence, be a big barrier to communication. To overcome this barrier, one should be open to new ideas and should learn to listen with thoughtfulness and an open mind.

3. **Halo/Horn effect:** Sometimes the listener may be too much in fear of or may completely distrust a speaker. In both these situations, the chances of success of communication are very less.

4. **Generation Gap:** The age, maturity, educational background and the eras in which a person grows up make a generation, which necessarily comes in the way of human communication. The generation gap becomes obvious in the use of language and style of speeches and the values of life to which they hold on to. Considering his age and maturity, an individual tends to apply different standards of judgement to judge the statements of the speaker. For example, in an organisation older workers gradually form their social group, which often remains apart from the younger workers. Their likings and interests are different and they take less interest in sports, parties and movies. Gradually, the older workers feel socially isolated and insecure because of the widening gap of communication between the older workers and the younger workers. This gap in itself sometimes causes conflict and poor communication because each generation is dissimilar and prefers its own mode for communicating. However, when employees learn to work together as a team, they just might be able to close the communication gap in the workplace.

5. **Phobia:** Feeling nervous in social situations is something everyone experiences from time to time. A racing heart, nervous muscle spasms and/or sweaty palms can affect adults and children alike. Extreme shyness, self-consciousness and a growing fear of embarrassment all restrict personal action and expression, and social phobia therefore can have a long-standing negative effect that limits emotional and social response.

Social phobia is a psychological difficulty that restricts personal experience, exchange and interaction with others. It is often misdiagnosed as depression and/or panic or personality disorder. For some people social phobia may present itself in situations that are public – for instance, as a fear of eating or drinking in the company of other people. Symptoms of social phobia may restrict the ability to hold down a job or influence behaviour around others.

Social Phobia, or Social Anxiety, is an automatic illogical thought process that triggers feelings of inadequacy that feeds negative action, expectation and appraisal.

Someone experiencing social anxiety may display self-conscious behaviour patterns or become restriction in social interaction with others. Sensitivity to public embarrassment, criticism and fear of making a mistake all add to the overwhelming sense of inadequacy that individuals experiencing social phobia may endure.

Social Anxiety may also limit personal experience by compounding feelings of loneliness and disappointment and of missing out of opportunities to interact in the way other people find it easy to do. Therefore everyday situations like chatting to colleagues at work can become an issue because social phobia fears begin to take hold.

6. **Status barrier:** Status is a position or social rank of a person in a group. It depends on the person's abilities, amount of pay, job-skills, seniority; type of work assigned, age, etc.

   Status reflects the degree of power, authority, importance and responsibility placed on an individual by the other people in the society. The people at the lower status are usually afraid of communicating unpleasant and unfavourable information to the high status people. They become obvious of their own status in relationship with the status of their superiors.

   This status consciousness is harmful in the process of upward communication. People fear that the unpleasant facts communicated to their superior might have unfavourable effects on them, if the information displeases the superiors. They are unwilling to communicate their problems, shortcomings, mistakes and other unfavourable information to the higher-ups because of their fear that the superior might consider them incompetent and unworthy to do their jobs. They do not show courage of offering suggestions and plans of improving the organisation and its procedures for the fear of being called arrogant by their superiors.

   The high-ups too are strongly conscious of their status. In order to protect the dignity of their status, they avoid accepting suggestions from the subordinates and presume that their higher status stands for better knowledge and competence than any of their subordinates. These assumptions prove serious barriers to communication between them.

7. **Attitudes and values:** The attitudes serve the personal needs of the people. They provide need satisfaction to the individuals. When the message is unfavourable to the receiver, he cannot be easily persuaded by it. The people in terms of their attitudes and values interpret the message. Their attitudes and values are different not merely because they are physically different but also because they have different backgrounds. They deal with the individuals and events according to their attitudes and assumptions. Their personal attitudes, values and opinions are the barriers to an effective communication. The most agreeable information for anybody of us is the one, which is favourable and palatable to our opinions, values, norms and attitudes. The message, which runs contrary to our views and beliefs, is not easily acceptable to us even when it is factual and true. We promptly accept the government policy if it is favourable to our business, but we express our strong dislike towards it if it adversely affect our business. Even the process of interpreting the message is consistent with the existing attitudes and values. It is due to the fact that our

thinking is colored and characterised by our attitudes and values. Sometimes, these attitudes and values can have emotional basis. Such attitudes are extremely difficult to change.

8. **Abstracting:** We use language to communicate our experiences and feelings, but we cannot communicate every detail of it. We cannot communicate every detail of our experience to others. Also, we focus our attention on some details and do not bother about the rest. We prepare a business report on our observations of the various events in the market. While preparing it, we abstract the reality and report only the valuable characteristics of the market. We observe partially and communicate partially because our experience of the event is also partial. When we try to convert our experiences and observations into words, we further abstract it by using selected words, which involve leaving out the details. If we try to completely describe a simple object like a 'shoe' we would require several volumes for it, which would still be insufficient to describe the object.

9. **Closed-Mindedness:** A person may close his mind to communicate receptions, if he considers himself to be a person who knows 'all' about a particular subject. It is very difficult to communicate with a man who has a deeply rooted prejudiced mind. Such a person is not prepared to receive any message on a subject about which he assumes to know everything. His mind is closed to new ideas, facts and suggestions. If an employee approaches his closes-minded boss with some suggestions to improve the work of a business unit, the boss would angrily reply to the employee by saying that he knows better than the latter regarding what should be done for the betterment of the organisation. Perhaps, he may further warn the employee that the latter should never try to teach him again. Thus, he completely rejects the information and recommendations of the communicator even before he knows the real facts. The reason behind his closed-mindedness is his deeply rooted prejudices.

10. **Distortion, filtering and editing:** When a message is transmitted through translations, interpretations, explanations and simplifications, some part of it gets distorted or lost. The accuracy of the message is lost and the transmission becomes imperfect as the message goes through the filters of translations and simplifications. The upward communication also tends to be distorted and filtered. The negative effects of the informal channel like grapevine are due to distortions and filtering. The message in grapevine receives fresh additions with every repetition until it gets worst. Thus, often the original information communicated through formal and informal channels gets lost or distorted to a large extent and very little of it is retained.

11. **Bad listening:** Bad listening is one of the major communication problems. Misunderstanding and conflicts can be reduced if people would listen to the message with enough attention. Most people do not listen very well due to various distractions, emotions, excitement, indifference, aggressiveness and wandering attention. One of the major reasons for bad listening is an individual's continual thinking about his own problems and worries. The poor listeners always feel that the thought in his mind is more interesting than what the speaker is saying. Bad listening can also be due to some strong reason for worrying. An employee may get engrossed in worrying about the sickness of his daughter rather than listening to the instructions given by his manager. Some listeners mentally argue with the speaker before comprehending the complete message. This usually leads to misunderstanding and conflict. Their impatience to talk out their thoughts and their lack of interest in the message contents are strong barriers to communication.

12. **Emotions:** Negative emotions are obstacles in the communication. Emotions are our feelings about the world around us. Usually, the positive emotions such as joy or affection do not interfere with communication, but the negative emotions act as strong barriers to

effective communication. Emotionally excited communicator is unable to organise his message properly. His excited or nervous state of mind does not allow him to think clearly. He expresses his blurred thoughts with gesticulations and keeps on repeating the same words. He cannot even grasp the message sent by the communicator in its true sense. This is especially true when one's negative emotion is uncontrolled and misdirected. It makes him blind to reason.

13. **Resistance to change:** If we receive a message, which proposes a new idea, we tend to be inattentive to it. The new idea is discarded consciously or sometimes unconsciously if it conflicts with our beliefs, morals, values, attitudes and opinions of the receiver. The average adult human mind ignores the new idea, especially when he feels insecurity and uncertainty about its aftermath. He feels that the things go along just fine with him and he would be insecure if the changes are introduced. He is also suspicious about its success in future. Because of its uncertainty, he quickly concludes in his mind that the proposal would not be successful. He even further feels that the proposal would make things worse for him. The new idea is considered as a proposal, which is not needed. Thus, the average human mind, which resists change, does not accept the new ideas from the communicator.

(d) Semantic Barriers:

An important means of communication is language and it must be used properly. Words having different shades of meaning and pronunciation should be used very carefully. For example, the word "record" can be used as a verb as well as a noun with a difference in stress but without any difference either in pronunciation or in spelling. The words like 'site, cite, sight' can cause misunderstandings. In certain cases, technical terms may not be understood particularly by those who are not associated with that field. All such factors are included in the term semantic barriers.

(e) Cultural Barriers:

Any language is the expression of thoughts, feelings and experiences in terms of the cultural environment. The same language when used in different cultures, takes a different colour and conveys a different meaning. Sometimes, language is modified according to the person with whom one is speaking. There is a difference in the use of language for discussions, formal talks, informal talks, etc. Language influences the behaviour of individuals and groups.

1. In certain cases, language itself becomes a barrier in communication.
2. Words are mere symbols and frequently convey more than one meaning, both in the mind of a sender and that of a receiver.
3. Symbols represent subjective and objective meanings.
4. Favourable and unfavourable arrangement of words depends upon the context in which it is used.
5. Communication difficulties arise when
   (i) One does not/cannot express what one means.
   (ii) One uses language, idioms, phrases etc. one may not mean.
   (iii) One speaks at the wrong moment.
   (iv) Words carry different meanings for different people, i.e. contextual meaning.

(f) Status Block:

A person, because of his status in a particular business enterprise, may develop a feeling that he knows everything about running that business. Such a person would not readily accept the fact that at times even his subordinates may have useful ideas which can contribute towards the success of the business. In such a situation, the useful ideas of the subordinates may go unheard on numerous occasions only because of the disparity in their social status. Thus, the status itself becomes a block and may be difficult to overcome from both sides unless an "open mind" policy is followed.

(g) Important barriers in communication can be summed up as follows:

1. Lack of clarity and precision
2. Semantic distortion
3. Premature evaluation
4. Inattention
5. Failure in communication
6. Fear or resistance to change
7. Distrust of communication.

Causes of Communication Barriers

- The barriers in communication have to be torn down or improved upon, in order to make communication effective. To do this, the causes of communication barriers need to be identified, understood and eliminated.
  1. Vagueness in communication is a major barrier in effective communication.
  2. Empty words, phrases, careless omissions, lack of coherence, repetition, failure to clarify implications, poor instructions are the main factors of bad communication.
  3. Lack of clarity and precision leads to costly errors and corrections.
  4. Managers receive different types of directives from superiors and also receive information from subordinates. A manager has to translate them both ways, i.e. he has to transmit superiors' directives to subordinates and the subordinates' feelings, thoughts and reactions to the superiors. At times, it is not enough to pass on the communication in its original form. It must be either put into words appropriate to the framework in which the receiver operates or else it would involve the risk of poor communication.
  5. **Transmission:** Successive transmission of a message generally reduces its accuracy. In case of oral communication, 30 per cent of the information is lost in each transmission. In case of a large organisation, it is not possible to rely upon oral communication from one level to another. Even written communication followed by interpretations are sometimes subjected to loss of partial meaning in transmission.
  6. **Inattention:** Sometimes carelessness in reading printed matter like notices, bulletins and minutes of meetings is a common feature in many organisations.
  7. Not listening, not paying attention to what is being said and not following directions seem to be the most important drawbacks of human behaviour.
  8. Overlooking assumptions made by a person without getting it clarified and non-communication of critical information may become a barrier in the way of effective communication.
  9. **Insufficient Time for Adjustment:** Some announcements have a vital bearing on the working conditions of subordinates. For example, the declaration regarding changes in shifts, place, orders etc., are very important. The subordinates require time to think and understand the full meaning of the message. It is not appropriate to force changes on the workers before they are able to adjust to such changes. This aspect is very important from point of efficiency.
  10. **Distrust of Communicator:** Sometimes superiors modify the messages that follow original communications believing that to be the right thing to do or due to inability to comprehend the message.. A superior might even alter the meaning of the message received by him due to wrong interpretation or poor judgement, thus corrupting the original communication. This leads to lack of enthusiasm, distrust and delays among subordinates.

11. **Premature Evaluation:** Those who wish to communicate must be listened to in an unprejudiced fashion before any response is generated. Otherwise, it becomes a barrier.
12. **Fear:** Managers must accept that they have to depend on their subordinates for information. Fearing the consequences of disclosure, subordinates sometimes mislead the superiors. This may become an important barrier in the way of communication.
13. **Failure to Communicate:** Laziness in transmitting messages becomes the main barrier in effective communication. One cannot communicate everything that is obviously necessary.

Effective communication is a process of meaningful interaction between human beings. It is fundamental for achieving successful co-operation and co-ordination in an organisation.

In brief, the main obstacles in the way of effective communication are the different types of barriers. They must be overcome by keeping oneself alert in all possible ways. Each type of barrier such as mechanical, physical, psychological, semantic or cultural has to be deliberately avoided.

How to Overcome Communication Barriers

- So far we looked at the different types of communication barriers. Each of these communication barriers can be easily eliminated, with the proper approach of the sender and the receiver.
- **Mechanical barriers** are mainly in the form of defects, failures or improper use of or failure in the mechanical devices (such as telephones, fax-machines, cell-phones, computers, radios, loud speakers, microphones, TV sets, mobile phones, etc.) that are used for sending messages.

  Solution: The sender has to take proper precautions to see that particular mechanical device being used to communicate a message is in proper order and will not create disturbances while in use. Factors like traffic noise, loud conversations etc. should be eliminated as far as possible. Ignorance about the proper use and handling of mechanical devices also acts as a hindrance. Hence the sender should have sound knowledge of the device he/she is using. Proper maintenance of devices ensures efficient and timely transmission of messages.
- **Physical barriers** are mostly related to physical shortcomings or drawbacks on behalf of the sender and/or the receiver. Ill-health, sleeplessness, fatigue, overworking etc. may prohibit the sender of the message to convey what he desires; it may even prohibit the receiver from understanding the received message properly.

  Solution: The sender of the message should ensure that he is in good health to function as the sender of the message. In addition to physical fitness, the sender should be careful enough about the language he uses, correctness of the terms he uses, precision in putting his thoughts, choice of words, capacity and intelligence level of the listeners and their limitations, use of theoretical jargons etc.
- **Psychological barriers** are large in number and are concerned with social, mental and personal environment. Self-image of the sender and receiver may also affect communication.

  Solution: the sender of a message has to keep a clear and open mind while transmitting messages, as should the receiver while receiving the message. Both should ensure that transmission and reception of the message does not get affected due to psychological factors, biases or prejudices.
- Appropriate use of language by the sender would help to avoid **semantic barriers** in communication, since cultural barriers also reflect on the quality of communication.
- Sending a message and making it effective is not an easy job. It requires study, concentration, patience, and proficiency in many aspects. Only then is it possible to make communication effective. The best test of effectiveness in communication is the quality of feedback one receives. To be very precise and brief, care must be taken to follow and adhere to the principles of effective communication and avoid different barriers in order to make communication quite effective.

System of overcoming barriers in communication:

1. Proper information to employees
2. Proper listening
3. Developing interpersonal relations
4. Communication through action and deeds
5. Use of proper language
6. Use of the grapevine (informal) communication
7. Feedback of information.

3.3 DEVELOPING EFFECTIVE MESSAGES

- The process of communication can be said to be complete when the sender gets feedback from the receiver and the feedback conforms with the transmitted message. This process is not as simple as it appears. It is affected by numerous aspects about the sender of the message, form of the message, content of the message, media chosen for sending the message, the nature and quality of the receiver i.e. listener or reader, group of people, audience, their knowledge, understanding and so forth. The effectiveness of the communication can be assessed only after the above cycle is completed. It will thus be seen that the following aspects are important in developing effective messages:
  1. Purpose of Communication
  2. Knowledge about the Audience
  3. Structuring the Message
  4. Selecting the Proper Channel
  5. Avoiding Barriers in Communication
  6. Facilitating Feedback.

(a) Thinking about the Purpose

Once a thought, idea or feeling occurs in one's mind, one thinks of making it known to others. Depending upon the nature of thoughts, ideas or feelings, one has to decide whom to communicate it to. A person may be required to address his feeling to an individual. For example, if he has seen a movie or play, read a great story or a novel, or had a unique or interesting experience, he may feel that he should share his pleasure with his friend, brother or sister, or a family member. He will tell his feelings of pleasure in a different way to each one. Supposing a leader in the society has certain ideas about public welfare, he may like to address a gathering of a group of people to make them aware of his plans. For instance, a company manager who intends to make his staff know about the new factory rules, working of newly installed machinery, financial benefits to the workers, etc. will see that the relevant people in the company are informed about his thoughts, plans or ideas.

(b) Knowing the Audience

The method to be adopted to communicate is decided by sender according to the standard, status, knowledge, temperament of the receiver. Thus, before one gets involved in the process of communication one must understand to whom one is to address to, so that the success of one's message can be assured. If this is not done, it is likely that a person's efforts or intention to give vent to his thoughts, feelings or ideas would become futile. It may lead to waste of time, energy and money of all concerned. To achieve success in sending any communication, one must carefully think of the audience. He should choose the correct audience for listening or receiving his communication.

(c) Structuring the Message

Once if the sender is clear about the message to be communicated, and the person he wants to communicate it to (i.e. the receiver), it is necessary to choose the proper structure, format or framework for the message. This structuring of the message decides the success of the communication. The messages can be conveyed in oral or written form.

Oral form takes the nature of face-to-face conversation, telephone or cell-phone talk, audio-visual conferencing, dictation in a dictaphone, etc. These conversations should be clear, precise, appealing, courteous, mannerly and pleasant. If the conversation is appealing and convincing, it will have a positive outcome. Half the work is done if the conversation is successful.

The other format is the written one, which includes such written formats as circulars, memoranda, notes, drafts, letters, reports, notices etc. These formats are more or less decided by conventional methods. The content of these documents has to be very clear, precise and understandable. Language used should be such as would not hurt the feelings of the reader. They should be in a simple style and in an organised manner. They should not be monotonous or the reader may lose interest in the message. A properly and clearly worded, brief, precise, to the point message, helps to create goodwill and does not create doubt or confusion in the mind of a reader. Such carefully drafted documents effectively influence readers. Success of any organisation and its image in the public largely depends upon the quality of correspondence. Progressive companies believe in getting positive and favourable responses if they write in a personal and informal manner rather than in old fashion and pedantic jargons.

(d) Selecting the Proper Channel

Unless one chooses the proper channel of communication to convey a message or an idea, the purpose of communication would not be served. Communication can take place by adopting different channels. It may be verbal or written.

Electronics has brought a revolution in information technology. There are many media which can easily pass on messages. Each one has its own merits and demerits in a given situation. Selecting a suitable medium of communication depends upon the nature and urgency of sending messages.

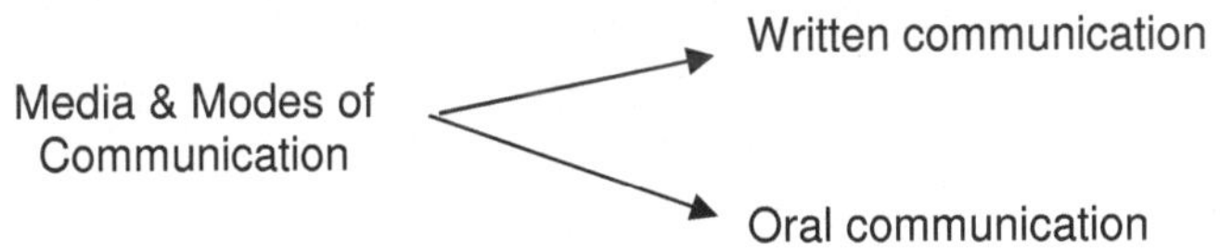

Written communication is transmitted by mail, telegraph, telex, fax, e-mail, newspapers, notice boards etc. Oral communication can be passed on by air vibrations when it is face to face or by microphone, loudspeaker, telephone, radio, cinema, TV etc.

(i) Postal service: This is a government owned network service which usually uses rail, road and air passages to link people within and outside the country. Various types of services are available from the postal departments; for example, mail, registered mail etc. It also includes Quick Mail Services like express delivery, speedpost etc.

(ii) Telegraph: In most countries it is a government owned network. It works by transmitting different letters of the alphabets. It can be sent by express, ordinary reply-paid telegrams. It gives the impression of urgency and is used when urgent action is expected.

(iii) Telex: It is an electronic device in which the message at the sender's end is typed exactly the same at the receiver's end. A machine is fitted with a roll of paper and messages can be transmitted continuously. The main advantage of this machine is that no person is required to attend to it. It types the messages automatically. The received message can be acquired and read even after a lapse of time. In certain cases, telex messages can be communicated through the screen, when moment to moment information is required to be transferred, for example, in the newspaper office, share markets etc.

(iv) Telephone: It is the most useful medium of transmitting information for oral communication. It allows two or more persons to exchange their views even though they are not at the same place. Intercom and STD facilities are now available in all major cities of the world. It is an instrument which can be used for instant communication. In certain advanced countries, conference facilities are also available. Cordless and wireless telephones and cell-phones (mobile phones) enable people to contact one another even while they are in transit. It transmits not only the words expressed but more significantly it transmits the tone, and the pitch of the voice of the speaker. Telephone communication is very personal. In certain cases, it is cheaper than written communication. In this case, immediate feedback is available and a lot of valuable time can be saved. Telephonic communication does not, however, provide a record of what is transacted and, therefore, the chances of misinterpretation are high. The receiver in telephonic talk gets a chance to listen and interpret it. It is however, possible to know the right frame of mind of those in conversation on the phone. In several cases, there is a possibility of breakdown in communication channels which is more frequent particularly in rainy seasons. Disturbances can also occur due to cross connections because of environmental distractions. For some people, conversation on the phone is a skill, very few people can master it and for some it serves as a life line.

(v) Telegrams: In this case the use of language in the printed form must be precise and accurate. Unnecessary words and descriptions are avoided to make it quite brief. The word 'STOP' is used to avoid confusion. All these are managed by Telecommunication Department and they are charged according to the number of words used in the message. Fax messages are charged according to the transmission time it covers. Telex and fax facilities are now available throughout the country and outside the country as well. In this case, the receiving operator's presence is not required. Telex messages can be received at any time of the day and can be collected later. Hence it is most convenient for organisations to send actual information where immediate feedback is expected. The main advantage of Telex and Fax are their quickness and accuracy. Fax is a machine which produces a photo copy of a document on the receiver's machine when it is inserted at the sender's end. It reproduces the entire document from one end to another end and it is done almost instantaneously.

(vi) Radio: Radio is a medium of mass oral communication. News, notices, announcements, advertisements, songs, conversations etc. are transmitted to the public by radio communication. It permits only a one WAY mass media communication.

(vii) Television (TV): TV has become a medium of mass communication. It has the advantage of audio as well as video facilities. It has greater impact on people than that of radio. It, however, allows only one way communication.

(viii) Film: This has become the most powerful medium of communication these days. It is an audio-visual medium. It combines all possible forms like written, oral and visual. A film can be shown either in a theatre or on video.

(ix) Newspapers and magazines: These are mass communication media used by the reading public and also by organisations for inviting tenders, applications for recruitments, notices etc. It keeps the public in touch with various activities of day to day life.

(x) Posters and bill boards: These are used to convey simple ideas. They convey short messages. Location of posters plays an important part. They are displayed at places where it can easily catch the attention of the public. They are used for mass publicity and mass communication.

(xi) Notice boards: These are used in organisations where internal communication is important. Notice boards are required to be properly located so that the people concerned can conveniently read them. The notices displayed should be simple and properly worded so that the matter is grasped quickly and easily.

(xii) Courier service: Courier services these days are under private sector agencies. They collect the matter and deliver it to the person at their exact destination. Although the cost of providing such service is a bit high, people prefer it, due to the speed and certainty of delivery.

(xiii) Hand delivery: In certain cases business organisations make their own arrangements to deliver their letters and articles through their own employees. It ensures prompt delivery. Even though cost(s) may be little high, it is used when reliable and quick delivery with proof or acknowledgement is required.

(xiv) Photographs: Photographic mode of communication is used to recognise the person and to verify his identity. It is a visual presentation which is often more effective in representing information.

- Oral communication, as discussed earlier, may be face-to-face conversation i.e. between person to person, person to group or through different modern devices such as the telephone, cell-phone, audio-conferencing etc. For all these, there must be something common between the sender of a message and the receiver of the message. They must be using common devices with which both are conversant. The choice of channel also depends upon the purpose for which it is used. Choice of a wrong channel can defeat the purpose of communication. For example, if detailed and accurate information is to be passed spoken words, face-to-face conversation or conversation through telephone etc. would not be useful. It has to be communicated through circulars, bulletins etc. In modern days, technical papers or presentations are necessary to be supported with visual aids. Without the knowledge of these devices the receivers of the message would be confused. Time element also plays a significant role in the choice of communication. Sending a letter by the postal channel may not serve the purpose because of the time taken by postal delivery. On the other hand, sending the same message by e-mail would save time as the message reaches the receiver almost instantaneously. However, both the senders and receivers must be connected by internet devices. Hence it would be an important factor to choose an appropriate channel of communication to serve the desired purpose.

(e) Minimising Barriers

It has already been seen that there are many barriers in communication which come in the way of effective communication. In order to make communication effective and useful for the achievement of objectives of an organisation, efforts should deliberately be concentrated on avoiding or controlling these barriers. This can be done through several measures, which can be briefly discussed as follows:

Remedies to overcome barriers are briefly enumerated in the following table:

Types of barriers	Nature or forms of barriers	Remedies to overcome the barriers
1. External barriers	• Not in control of the receiver and sender of the message. • Occur due to noise. • Time gap. • Physical distance.	It can be overcome by making messages brief and precise. Proper and timely maintenance of the devices used.

contd. ...

2. Physical barriers	• Occur due to distance. • Overload of information. • Inability to read maps, graphs, charts etc.	
3. Psychological barriers	• Attitudes and opinion disturb the communication. • Emotions block the mind • Occurs due to close mindedness and in- attention. • Distrust and poor retention also leads to it.	Be open minded. Encourage subordinates to express. Give complete attention. Listen patiently.
4. Linguistic and cultural barriers	Different cultures give different meaning to the same word. One may not also be able to express what one means.	Proper judgement of the audience or listeners.
5. Organisational barriers	Occur due to long chains of command, poor layout, unsuitable medium.	• Reduce chains. • Proper layout. • Reduce fear in the mind of subordinates.
6. Semantic barriers	• Wrong interpretation of words. • Can occur due to different meaning and shades of words. • Poor expression. • Defective interpretation.	• Use words which are familiar to both sender and receiver. • Clarify the shades and meanings of the words. • Use words with exact meanings. • Translate messages correctly. • Give sufficient practice or training.

(f) Facilitating Feedback

Feedback can be taken as response from the receiver to the sender of the message. The process of decoding commences when the message reaches concerned person. At that time, the sender expects that the receiver of the message gives some kind of response. This response in the form of the receiver's reaction helps the sender to assess how much part of the message is understood by the receiver. The sender certainly wishes that the receiver fully responds in the desired way. It is possible to know only by the way of feedback. It can be stated that feedback is the knowledge of the result of communication, which is essential for knowing or modifying further communication. Hence, provision should be made to facilitate the process of feedback so that the sender can adjust further communication without much difficulty.

- Following are the important points which can be useful for facilitating the process of feedback :
  1. There should be face to face conversation which can give immediate or continuous feedback.
  2. Facial expressions of the receiver can provide immediate feedback to indicate whether the message is fully understood or not.
  3. The speaker can modify his further communication according to the need of the feedback.
  4. The speaker can change the ways and means by which the receiver can get the complete information.

5. In certain cases the speaker can ask certain questions to facilitate the feedback process.
6. The receiver should also be given an opportunity to get clarifications so that the feedback process can be easily accelerated.
7. In face to face contact, the speaker can influence the listener and can get the required feedback spontaneously.
8. In certain cases, during a conversation, nodding the head, clapping, smiling etc. can help in the process of feedback.
9. Inattentiveness, yawning, boredom etc. can be taken in the negative sense but it is also a kind of feedback in conversation.
10. Facial expressions of the receiver plays an important role in the feedback system.
11. In case of written communication feedback is a bit slow, as it takes time to get back written reactions.
12. In mass communication, feedback is still slower.
13. The receiver can check up on his understanding if the perfect feedback system is provided.
14. Summarising the message can be tactfully applied in the feedback system facility.
15. The other method is to encourage the receiver to write questions. As it is conditional here, the receiver can also ask the question in case he understands them.
16. Persons working on higher posts cannot get immediate and correct feedback from their subordinates because of fear.
17. One should be a patient listener so that one can get feedback in due course of time.

Writing Articles

- The basic objective of writing articles for newspapers and magazines is to communicate views, ideas and opinions to a large number of readers. The practice of writing articles gives an opportunity to students to develop and improve the power of expression. People engaged in the technical field are no exception to this rule. They are constantly required to communicate to others in the form of technical articles, notes, reports, processes, functions and tools and machines, preparation of work projects etc.
- Although there is no precise definition of article writing, it is generally considered as a group of sentences and paragraphs which deal with one unit of thought forming a definite state in the development of the subject or the theme. Article writing, technical or general as the case may be, involves two important points viz.
  (i) An article deals with only one topic; and
  (ii) The topic is a significant and important part of the subject or theme as a whole.
- One should be aware that while writing, one is concerned with different units of thoughts which vary considering the degree of their importance. In fact, the theme or subject is itself a unit of thought. However, it consists of different paragraphs bringing out the important features of different aspects of the theme. Similarly, each paragraph is also a group of sentences, each one dealing with a still smaller division of a thought. The paragraph of an article therefore differs from a group of sentences in that it covers a large topic that carries discussion forward through the definite stage.
- While standing to write an article, the topic of the article is often stated in a brief sentence, called the 'topic sentence' which sums up the entire content of the article. The topic sentence is generally written at the beginning. However, there is no hard and fast rule about its place. Sometimes, it can also come at the end or may find place in the midst of the article. The topic sentence is found in the beginning so that there is no possibility of wandering away from the main subject of an article. The topic sentence should give an exact statement of the subject matter it covers. The technical article must contain a discussion of the subject stated in the topic sentence.

- An article is a group of sentences and paragraphs. It follows that every sentence in a paragraph must bear a relevance to the topic and must contribute something worthwhile to its developments. Thus, the thoughts are presented as a unit which can be readily understood by the reader while reading an article. This is possible only if the development of an article occurs in a logical order, where one thought leads to another so that the flow of the thought is kept disturbed. For achieving coherence or continuity of thought, certain conjunctions such as hence, but, and, then, therefore, so are very useful. For bringing variety, liveliness and making the article interesting, the sentence pattern must vary. Technical students must try to write simple structural patterns and achieve as much variety as possible.
- As a general principle, one should avoid frequent use of short paragraphs, each consisting of only a few lines. This leads to the possibility of dividing a significant topic into two or more groups, each one dealing with comparatively unimportant details. As a result, the pieces appear to be a series of jottings or notes rather than a well-organised discussion on a significant topic. Hence, the technical student should try to combine and bring together the sentences which logically belong to one another forming one article. All those groups of sentences which logically belong to topic should appropriately amplify it. This will also help to maintain a logical sequence of the development of the theme or subject.
- The end of a paragraph, like the end of the whole article, should give the impression of completeness. An ending makes a good impression of the article keeping its impact on the minds of the readers for a longer period. Hence, the concluding sentences of an article are as important as the opening sentences. The first sentence should arouse the reader's interest and the last sentence should satisfactorily conclude the topic. The last sentence should sum up the contents of the article or make some brief commence on what has gone before. It must make the reader feel that the article is properly concluded.

Hints for good article writing:

1. Select the proper subject of an article.
2. Think about the topic of an article from all angles.
3. Ask yourself the questions – what, why, where, when, which, whom and collect certain answers.
4. Write answers to these questions.
5. Arrange them in a logical order.
6. Connect all sentences with proper conjunctions.
7. Think about the beginning and ending sentences of a paragraph to complete the article.

- Start writing article keeping in mind the following points:
  (i) Unity of thought
  (ii) Statement of topic
  (iii) Logical order of thoughts
  (iv) Coherence
  (v) Arrangement of article
  (vi) Length of the article
  (vii) End of the article

Important Points

- **Introduction**
  1. The main objective of communication is to make others know what one thinks or feels.
  2. Purpose of communication differs from person to person.
  3. Personal communication differs from organisational and business communication.
  4. Business and organisational communication should be exact and not vague or random.
  5. Objectives in business can be achieved through communication.
- **Definition of Effective Communication**
  1. Effective communication is a vital process.
  2. Effective communication is the life blood of any business.

3. Attributes of effective communication:
   (i) May be intentional or unintentional
   (ii) Dynamic
   (iii) Systematic
   (iv) Interactive and transactive
4. Effective communication requires proper planning and thinking.
5. Conditions of effective communication:
   (a) Clarity
   (b) Completeness
   (c) Conciseness
   (d) Consideration
   (e) Correctness
   (f) Courtesy

- **Principles of Effective Communication:**
  1. Clarity
  2. Precision
  3. No vagueness/should be precise.
  4. No confusion
  5. Understandable by all
  6. Completeness
  7. Sincere participation of receiver and communicator
  8. Proper channel
  9. Suitability of time
  10. Integrity
  11. Co-operation
- **Principles of Effective Communication are:**
  (a) Flexibility
  (b) Informality
  (c) Feedback
  (d) Consciousness
  (e) Attractiveness
  (f) Correctness
  (g) Consideration
  (h) Courtesy
  (i) Body language.
- **Communication Barriers**
  1. Communication is a complex interaction between a sender and a receiver.
  2. There is possibility of misunderstandings.
  3. There are many obstacles in communication.
  4. Barriers can be
     (a) Physical
     (b) Mechanical
     (c) Psychological
     (d) Cultural.
     (a) Physical barriers which disturb the conversation (traffic, noise etc.)
     (b) Mechanical barriers which are not under the control of a sender or a receiver (defective telephone and so on)
     (c) Environmental and other barriers like moods, etc.
     (d) Ignorance about the medium
     (e) Sender's and receiver's limitations

- (f) Psychological barriers (frame of mind, desire, dislike, fear etc.)
- (g) Self-image (self concept)
- (h) Wrong assumptions
- (i) Tendency of generalisation
- (j) Interpretation of message
- (k) Influence of groups and others
- (l) Self -image which goes against a listener's will
- (m) Status and feelings
- (n) Failure in understanding human nature
- (o) Inefficiency in reading and writing
- (p) Use of words with different meanings
- (q) Cultural barriers: Language influences behaviour of groups and individuals
- (r) Feedback helps in reducing barriers

- **Important Barriers:**
  - (a) Lack of clarity
  - (b) Semantic distortion
  - (c) Premature evaluation
  - (d) Inattention
  - (e) Failure in communication
  - (f) Fear of change
- **Overcoming the Barriers in Communication:**
  - (a) Proper information
  - (b) Proper listening
  - (c) Development of interpersonal relations
  - (d) Communication through actions and deeds
  - (e) Use of proper language
  - (f) Use of grapevine (informal communication)
  - (g) Feedback.
- **Barriers in Effective Communication**
  1. Vagueness in communication
  2. Empty words
  3. Lack of clarity
  4. Improper transmission of message
  5. Non-listening
  6. Insufficient time for adjustment
  7. Distrust of communicator
  8. Premature evaluation
  9. Fear in the minds of subordinates
  10. Laziness in transmitting messages.
- **Developing Effective Messages**

  Thinking about the purpose
  1. Man is a social animal involved in the act of communication.
  2. Feelings are transmitted.
  3. Feedback confirms thoughts.
  4. An effective message must have:
     - (a) Purpose
     - (b) Knowledge of audience
     - (c) Proper channel
     - (d) No barriers
     - (e) Feedback

- **Knowing your audience:**
  1. Speaker must understand his/her audience.
  2. Selection of proper subject.
  3. Beginning of the process of communication.
  4. Success of message can be assumed.
- **Structure of the message**
  1. Speaker must choose proper structure, format and framework.
  2. Message can be conveyed orally or in the written form.
  3. One has to make the conversation successful.
  4. Speaker should not hurt anyone during conversation.
- **Selecting the proper channel:**
  1. Basic object is to make communication understood by others.
  2. Verbal, oral or written channels are to be selected.
  3. Electronics has revolutionised information technology.
  4. Information technology has revolutionised the transmission system.
  5. Postal services, Telegraphs, Telex, Telephone, Telegrams, Radio, T.V., Films, work as powerful media of communication system.
  6. Newspapers, Posters, Notice boards, Courier services are also important means of communication with respect to the (educated) public.
- **Minimising Barriers**
  1. There are many barriers in communication which come in the way of effective communication.
  2. They can fall in one of the following groups:
     (a) External or Mechanical
     (b) Physical
     (c) Psychological
     (d) Linguistic and Cultural
     (e) Organisational
     (f) Semantic
  3. Vagueness in communication is an important barrier.
  4. Language skills reduce barriers.
  5. Lack of precision leads to failure in communication.
  6. Effective communication depends upon the method of transmission.
  7. Thirty percent information is lost in each transmission.
  8. Encourage attentive listening.
  9. Avoid laziness in transmission.
- **Facilitating feedback**
  1. Facilitating feedback can be considered as responsiveness.
  2. Feedback is the result of communication.
  3. The process of feedback is to be facilitated and encouraged.
  4. Face-to-face conversation is always more effective.
  5. Certain questions can be asked for receiving the correct feedback.
  6. In face to face conversation, the speaker can influence the listener.
  7. Clapping, smiling, nodding etc. are expressions of feedback.
  8. A negative mood, boredom, yawning etc. indicate a negative attitude.
  9. Facial expressions are also signs of feedback.
  10. Fear in the mind hampers the process of immediate feedback.

Practice Questions

1. State the importance of effective communication in the engineering career.
2. Define effective communication and state the importance of communication for the development of your career.
3. What is effective communication ? What are the different meanings of the term ?
4. Explain in brief the major objectives of effective communication.
5. State, in brief, the different features of effective communication in fulfilling the objectives of an organisation.
6. What are the important attributes of effective communication ?
7. Explain, in brief, the different basis of effective communication.
8. State the different types of barriers of communication and how they can be overcome.
9. What are the different barriers to communication ? Explain mechanical barriers and physical barriers.
10. Explain, in brief, physical barriers and psychological barriers.
11. Explain, in brief, the concept of 'self-image'.
12. Explain, in brief, psychological and semantic barriers.
13. Explain, in brief, cultural and semantic barriers.
14. Explain the system useful for overcoming barriers to communication.
15. Describe how the barriers to communication can be overcome in order to make communication effective.
16. What are the different barriers in effective communication ?
17. Enumerate the barriers of communication in a classroom situation. Find out how they can be overcome.
18. State the importance of effective communication.
19. Explain the idea of 'Developing effective messages'.
20. Explain the concept of "Knowing the audience".
21. How can the proper channel and structure of the message be selected while communicating a certain message ?
22. What are the various channels used in today's modern age to make communication effective ?
23. How can the various barriers of communication be minimised ?
24. Explain the idea of feedback in communication.
25. Write short notes on:
    - (i) Effective communication
    - (ii) Basis of effective communication
    - (iii) Communication barriers
    - (iv) Physical barriers
    - (v) Mechanical barriers
    - (vi) Psychological barriers
    - (vii) Semantic barriers
    - (viii) Cultural barriers
    - (ix) Self-image
    - (x) System of overcoming barriers
    - (xi) Failure of communication
    - (xii) Effective messages
    - (xiii) Knowing the audience
    - (xiv) Selecting the proper channel of communication
    - (xv) Minimising barriers
    - (xvi) Feedback.

■■■

Chapter **4...**

Non-Verbal Graphic Communication

Contents

4.1 PRINCIPLES OF NON-VERBAL COMMUNICATION

- As can be seen from the above, it is necessary to understand the principles of non-verbal communication, as it accentuates, reinforces or even replaces the verbal message, is visible and can be subject to interpretation by the receiver, either as supporting the verbal communication or not. Hence to use this mode effectively, the communicator has to understand these principles and avoid giving wrong cues. Briefly stated, they stem from the basic propositions that it is necessary to understand the significance of non-verbal communication as the second communication and align it with intent, thought and words. The key characteristics of an effective communicator are openness, connection, passion, and listening. It is necessary to remember that "Actions speak louder than words."

1. It is necessary to **align body language to verbal communication** to make the communication effective. If the two are not aligned, or they give contradictory messages, the receiver will focus more on the non-verbal aspect and believe those cues than what is being said. The receiver will interpret the intent and emotions of the sender through his/her facial expressions, posture, his kinesics etc. These are extremely transparent and it is not easy for the sender to pretend that the situation is otherwise. A speaker may begin his speech with impressive words, but if he is nervous or is not sure of himself, his expressions will betray him easily. That is why, before a speech, effective speakers take some time to take a good look at the audience, focus on the emotions they want to convey and form their words and their body language accordingly. To catch the imagination, interest and attention of the audience, they start with a joke or a story or an anecdote, some statistics etc.
2. Before beginning the speech, it is necessary to **establish an open relationship with the audience**. It is natural for people to start evaluating the person in front of them and developing a relationship even before any words are exchanged. They do it very fast, spontaneously and unconsciously. The two parties to the communication may not be aware of this exchange of non-verbal communication, but even before the speaker utters a single word, his success or failure as a speaker is determined. Hence it is necessary to develop an open relationship with the audience, whether it is a single person or a large group. Only then can he be able to convince the audience and persuade them to think in the way he wants them to. This can happen only when verbal message and body language are in alignment. If you send out mixed messages, your audience will pick up on them unconsciously and resist you because you do not appear honest to them.

3. ***Be authentic*:** By aligning body language with verbal communication, the speaker conveys authenticity and people will listen to him. People do not trust those who tell lies or pretend in any manner. If there is basic lack of alignment between content and body language, it does not inspire trust and does not get read as authenticity. The door to the audience is closed.
4. ***Decision making process is basically an emotional process and so emotions should be communicated and not hidden.*** Human brain links memories to emotions, depending on how significant the memories are. These emotions are experienced non-verbally. The speaker tries to influence the audience by appealing to their emotions. Thus, the source of non-verbal communication is emotions. When we evaluate the environment and the speaker, it is done non-verbally and unconsciously. We look for signals that indicate comfort level in the environment and try to connect with it. We unconsciously size up the speaker. The conscious part comes in much later.
5. ***Non-verbal communication takes precedence over speech.*** This can be seen from the sequence in a speech: intent (what the speaker wants to communicate), gesture, thought and lastly, speech. But if the speaker consciously thinks about what gestures he will use, it becomes totally unnatural, as the gesture will happen too late. The sequence then will be: thought - speech - gesture. Since people evaluate gestures unconsciously, they will unconsciously notice that the speaker is not authentic or is putting up a false performance, or is foolish. (Students could check this out in front of the mirror and notice the difference.) The fluidity of the speech is lost.
6. ***The speaker should be conscious of his body language and try to master it.*** In an informal meeting where comfort levels all around are very high, such care is not necessary. But if the stakes are high, it is not just about the speaker, but the audience as well. If the speaker's body language conveys tension or fear, the emotions are picked up by the audience as well. While facing the superior or the company board or while negotiating which makes or breaks the company, the speaker has to be mindful of what his body language communicates. As the saying goes, 'see yourself as others see you'.
7. ***It is necessary to be aware of the body language of others*:** People use body language to signal their intentions much before they express them verbally. Sometimes, only the body speaks. They speak first and they rarely lie. Words may fail their speakers. An effective speaker does not depend on words alone. Hence it is necessary for a good communicator to consciously learn to read and understand one's own and others' body language and react accordingly.
8. ***It is necessary to have control over one's own emotional intent and the appropriate gestures will follow.*** The speaker should understand the emotions he wants to convey and express it forcefully. This applies to business speakers as well. Emotions are literally infectious. They are picked up from the people around us. While promoting a product, the appeal is not just to the logic but to the emotions of the consumers as well. The enthusiasm, anger, joy or sorrow expressed by the speaker spreads rapidly to the audience as long as the emotion displayed is not a fake. A successful speaker will think of the emotions he wants to convey and rehearses the speech along with gestures, movements, effective pauses etc.
9. ***The emotional intent has to be matched by conscious thought and appropriate words.*** Very often in the organisational environment, out of respect or as a norm, people may not show their real emotional intent in the words they express. But if the stakes are high and the speaker is being watched by many people, the two need to be properly aligned. It may be a crucial business meeting, where a lot of funding decisions are to be made. It may be a major negotiation process. It may be a meeting with employees to

congratulate their team on their significant achievement. There may be many such situations. The speaker will have to focus on his emotional intent and choose appropriate words. The speaker will have to be alert to both emotion and word choice for a longer period. All this takes a lot of practice and distinguishes a man of experience from a novice in this area. Terms like 'charismatic speaker' refer to people who can be passionate, open and capable of influencing the thoughts and behaviour of the others.

10. ***An effective speaker shows that he is open, authentic, connected and at the same time passionate about what he is communicating.*** *He is a good listener as well.* As has been said in the previous sections, all this comes from effectively aligning the emotional intent, body language and words. In such a person, the emotions and the message are perfectly aligned. He believes in what he is saying. Such a person listens to the other party carefully and responds properly.

4.1.1 Non-Verbal Codes

(A) Kinesics

- The study of body motion and facial expressions as related to speech is called *kinesics*. It is an important factor in non-verbal or oral communication.
- Kinesics is non-verbal behaviour related to the movements of a part of the body or the body as a whole. Generally, all communicable body movements are classified as kinesics. Kinesic communication is one of the most important non-verbal communication forms. Communication by way of body movements is endless. For example, when the forefinger and middle finger are indicated in the shape of the letter 'V' it becomes the sign for victory. In certain cases two fingers are used to symbolise number two. Sometimes '0' may be indicated as OK, in other cases it is taken as a zero.
- There are some non-verbal signs that regulate and maintain the flow of speech during meetings or conversations. These can be both kinesic, such as nodding of the head, as well as non-kinesic, such as eye-movements. The information which flows as a part of feedback to indicate whether the receiver has understood the message can be highly confusing.
- Kinesics is an important part of non-verbal communication. The movement of the body conveys a specific meaning and interpretation. There are movements which carry a significant risk of being misinterpreted in certain cases of communication.
- Body language is a comprehensive term used for the method of communication through different parts of the body other than the tongue. It includes, all sorts of body movements, gestures, facial expressions, postures, miming, touch etc. depending upon the physical space between the communicators. Silence as a body communication also plays an important role in certain exchanges. Body language, by and large, is unintentional and more truthful than verbal messages. It is easy to deceive a person by expressions.

Body Language

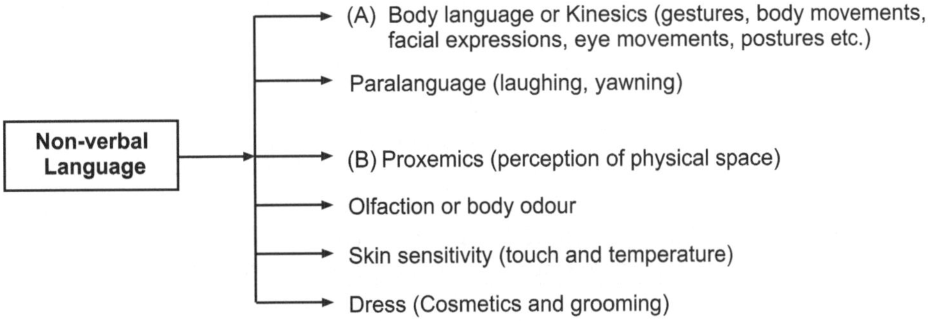

(B) Proxemics

- There are several non-verbal signals which are understood by the sender and the receiver. One such source is space. It communicates in its own way. *The degree of perceived physical or psychological closeness between people is called proxemics.* The term proxemics was introduced by anthropologist Edward T. Hall in 1959 to describe set measurable distances between people as they interact. Communication experts call this dimension of communication, proxemics. It can also be called as 'space language'. In this context, space means distance between the sender of the message and the receiver of the message in oral communication. Space can be classified into various categories according to distance.
  1. Intimate (soft whispers, pat on back etc.)
  2. Personal (close circle, shake hands etc.)
  3. Social (used in official relationships)
  4. Public (formal and objective)
  5. Fixed and Semi-fixed (arrangement of furniture).
- Broadly, Intimate (0-2 ft.), Personal (2-4 ft), Social (4-12 ft.) and Public (more than 12 ft.). When strangers enter the wrong zone, we feel uncomfortable. E.g. In an airplane / elevator, this space gets compromised and we deal with it by "dehumanizing" (making no eye contact and acting busy) those around us. Even a loud cell phone conversation trespasses this space!
- Personal space and interpersonal distance, i.e., physical distance between the communicating partners signals their level of intimacy and comfort. **If an unknown person stands too close or touches too often, it will make the other person feel uncomfortable**. Sitting on the table in the office, standing too close to the superiors etc., may not be appreciated. There are unspoken and invisible rules governing personal distance. People stand too near when they are more intimate. When they are not so close, they should be at a distance.
- Edward T. Hall is most associated with proxemics, the study of the human use of space within the context of culture. In *The Hidden Dimension* (1966), he argued that human sensory apparatus is the source of all perceptions of space, and these perceptions operate at the subconscious level and are moulded and patterned by culture. These perceptions can impede communication and understanding in cross-cultural settings. He analysed both the personal spaces that people create around their bodies as well as the macro-level sensibilities that shape cultural expectations about the organisation of streets, neighbourhoods and cities.
- Edward Hall's theory of proxemics suggests that people will maintain differing degrees of personal distance depending on the social setting and their cultural backgrounds.
- Hall's most famous innovation has to do with the definition of the informal or personal spaces that surround individuals:
  - **Intimate space -** the closest "bubble" of space surrounding a person. Only the closest persons can reach this level.
  - **Social and consultative spaces -** the spaces in which people feel comfortable conducting routine social interactions with acquaintances as well as strangers.
  - **Public space -** the area of space beyond which people will perceive interactions as impersonal and relatively anonymous.

Hall describes the handling of space during conversations: "A U.S. male...stands 18 to 20 inches away when talking face to face to a man he does not know very well; talking to a woman under similar circumstances, he increases the distance about four inches. A distance of only 8 to 13 inches between males is considered...very aggressive. Yet in many parts of Latin America and the Middle East, distances which are almost sexual in connotation are the only ones at which people can talk comfortably." Hall (1955) concluded: "If you are a Latin American, talking to a North American at the distance he insists on maintaining is like trying to talk across a room."

- Hall argues that cultural expectations determine these spaces. Americans traveling to Europe would tend to back away instinctively, as the distance maintained there almost half of what Americans do.
- Every culture has similar internalised expectations about how fixed and semi-fixed features, such as furniture, buildings and cities should be organised. United States cities, for instance, are customarily set out along a network, a preference inherited from the British, but in France and Spain a star pattern is preferred.
- In general, distances differ according to culture and the closeness or otherwise of a relationship, such as friends, co-workers, sales people etc. and very often, this is expressed intuitively. Again it depends on the place of interaction, such as public places like parks, restaurants, or on the street and in private homes. It also depends on the gender of the communicating parties. It is necessary to study the cultural norms and look for signals of discomfort caused by invading the other person's space. It is necessary to be aware of the other persons' body language as well as their attitude and respond accordingly. It is always correct to maintain a dignified distance from the receivers.

(C) Haptics

- Haptics is a scientific term that indicates the sense of touch. Haptics is the science concerning human interaction by means of touch. The word haptics is derived from Greek, meaning *"to grasp, touch or perceive"*. Haptics is often accomplished through electrical motors. There are other methods to create pressure sensations such as those with devices that are hydraulic or pneumatic etc. In feedback, a user can feel forces applied directly to the skin, which are detected by the user through sensors within the skin. Tactile feedback can also be applied by the user through electrical currents applied directly to the skin or objects that can vary in temperature touching the skin. Haptic devices have varying complexities and can work in different ways. Pressure devices are often described by their degrees of freedom. Haptic devices are available for commercial applications.

(D) Vocalics

- The dictionary meaning of vocalics is 'concerned with a vowel/vowels'. Vocalics is the study of non-verbal cues of the voice. Vocalics means such features as are related to voice, including intonation and pitch. It is the area of language related to the way voice and intonation are used. Voice has characteristics like tone, volume and pitch. Tone can be considered as the quality of the voice. Volume is the loudness or softness of speech which can be suitably modified according to space, distance, number of listeners, nature of audience and self-confidence etc. A high pitch voice is not preferred because it gives an impression of immaturity. Voice generally becomes high pitched when a person is encompassed by a sense of fear. Rapid speech shows some kind of excitement. Stress on a particular syllable can alter the meaning of a sentence.

(E) Physical Appearance

- *Objective language* is non-verbal message communicated through appearance. As a medium of non-verbal communication, it includes arrangement and display of material things. This method may include intentional or unintentional communication through material things like clothing, make-up, ornaments and other accessories, books, buildings, room furniture, interior decorations etc. These offer signals relating to the context, such as formal and informal settings.
- Dress and decoration communicate a great deal about the speaker's rank, status, personality, feelings, emotions, attitudes, opinions etc. What they reveal is something special about the person. For instance, when a professional bathes every day, wears clean clothes and his socks are not stinking, it indicates his outlook and concern for others. On the other hand, if he has

body odour and wears clothes and socks which have not been washed for a couple of days, he will be looked down upon right at the outset. First impressions count and whether one likes it or not, people are immediately judged by their appearance.

- *Professional dressing* conveys a non-verbal message as it creates an impression about the person, his background, his tastes and choices. Obviously, the person is instantaneously assessed by the audience. Hence the emphasis is on choice of dress and accessories.
- Professional dressing includes dressing appropriately for a job, an interview, an internship, a networking event, etc. Each company and event is different. Secondly, there are many other variables, like time of event, season, the city of residence and work, mode of transport being used, and whether person will be sitting or standing most of the time, dress code announced, what the others are likely to be wearing etc.

(F) Chronemics

- One of the most immediate features of chronemics is the attitude of different cultures to punctuality. Different cultures give different weightage to punctuality. For example, consider a multinational organisation where, managers from different countries are required to attend a meeting at a specified time. There will be cross cultural communication in such a gathering. Since each country has a different idea and/or concept of punctuality, all the participants would arrive at different times, some would arrive before time and some even after the specified time. None of them would pay much attention to either arriving early or late probably because of the different cultural attitudes to the time factor or punctuality. As in the case of punctuality, there is also a varied cultural attitude towards management of time and orientation to time with respect to past, present or future. Therefore, there is a problem about how cultures formulate their approach to time. A skilful management of activity, therefore, depends on one's attitude towards synchronic and sequential time.
- Under time language, indications are given about how important time is. owadays, time management has become an important part of business transactions. Organisations adopt various means and methods to save time because after all 'Time is money'. All communication should be appropriately adjusted to be effective.
- Use of time is a very subtle factor in non-verbal communication. Time is non-verbal communication. Having someone give you time to speak, or giving you their time and listening to what you are saying, can greatly boost self-esteem. A person who wastes his own time and that of others can certainly be considered inefficient. Chronemics includes a sense of time used for presentations, conducting meetings, giving good or bad news etc. as it creates good and bad impressions. In brief, it can be stated that chronemics is an important non-verbal method of communication as time also conveys certain messages.

(G) Artifacts

- Artifacts are concerned with the environment and objects around us. The physical environment around us conveys its own non-verbal language. Surroundings consist of different objects. There are two important parts of surroundings - colour, and layout. Colour suggests different attitudes, behavioural patterns and cultural backgrounds. For example, white colour stands for peace and surrender and red suggests danger. Decorative articles are also useful in effective communication. Artifacts are also personal objects we use to announce our identities and personalize our environments. Artifacts are thus material objects as an extension of oneself. For example, clothing has the power to influence. Change left in a phone booth was returned to well dressed people 77% of the time, poorly dressed people only 38% of the time.
- Several studies show that fancy suits, uniforms and high-status clothing are related to higher rates of compliance.
- Artifacts announce professional identity. Nurses and doctors wear white and often drape stethoscopes around their necks; executives carry briefcases, whereas students more often backpacks. White-collar professionals tend to wear tailored outfits and dress shoes, whereas blue-collar workers often dress in jeans or uniforms and boots. The military requires uniforms that define individuals as members of the group. In addition, stripes, medals, and insignia signify rank and accomplishments.

- We also use artifacts to define settings and personal territories. When the president speaks, the setting usually is decked with symbols of national identity and pride such as the flag. At annual meetings of companies, the chair usually speaks from a podium that bears the company logo. In much the same manner, we claim our private spaces by filling them with objects that matter to us and that reflect our experiences and values. Lovers of art adorn their homes and offices with paintings and sculptures that reflect their interests. Professionals may decorate their offices with expensive furniture and framed awards to announce their status or with pictures of family to remind them of people they cherish. Web sites, too, are defined by artifacts such as photos, animations, and colours that express the creators' personalities and interests
- Although the human body transmits a number of messages, these messages do not carry any sense until they are translated by someone. The process of translation is important where artifacts are concerned.
- In order to communicate by means of embodiment media, both the sender and receiver of the message must assign the same meaning to the common embodiment media. The assignment emerges through interaction between humans and artifacts. The designer of an artifact extracts essential characters of human communication by observation and it is to be implemented in the artifact. All artifacts do not have the same character (embodiment) as humans.
- "No" is often represented with a shake of the head. In order to express a negative message, the sender must have appropriate embodiment media. It means that the sender must have a head, neck, body etc. and the ability to realise the action. Lifeless articles like a chair, table, briefcase do not have such embodiment. Human beings can adopt various other methods to understand embodiment media.

4.2 ASPECTS OF BODY LANGUAGE (TYPES OF BODY LANGAUGE)

- There are various ways in which non-verbal communication can be classified. There are four most common types:

Types of Non-Verbal Communication

Body Language	Space Intimate (18") Personnel (18" – 4') Social (4 to 12 feet) Public (12 feet to seeing and hearing)	Time (Time orientation becomes part of message)	Para language (How we say some things)

- These four types are important in speaking and listening. The human body sends non-verbal messages through arms, fingers, expressions, postures and so on. Face and eyes are most important. Gestures are useful for sending non-word messages through body parts. Another area of body language is physical appearance; for example clothing, hair style, jewelry, cosmetics etc. can affect body movements. Yet another type of non-verbal communication includes space and the way it communicates, meaning in speaking and listening.

Kinesics:

- The study of body motion and facial expressions as related to speech is called *kinesics*. It is an important factor in non-verbal or oral communication.
- Kinesics is non-verbal behaviour related to the movements of a part of the body or the body as a whole. Generally, all communicable body movements are classified as kinesics. Kinesic communication is one of the most important non-verbal communication forms. Communication by way of body movements is endless. For example, when the forefinger and middle finger are indicated in the shape of the letter 'V' it becomes the sign for victory. In certain cases two fingers are used to symbolise number two. Sometimes '0' may be indicated as OK, in other cases it is taken as a zero.

- There are some non-verbal signs that regulate and maintain the flow of speech during meetings or conversations. These can be both kinesic, such as nodding of the head, as well as non-kinesic, such as eye-movements. The information which flows as a part of feedback to indicate whether the receiver has understood the message can be highly confusing.
- Kinesics is an important part of non-verbal communication. The movement of the body conveys a specific meaning and interpretation. There are movements which carry a significant risk of being misinterpreted in certain cases of communication.
- Body language is a comprehensive term used for the method of communication through different parts of the body other than the tongue. It includes, all sorts of body movements, gestures, facial expressions, postures, miming, touch etc. depending upon the physical space between the communicators. Silence as a body communication also plays an important role in certain exchanges. Body language, by and large, is unintentional and more truthful than verbal messages. It is easy to deceive a person by expressions.

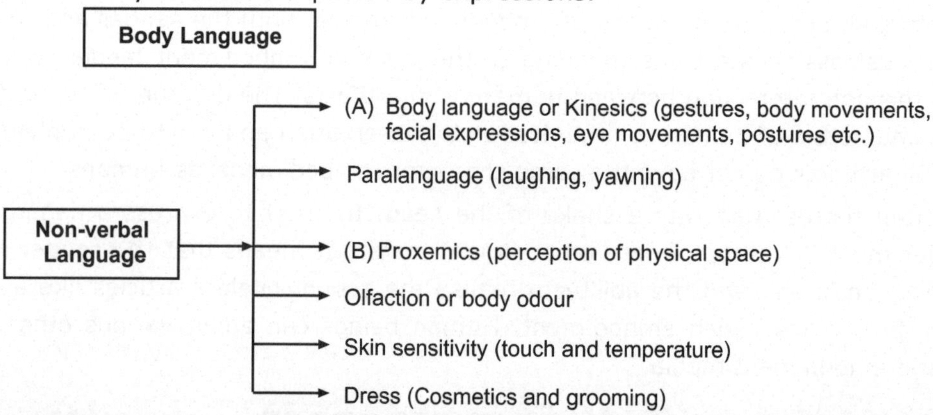

Presentation Plan

- The art of public-speaking is the fruit of constant efforts. There is no definite formula to be adopted to become an effective speaker. Enough planning and preparation is essential for a successful presentation. Preparation is the key to overcome nervousness.
  1. As far as possible the speaker should not read out a written speech word by word.
  2. Written speech seldom sounds fresh.
  3. It is not necessary to memorise the speech either. It never exhibits spontaneity.
  4. Memorisation of speech would hamper flexibility. Consequently, communication will suffer.
  5. Face to face conversation or interaction involves thinking and speaking and not mere repetition.
  6. Even while reading the speech, one must lift the head occasionally and observe the audience.
  7. The speaker must make adequate notes and points and arrange them in a proper order before actually starting the speech. It should bear proper headings and sub-headings.
  8. Appropriate time should be devoted for introduction and to the main body of the talk.
  9. Notes and points should be written neatly and in bold letters.
  10. A positive attitude needs to be generated while speaking.

(a) Facial Expressions:

- Facial expressions exhibit attitudes. A smile expresses friendliness and affection. Raising of eyebrows conveys surprise, while a furrowed forehead expresses worry and anxiety. Various parts of the human body, particularly eyes, are used for communicating information. Eyes are the most significant organs which communicate subtle messages. Really speaking, eyes by themselves do not show emotion; but when they are associated with other parts of the face, they are windows to the soul. Squinting, winking, staring and gazing express a world of meaning. Looking at the eyes, face and other verbal signals during conversation perform certain important functions:

1. Looking helps with the synchronising of utterances.
2. It provides definite feedback.
3. It provides additional non-verbal information which elaborates verbal messages.

(b) Eye Contact:

- Eye contact is an instantaneous and effective means of establishing rapport with an audience. A good speaker first looks at the audience and takes a pause before beginning his speech. It helps him to make a good impression on the audiences. In case the speaker gives breaks and puts his ideas into small units rather than in a lengthy narration, it allows listeners some intervening time to grasp those ideas. Once this is done, the speaker can proceed to explain the subsequent points or ideas. While speaking it is also necessary to maintain eye contact with the audience. If the listeners feel that the speaker is ignoring them, they are also likely to ignore the speaker and what he says. Through eye contact, the speaker gets signals whether the channel of communication is open and operating on right lines. This sense itself works as the feedback.

(c) Vocalics (Aspects of Speech or Para Language):

- The dictionary meaning of vocalics is 'concerned with a vowel/vowels'. Vocalics is the study of non-verbal cues of the voice. Vocalics means such features as are related to voice, including intonation and pitch. It is the area of language related to the way voice and intonation are used. Voice has characteristics like tone, volume and pitch. Tone can be considered as the quality of the voice. Volume is the loudness or softness of speech which can be suitably modified according to space, distance, number of listeners, nature of audience and self-confidence etc. A high pitch voice is not preferred because it gives an impression of immaturity. Voice generally becomes high pitched when a person is encompassed by a sense of fear. Rapid speech shows some kind of excitement. Stress on a particular syllable can alter the meaning of a sentence.

(d) Gestures:

- In certain cases, gestures are used extensively when a person does not know the local language. The deaf mainly depend upon gestures for their sign language. In certain discussions, gestures of a person convey much more than what he speaks, for e.g. enthusiasm, attitudes, emotions, etc. Gestures like shaking the hand display friendship and warmth. Other such examples are holding of hands, patting on a shoulder etc.

(e) Postures:

- Each movement or posture of the body has expressive and defensive functions. The way in which one sits or stands, walks in or walks out tells a lot about the person's emotions and attitudes. But no body movement itself has any exact meaning. Just as spoken words are assisted by gestures, posture is also taken together with other verbal and non-verbal clues. For example, lowering of the head indicates the end of the statement, or raising of the head indicates the end of the question. A large number of postural changes indicate the end point in interactions; for example, the end of a thought or the end of a statement. A shift in posture shows that something is happening. This aspect of body language is known as "Proxemics."

(f) Dress and Appearance:

- The kind of dress that is worn by a person and his grooming shows his status and attitude. Most people are influenced by how the speaker looks and the clothes he/she is wearing. Physical attractiveness plays an important role in one's assessment of people. Attractive people are generally considered to be more successful, more intelligent and more competent than those who do not look or appear attractive. A person's physical appearance and dress conveys a great deal of information about him. One's judgement or evaluation of others is greatly influenced by physical appearances. It has a definite impact on the communication process.

(g) Haptics (Communication by Touch):

- Haptics is a scientific term that indicates the sense of touch. Haptics is the science concerning human interaction by means of touch. The word haptics is derived from Greek, meaning *"to grasp, touch or perceive"*. Haptics is often accomplished through electrical motors. There are other methods to create pressure sensations such as those with devices that are hydraulic or pneumatic etc. In feedback, a user can feel forces applied directly to the skin, which are detected by the user through sensors within the skin. Tactile feedback can also be applied by the user through electrical currents applied directly to the skin or objects that can vary in temperature touching the skin. Haptic devices have varying complexities and can work in different ways. Pressure devices are often described by their degrees of freedom. Haptic devices are available for commercial applications.

(h) Voice Modulation:

- The speaker must know how he should use his voice effectively. To want to become an effective speaker, a good voice is a must. A good voice is no doubt a natural gift; but one can make efforts to improve the quality of one's voice with skilful training and practice. Proper attention to this aspect would lead to the acquisition of confidence and power.
  1. Variations in the pitch and tempo of the voice are essential to transmit the message effectively.
  2. Fast delivery of speech indicates lack of confidence, and betrays the object of the speech and the image of the speaker. The speed of delivery must be such that the audience can keep pace and understand the matter properly. One should not go beyond the speed of 125 to 150 words per minute. One has to bring the speed of speaking within this range by proper practice.
  3. Pronunciation of the words used must be correct with due stresses wherever required.
  4. One should speak loudly enough so that everyone including even the last man in the audience is able to listen to the speech clearly.
  5. Do not use repeatedly words like "You see", "I mean," "Do you understand," "Is it clear" etc.
  6. The microphone should be kept eight to ten inches away from the mouth to ensure proper transmission of ones voice to the audience.

(i) Audience Awareness:

Before starting the speech the speaker should first:

1. Size up the listeners.
2. Consider the age, sex, background and interest of the listeners.
3. Ascertain whether the audience is friendly or hostile.
4. Use common sense to become a good speaker.
5. Select an approach that suits the audience.
6. Make the listener feel that he/she is being addressed individually.
7. No verbal fireworks are necessary to arrest the attention of the audience.
8. Make the audience feel that the speaker is sincere in gaining their interest.
9. Dramatise certain ideas to overcome the barriers of communication.
10. Create an impression that you want to share the views and ideas of the audience.
11. A 'joke' should be delivered in such a way that it appears to be part of the speech. The audience should receive it almost unprepared.
12. Do not get distracted if a listener smiles or whispers to a neighbour.
13. Concentrate on ideas.

(j) Silence:

- *Speech is silver, but silence is golden.* It means silence speaks louder than words. It is a non-verbal language par excellence. It establishes the relationship between communicators and their attitudes towards one another. Moments of silence show that the parties involved are either confused or do not know how to continue the conversation. In case of public speaking, when the speaker punctuates his remarks with periods of silence, he could be using it for stressing the importance of his message. In some cases, silence is generalised. For example, when a worker fails to complete his assigned job and the supervisor makes enquiries about it, his silence speaks for his failure.

(k) Physical Appearance:

- How one looks and what one wears affects listeners. Thus some organisations are more causal in dress, others are not so. Learning an organisation's culture includes learning its dress code. Most organisations encourage men to wear a shirt, a tie and a suit and women to wear a white blouse and a dark skirt and a jacket. But these rules are continuously changing. Terms as 'smart casual' or 'business casual' indicate symbols for appearance, and it is important that one knows what such terms mean is one's particular surroundings or organisation.
- Clothing, hairstyles, neatness, postures, stature are part of personal appearance. They convey impressions regarding occupation, age, nationality, social and economical level, job status and good or poor judgement, depending on some conditions. Aspects of surroundings include room size, location, furnishings, machines, architecture, wall decorations, floor, lighting, windows, view and other related features wherever people communicate orally. Surroundings will vary according to the status and according to the country and culture.
- Whether one is speaking to one person face to face or to a group in a meeting, physical appearance and surroundings convey non-verbal stimuli that affect attitudes towards spoken words.

4.3 GRAPHICAL COMMUNICATION: INTERPRETING VISUALS AND ILLUSTRATING VISUALS LIKE TABLES, CHARTS AND GRAPHS

Visual Communication:

- Before primitive man learnt to use verbal communication, he might have used non-verbal and visual signs for communicating. He might have used pictures, images, paintings etc. Pre-literate man might have expressed himself visually. The discovery of letters of alphabet led to a form of writing and abstract symbols replacing visual representation of objects. Following are the important visual forms of communication today:

Communication Today:

1. Charts and graphs
2. Pictograms
3. Maps and plans
4. Tables
5. Slides
6. Film strips.

- In visual communication, very elementary and simple ideas, orders, warnings are used. It can be effectively used only in combination with other media. Diagrammatic representation combines the written with the visual. Pie charts and pictograms also combine the two, because there is always a written indication of what is shown in them. Such forms are very useful in the presentation of statistical data.

Advantages/Merits of Visual Communication:

1. To overcome the language difficulty, charts, graphs etc. can be effectively used for visual communication because they are easily grasped and understood by the viewers.
2. To make people remember the matter for a longer period, pictures and graphs can be used.
3. Visual communication makes comparisons easy.
4. To make publicity more effective, charts and graphs are introduced. They are very effective in changing people's attitudes.

Disadvantages/Demerits of Visual Communication:

1. Charts, graphs etc. cannot be used for conveying all kinds of information.
2. However, they are not suitable for discussions.
3. They are also not suitable for explaining 'ideas in detail'.
4. Charts, graphs etc. may sometimes convey misleading information.

Introduction to Graphic Communication

- **Importance of Graphic Language:** The importance of graphic language can be understood by comparing it with word languages by all those who are educated and attend school or college, learn to read, write and speak with some degree of proficiency. Language is a highly developed system of communication. But any single word in any language is not enough to describe the size, shape and its relation with a physical object. Engineering being an applied science, the communication of physical facts must necessarily be complete and accurate or else it may have disastrous results. *Qualities relationships* can be expressed mathematically. However, for designing machines and structures, graphic representation is necessary.
- Major engineering drawings are made in lines with different views arranged in a logical system of projection. This is the language of graphics which can be defined as the graphic representation of physical objects and their relationship. Engineers have to study the language of engineering graphics so that they can write it clearly for those familiar with it and can read it readily when written by others. It is essential for an engineer to master these specific modes of communication.

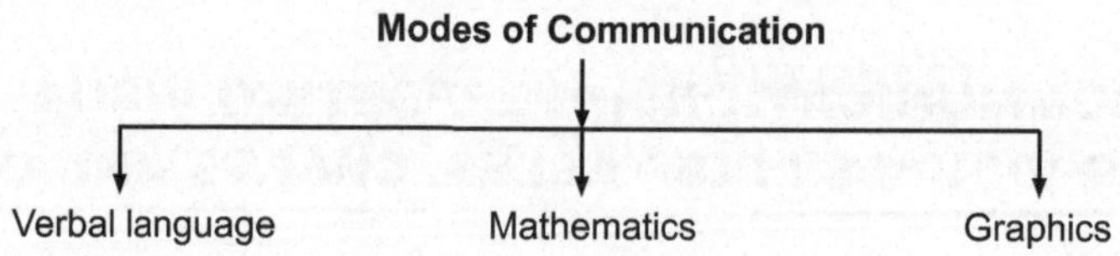

- Graphics is a method or language in which visual symbolism is applied for the description and representation of objects and concepts. Skill in each type of language enables an engineer to describe technical concepts accurately. Formal education is normally devoted to the development of language and mathematics. Graphics is the most important language with which engineers communicate.

Objectives of Graphic Communication

1. Graphic communication provides the ability to visualize.
2. Communicate design concepts and ideas.

- With a set of drawings, an engineer can represent his thoughts and ideas on paper. He makes possible structures of engineering projects such as bridges, machine tools etc. on paper.
- Business people are supposed to read numerous letters and reports every day. Writings that contain a lot of statistics will usually be difficult to read and grasp. Such information would be much easier to comprehend if it appears in the form of a graph or chart. One of the most valuable functions of a computer is that it can produce charts, graphs, and other visual aids accurately, quickly and easily. When statistical information is fed into a computer, a good graphics programme can automatically turn it into an appropriate visual aid. Not only are the reports containing visual aids easier to read but they also look more professional and appealing. Many pages of prose are difficult to read, even though the subject-matter may be fairly simple. A few well placed charts and graphs can make a report appear more helpful and arouse the reader's curiosity and interest. Most computer graphics programmes can reproduce visual aids in colour, making them even more attractive and appealing.

- Computer graphics are quite helpful to designers, engineers and research scientists as they plan new products and test new theories. Now new uses for computer graphics, both in and out of the office, are being discovered all the time. A computer can be used to produce maps, drawings, blue prints, a network for advertising and almost anything.
- **Pictograms:** Pictograms represent figures by pictures. The best form is that in which the value of a figure is shown by the size or number of figures shown. This form is an effective way of representing information to the general public.

Table 4.1: Tables are useful for presenting statistical information at a glance

No. of Total Labourers	Total Product (output in kg)	Marginal product (output in kg)	Average Product (output in kg)
A	B	C	D
1	100	100	100
2	250	150	125
3	375	125	125
4	450	75	112.5
5	460	10	92.00
6	460	00	76.66
7	450	– 10	– 64.28

- **Maps:** A map represents on paper the earth's surface or a part thereof showing countries, rivers, mountains, oceans, roads etc. Take for example, a map of India or a map of Maharashtra.
- **Signs and Signals:** Signs and signals are widely used to convey messages, warnings, requests etc. They possess certain universal characteristics, conveying the same meaning all over the world. The fundamental difference between a sign and signal is that while signs are always static, signals are dynamic. For example, mathematical signs (+ or –) are static while traffic signs are dynamic. A sign is, therefore, like a still photograph or picture, whereas a signal is dynamic, for example, a traffic signal.
- **Posters:** Posters are used for displaying advertisements. They are mostly displayed at strategic points so that they can easily catch public attention. Posters, nowadays, have become increasingly effective in attracting public appeal. A well designed poster has a great visual appeal. The message in a poster can be visualised in a simple form and can reach the literate and illiterate person.
- Certain kinds of information can be represented in the form of tables, charts, graphs, diagrams etc. However, they have certain limitations. They cannot be used to record discussions, instructions and legal arguments.
- **Tables:** A table is a part of the text. It may have three to four columns and may run over certain lines. An independent table is noticeably separate from the text. Usually a table consists of numerical data and gives statistical figures. When the given data is in the form of words, it is called a 'phrase table'. A table may have a number of rows and columns. A row runs horizontally across the page and columns run vertically. The first column is called a sub-head under which a number of items are listed. The other columns are called boxheads. Data which are not available or absent can be mentioned as dash. While preparing the table it is necessary to note the following points:
  1. First plan the data given.
  2. Prepare the form of the table.
  3. Select the position of the table.
  4. Show the figures in a clear manner.
  5. Each column should be of appropriate size.
  6. Logical arrangement of the given data is important.
  7. It is necessary to keep the size of the table suitable to the format.

Examples of a Table:

- The following Table 4.2 shows the age (X) of students in years and marks (Y) obtained by them in a certain intelligence test.

Age in years / Marks	16 – 18	18 – 20	20 – 22	22 – 24
10 – 20	2	1	1	–
20 – 30	3	2	3	2
30 – 40	3	4	5	6
40 – 50	2	2	3	4
50 – 60	–	1	2	2
60 – 70	–	1	2	1

- The following Table 4.3 shows the frequency distribution of the heights and weights of 300 students of a certain college.

Height (inches) / Weight (Pounds)	58 – 62	63 – 67	68 – 72	73 – 77	78 – 82
90 – 109	2	1	–	–	–
110 – 129	7	8	4	2	–
130 – 149	5	15	22	7	1
150 – 169	2	12	63	19	5
170 – 189	–	7	28	32	12
190 – 209	–	2	10	20	7
210 – 219	–	–	1	4	2

- **Charts**

  A chart shows the organisation of a process by representing its sub-divisions or the sequences of steps in that process. The three main types of charts are: (1) organisational charts (2) flow charts and (3) block charts.

- **Organisational chart**

  This chart shows the chain of command in an organisation. It can be observed that the chain of command moves vertically from top to bottom and the posts or equal ranks or status are aligned horizontally.

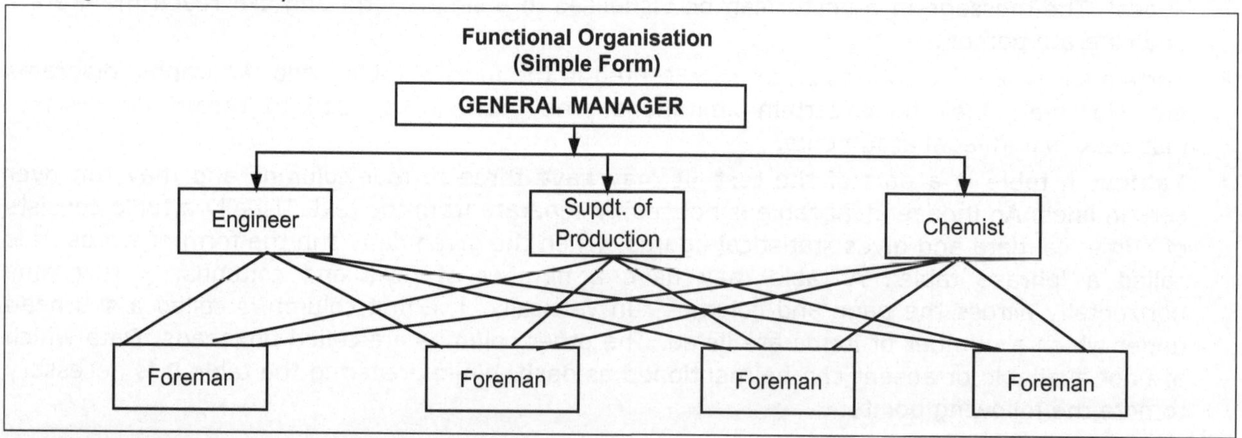

- **Flow charts:**

  A flow chart is generally used to indicate various steps in a process or sequence of operations.

Flow Process Chart

'Man'

Job: Writing a letter using a shorthand typist

Chart begins – typist in his own chair

Chart ends – typist puts letter in outward tray for dispatch

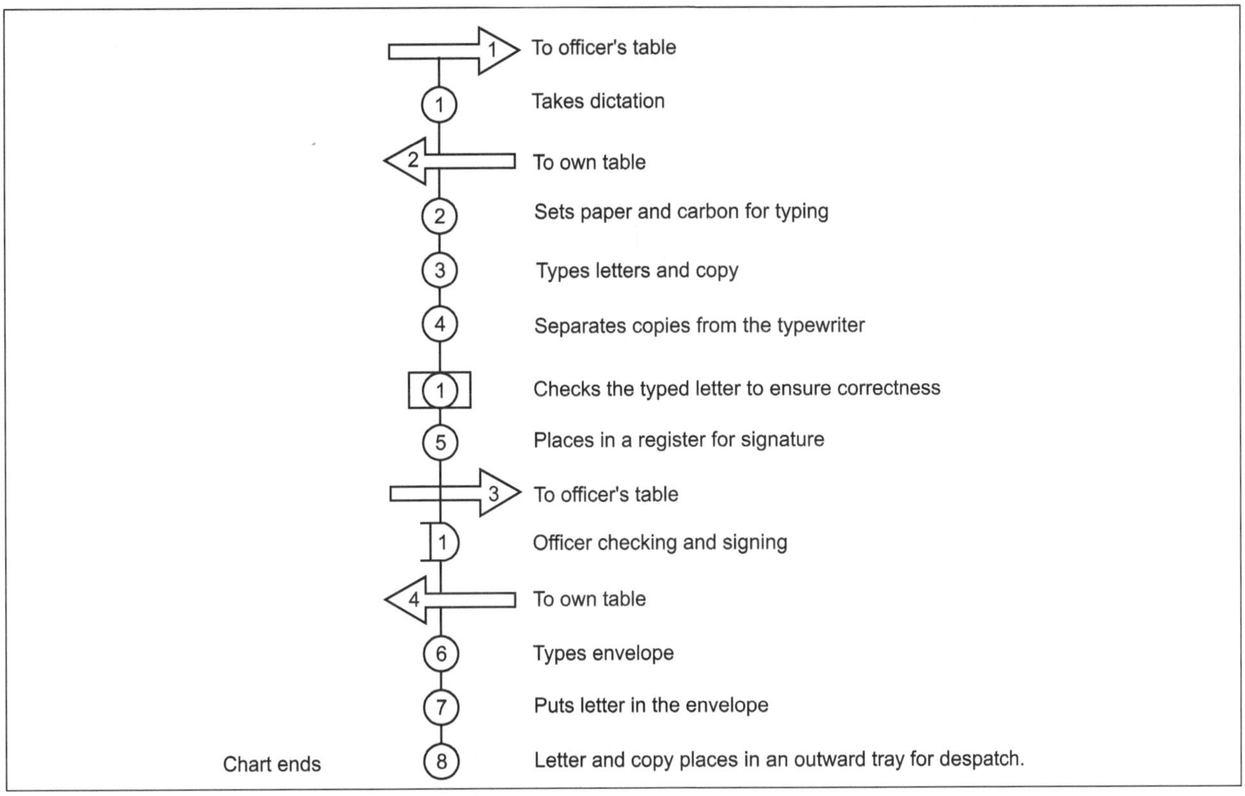

- **Summary:**

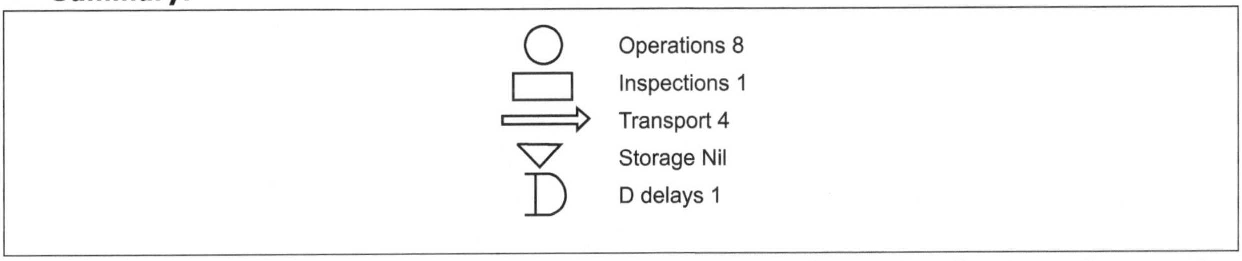

- **Block charts:** In this chart, blocks are used to indicate the major elements of a complex structure generally of a technical nature.

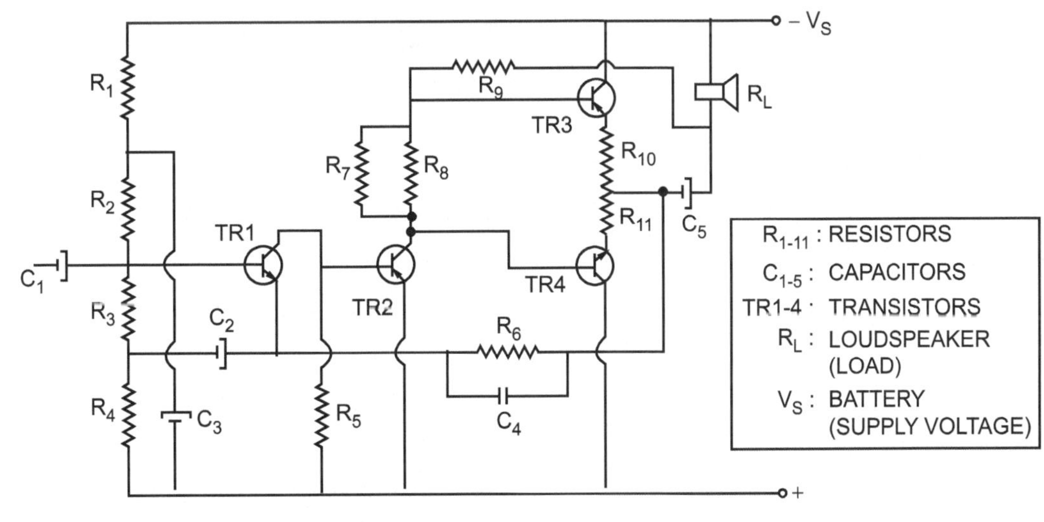

Fig. 4.1

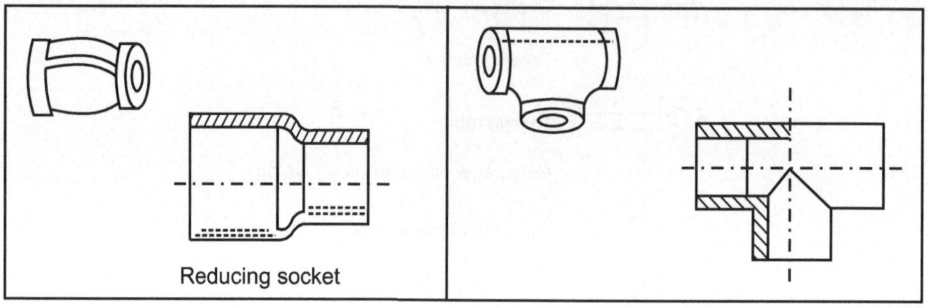

Fig. 4.2

- **Pie charts:**

An opinion poll was taken to find out how long the United Front Government would last

The following was the response:

Full term	17%
2 to 3 years	16%
One more year	09%
Less than a year	10%
Less than 6 months	11%
Less than 3 months	06%
Don't know	31%

The above information is presented below in the form of a pie-chart.

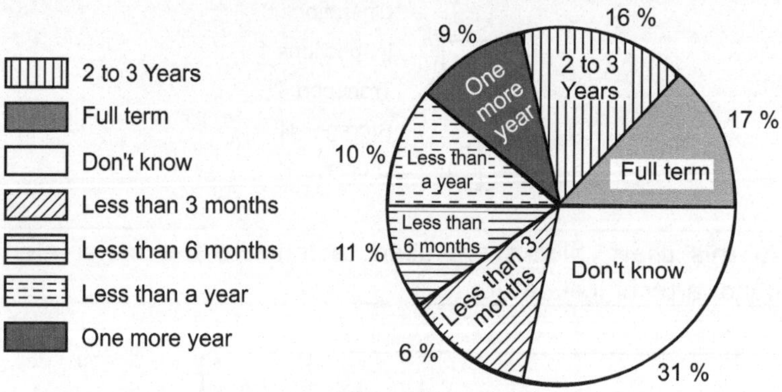

Fig. 4.3

- **Graphs:**

Bar Graph: This is another method of presenting data.

The data of incidence of fires in textiles factories in Mumbai during the year 1994, 1995 and 1996 are given below:

	1994	**1995**	**1996**
Short circuiting	84	61	52
Smoking	50	45	62
Debris/scrap burning	22	31	18
Unknown	5	9	6
Total	**161**	**146**	**138**

A bar graph to represent the above data is drawn below:

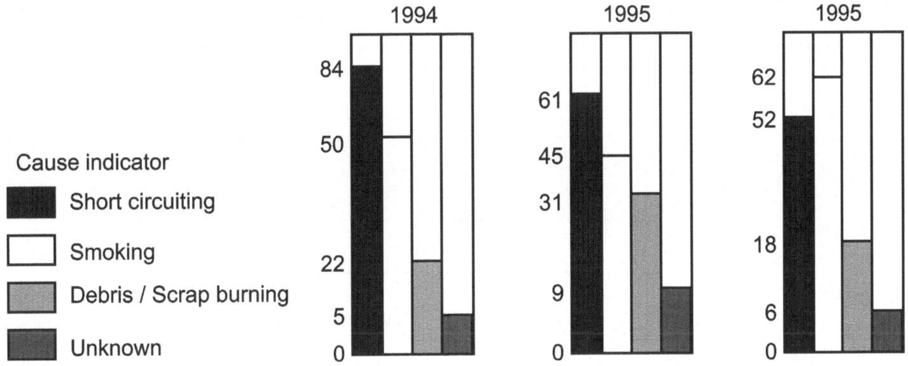

Fig. 4.4

- **Presentation of the information given in textual form can also be presented in the example given below:**

Flowering plants are classified into four categories on the basis of:

(i) Period of flowering,

(ii) Type of bark,

(iii) Method of storing of food material, and

(iv) The place where they grow.

In the first type, there are again three categories viz. annual, biennial and perennial.

Annual plants grow every year and their seeds die within that period. Aster, sweet pea are examples of this type. Biennial flowering plants grow once in two years while perennial plants grow without interruption. Daisy, rose, jasmine plants are examples of perennial flowering plants.

On the basis of the bark of a plant, there are two types viz. woody and semi-woody. The plants in this category are usually perennials. Plants like jasmine, rose belong to this type. The third type of plants is bulbous. They store food in a bulbous form. Tubers and Tulips are examples of this type. Lastly, on the basis of place of growth, there are also aquatic plants. These flowering plants grow in water. Lotus is a good example of this type. Lotus Is a perennial plant that grows without interruption.

The above textual details are presented in the following figure:

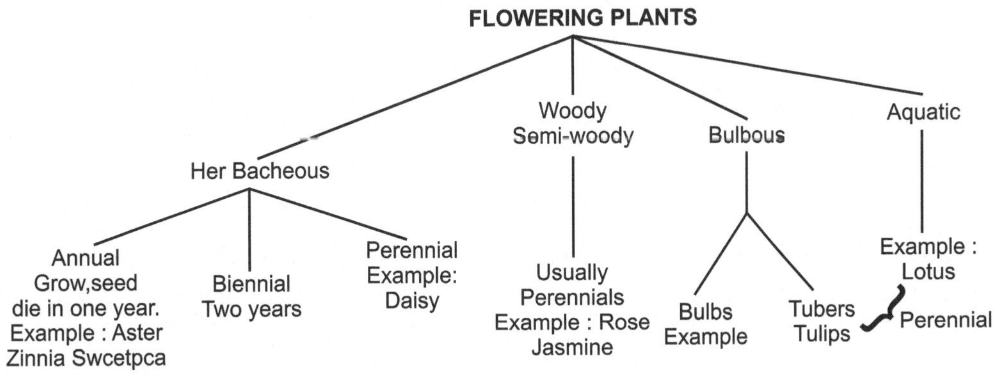

Fig. 4.5

- There are different types of graphs:

1. **Rectilinear graph:** This type of graph is used for showing the trend of two variables. An illustration of capital growth of an organisation over a certain period is given below:

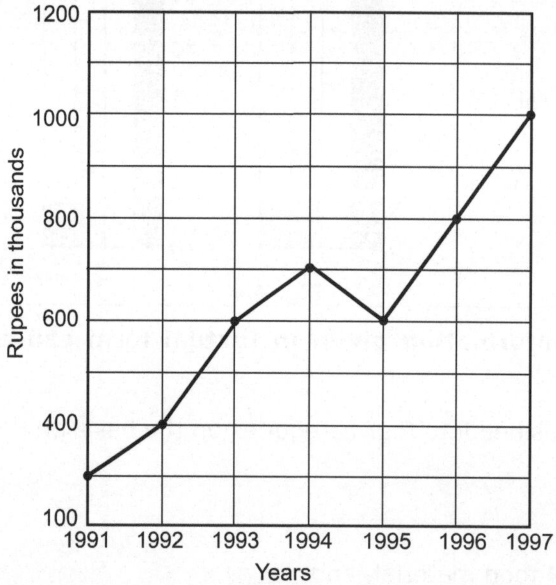

Fig. 4.6: Capital Growth during 1991 to 1997

2. **Multiple line-graph:** This type of graph is used to present the data of related units which have more than one indexline, or variables. In the following illustration, the data about admissions to Engineering courses related to three variables viz. number of applications received, candidates selected and actually joined are shown in a multiple line-graph.

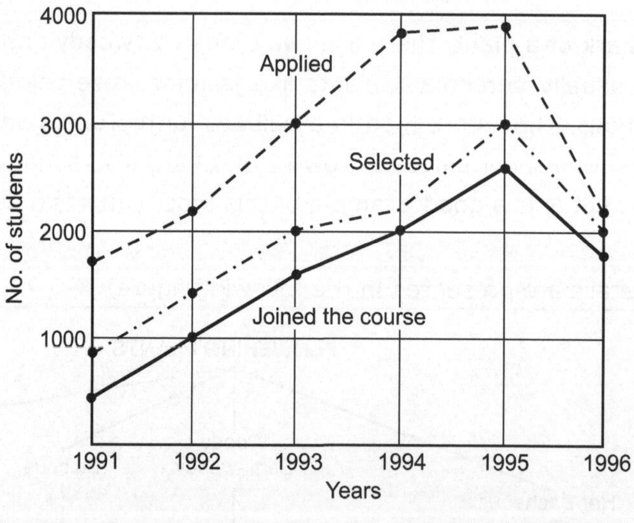

Fig. 4.7: Number of students admitted for Engineering courses in a city

3. **Semi-log graph:** In this type of graph the unemployed population has been shown on a logarithmic scale vertically, whereas years have been shown on a linear scale horizontally.

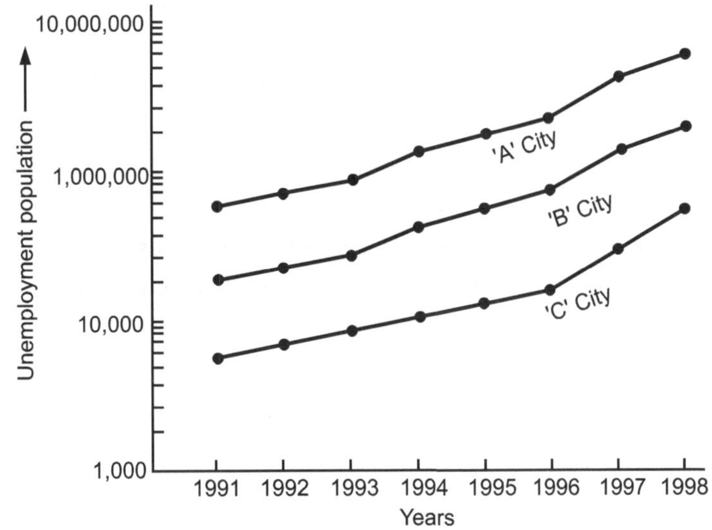

Fig. 4.8: Employed Population

It has the linear scale on one axis, X axis and the logarithmic scale on the other. This type of graph can predict future trends also.

Photographs: Photographs have limited use in emphasising important details. In this case, reproduction costs are high. They are, therefore, used particularly when actual appearance is to be shown.

Drawings: Drawings are mainly composed of lines. They may include shades, solids or cross hatched areas. They are used to represent objects as they appear to us.

Important Points

1. **Communication (Non-Verbal):**
   - Communication without the use of words is non-verbal communication. Messages conveyed by non-verbal methods are simple.
   - Maps, charts and graphs should be skillfully developed.
   - Communication by means of TV, cinema etc. is more powerful.
   - Non-verbal communication exerts an immediate effect on a receiver.
   - A non-verbal method takes less time.
   - It is more convenient.
   - People quickly react to pictures, colours etc.
   - Non-verbal communication provides data in a compact form.
   - Graphical representation provides information quickly and in brief.
   - Non-verbal method is more useful for comparison.
   - Warning can be given instantly through a non-verbal method.
   - TV is more useful than the radio.
   - Colour is a more useful and powerful means of communication
   - Traffic signals play an important role in cities.
   - A colour can have a psychological effect in motivation.
   - Colours represent moods and feelings.
   - In case of graphs, charts etc. additional information is given when a colour is introduced.
   - Pictorial method is useful in mass-communication.
   - Graphs and Charts give statistical information.

2. Non-verbal communication is more effective in representing figures

- Maps are used for showing routes.
- (+) or (–) in mathematics represent non-verbal communication.
- Signals become important when instant decisions are required to be conveyed.
- Visual signs and signals are used in communication when the receiver is not within hearing distance.
- Drum beats, sirens are used as auditory symbols.
- Tunes and sounds are also used for communicating secret messages.
- A number of non-verbal clues plays an important role in oral communication.
- Sometimes, non-verbal communication can become a substitute for verbal communication.

Body Language

(A) Kinesics: It is the study of body motion:

- It is related to speech.
- It is important in communication.
- Kinesics is non-verbal behaviour related to movement.
- All communicative body movements can be taken as kinesics.
- Body movements are endless.
- Some body movements can maintain the flow of speech during conversation.
- Kinesics is an important part of non-verbal communication.
- Kinesics is a comprehensive term where the tongue is not used.
- It includes various body movements, facial expressions, postures etc.
- Silence can also be taken as communication.
- Gestures are useful when a person does not possess knowledge of the local language.
- The hearing impaired depend on gestures.
- Facial expressions exhibit emotions and certain attitudes.
- Eyes are used for conveying certain information.
- Gazing, staring, winking, give different meanings.
- Eyes perform important functions during conversation.
- Looks convey different feedback.
- Postures express different functions.
- Lowering of head means an end of statement.
- Postures are a sign of different interactions.
- Dress and appearance reflect status and attitude of a person.
- Physical appearance plays an important role in judging people.
- Physical appearance and dress convey much required information. It has a different impact in communication.
- Sometimes silence plays a much more important role than what is expected.
- Speech is silver, but silence is golden.
- Eye contact is very important in communication.
- Eye contact establishes rapport with an audience.
- A good speaker first carefully observes his audience and then starts his speech.
- Eye contact conveys certain signals in communication.
- Eye contact works as feedback.
- Voice plays a vital role in communication.
- A good voice is required to become a good speaker.

- A good voice is nature's gift.
- Voice transmits messages effectively.
- There should be correct pronunciation of words.
- Common sense is required to become a good speaker.
- A speaker must consider views of the audience and include their interests in his speech.
- Speaker must concentrate on the ideas on which he is speaking.
- Speaker must have a definite presentation plan of his speech.
- Enough planning is required to become an effective speaker.
- Memorisation of speech hampers flexibility.
- Logical arrangement of speech impresses the audience.
- There should be a definite degree of nearness or closeness while communicating.

(B) Proxemics

- Physical or psychological closeness between people.
- It is a dimension of communication.
- It is also the distance between a sender and a receiver of message.
- It is called 'space language'.
- The physical environment around us gives its non-verbal language.
- Important surroundings are colour, design, layout, time etc.
- Smiling is a good non-verbal signal of communication.

(C) Haptics

- Indicates sense of touch.
- Concerned with human interaction.
- Executed through certain actions.
- Used in tactical feedback.
- Haptic devices have varying complexities.
- Haptic devices are available for commercial applications.

(D) Vocalics: Features related to voice –

- Includes intonation and pitch.
- Voice has characteristics like tone, volume and pitch.
- Loudness and softness of speech requires adjustment to space and distance.
- Rapid speech exhibits excitement.

(E) Physical Appearance:

- Clothing, hairstyle, neatness, stature etc. are part of personal appearance. Surrounding and other related features are also important in respect of physical appearance.

(F) Chronemics (Time Language)

- Chronemics is concerned with the importance given to time.
- Chronemics can also be inter-cultural communication.
- Use of time as a non-verbal method of communication is also called chronemics
- Time also conveys messages.
- Time is a mode of interpersonal communication.
- Time can be saved, wasted, given and taken.
- Punctuality gives pleasant or unpleasant feelings.
- Time is important in group activities.
- Receiving a telephone call late at night or early in the morning gives a different meaning and message.

- Time is considered as money in day-to-day life.
- All communication should be adjusted correctly.
- Use of time is an important factor in non-verbal communication.
- A person who wastes time can be considered inefficient.
- Chronemics includes a sense of using time.
- Chronemics is an important non-verbal method of communication as time also gives a certain message.

(G) Artifacts

Artifacts are concerned with the environment and objects.

- The physical environment around us gives an idea of non-verbal language.
- A person's surroundings consist of various objects.
- Surroundings consist of two parts - one is colour, another is layout.
- Colour suggests different attitudes.
- White colour suggests peace and surrender.
- Decorative pieces are also sources of different kinds of communication.
- Indifference towards colour reflects a poor cultural background.
- The artifact and the communicative ability of that artifact is important in non-verbal communication.
- A chair on its own when inactive does not communicate.
- Each article has its own function.
- The human body can transmit various functions but it has to be translated by someone.
- Sender and receiver must give same meaning to the common embodiment media.
- The designer of an article puts required characters in the article and they can be implemented.
- All artifacts do not have the same characters.
- Embodiment of media is more important in non-verbal communication.

Interpreting Visuals and Illustrating with Visuals like Tables, Charts and Graphs

- Primitive society learnt to use visual signals for communication.
- Pre-literate people used visual symbols for communication.
- Charts, maps, tables etc. are visual communications.
- Diagrammatic representation combines the written with the visual.
- Charts and pictograms are useful for effective communication.
- Visual communication overcomes language difficulty.
- Charts and graphs are easily understood.
- Charts and graphs are remembered for a longer period.
- Charts and graphs make comparisons easy.
- Charts and graphs cannot, however, provide all the required information.
- They have certain limitations.

Importance of Graphic Language

- Graphic language is useful for educated people.
- Language is a highly developed system of communication.
- Engineering drawings are made in lines with different views.
- Graphic mode of communication is important in the Engineering Science.
- Communication without the use of words is non-verbal.
- Maps, charts, graphs play an important role in non-verbal communication.

There are two methods in non-verbal communication:

1. Visual communication
2. Auditory communication

- Visual means are those that can be seen e.g. pictures, colours etc.
- Auditory means those that can be listened to e.g. bell, horn etc.
- Non-verbal is used to draw the immediate attention of a receiver.
- Pictures, colours and sounds are more effective in attracting people.
- TV and cinema are the most powerful devices in conveying messages.
- Pictorial representation is more powerful in mass-communication.
- Signals are more effective where instantaneous decisions are required to be taken.
- Non-verbal clues like body language, gestures, facial expressions, postures, appearance, silence play an important role in non-verbal communication.
- Charts, graphs, pictograms, tables, slides, films are the important features of visual communication.
- They are also more effective in the education system.
- A drawing is composed of lines.
- A drawing may include shades, solids or cross hatched areas.
- A drawing represents objects as they appear to us.

Practice Questions

1. State the importance of non-verbal codes.
2. Explain the importance of visual and auditory methods of communication in day-to-day life.
3. Non-verbal methods of communication have an instant effect. Explain the statement giving suitable examples.
4. Explain in brief the media of non-verbal communication.
5. Explain the importance of body language and state the importance of chronemics and artifacts.
6. Explain the following terms: Chronemics, artifacts, kinesics.
7. Explain the importance of kinesics, gestures, facial expressions, postures, dress appearance and silence in the communication process.
8. Eye contact plays an important role in establishing contact with the audience. Explain.
9. What is the importance of voice modulation in becoming an effective speaker ?
10. What are the important points a speaker has to take into consideration to become a good speaker ? Explain the term audience awareness.
11. Explain the following terms:
    (i) Proxemics
    (ii) Haptics
    (iii) Vocalics
12. What are the important visual forms of communication in modern times ?
13. What are the merits and demerits of visual communication ?
14. Explain the importance of graphic communication in Engineering Science.
15. What do you know about the following ?
    (a) Pictograms
    (b) Charts, graphs, tables and bar graph.

16. Write short notes on:
    (a) Media of non-verbal communication
    (b) Verbal communication
    (c) Body language
    (d) Artifacts
    (e) Kinesics
    (f) Gestures
    (g) Facial expressions
    (h) Postures
    (i) Dress and Appearance
    (j) Silence
    (k) Eye contact
    (l) Voice modulation
    (m) Audience awareness
    (n) Proxemics
    (o) Haptics
    (p) Vocalics
    (q) Charts and graphs
    (r) Graphic communication
    (s) Pictograms
    (t) Maps, signs and signals
    (u) Tables, Pie – chart
    (v) Physical appearance

■■■

Chapter **5...**

Formal Written Skills

Contents

5.1 OFFICE DRAFTING: CIRCULAR, NOTICE AND MEMO

(a) Notice:

A notice is also one of the forms of communication in the corporate world. Almost all business or industrial concerns have to draft and circulate notices when certain messages are to be communicated to the members of the organisation or instructions are to be given to the staff members, or certain information is to be given to the general public. Notices are either circulated by post when they are meant for people outside the organisation, or they may be displayed on the notice board if the instructions are meant for subordinates and people within the organisational set-up. Notices, sometimes, are required to be sent to the persons concerned within a certain stipulated time particularly when they are related to meetings such as General Body meetings, Shareholders meetings.

Time-bound activities are also required to be notified to the concerned people by adhering to the prescribed time. Some examples of notices are given for the guidance of students.

Notice for the First Meeting of the Board of Directors of a Company.

SIDDHARTH ELECTRICALS PVT. LTD.
Vakadewadi, Pune 411005

Date: 27th March, 2013

Telephone:
Fax:

Shri. XYZ
___________________(Name of the Director)
___________________ (Address)

Dear Sir,

The first meeting of the Board of Directors will be held in our Registered Company Office at 5.30 p.m. on Friday the 27th April, 2013 to transact the business shown in the enclosed agenda.

You are, therefore, requested to attend the meeting and participate in the business.

Yours Faithfully
Sd/-
Secretary

Encl: Agenda

1. Elect Chairman of the meeting
2. Elect Chairman of the Company
3. Produce certificate of incorporation of Company
4. Appoint Secretary
5. Appoint Legal Advisor
6. Appoint Chartered Accountant
7. Appoint Bankers
8. Appoint Auditor
9. Adopt common seal of the Company
10. Decide the date of next meeting
11. Any other subject with the permission of the Chair
12. Vote of thanks to the Chair.

Notice of General Body meeting of a Co-operative Housing Society

SIDDHI-VINAYAK CO-OPERATIVE HOUSING SOCIETY LTD.
Regd. No. PNA/HSG/7234

Date: 28th Jan., 2013

Notice

The Annual General Body meeting of the society will be held at the Society's Registered Office at 4.00 p.m. on Friday the 20th February, 2013, to transact business of the enclosed agenda.

Copies of the audited accounts statements such as Balance Sheet, Profit and Loss Account, Audit Report for the financial year 2012 - 2013 are enclosed.

All the members are requested to attend the meeting and participate in the discussions.

If the required quorum is not fulfilled at the scheduled time, the meeting will be postponed for half an hour and thereafter the meeting will be held at the same place. The decisions taken in the said meeting shall be binding on all the members.

If any of the members wish to move any subject in the meeting, the same should be communicated in writing to the Secretary so as to reach him before seven days of the scheduled meeting.

Chairman / Secretary

Agenda:

1. Elect Chairman of the meeting.
2. Read and approve the minutes of the last general body meeting of the society. (Already circulated).
3. Approve the accounts statements and the audit report.
4. Read the Chairman's annual report about the progress made by the society.
5. Read and discuss suggestions made by the members.
6. Any other subject with the permission of the Chair.
7. Vote of thanks to the Chair.

Public Notice

ABC COMPANY PVT. LTD.
3, Elphinston Road, Khadki, Pune 411003

Public Notice

It is hereby notified for the information of all people that Shri. R. V. Samant is no more in the service of the Company as a Company Co-ordinator and Chief Accountant. Nobody should, therefore, make any transaction meant for the Company with him. The Company will not be responsible for any such dealings with him by any person.

This notice is issued in public interest on Friday, the 28th March, 2012.

Sd/-
Advocate for
ABC Company Pvt. Ltd.

(b) Memo

The term 'Memo' is derived from the Latin word 'memorate' meaning thereby 'to mention' or 'to tell'. It is a short form of the word memorandum and its plural form is 'memoranda'. The literal meaning of the word memorandum is 'a note to assist memory'. A memo is generally used for internal communication between officers of the same level (equal ranks). Whenever an office or company executive wishes to issue certain directives or instructions to his staff, he writes a memo and sends it to the person concerned. Since a memo is meant only for internal communication, it is never sent outside the organisation. A memo is different from a letter so far as its form is concerned. It is written directly in a less formal style. A memo only has a heading, body and signature. It does not contain a salutation (Dear ...) and a complimentary close (Yours ...). Since it moves from one department or employee to another, it must contain words like 'From' and 'To'. A memo has to be very brief and to the point. It is one of the most convenient and inexpensive devices. A memo can be used:

1. To seek suggestions.
2. To communicate policy changes.
3. To give suggestions.
4. To issue instructions to the staff.
5. To request for help or information.
6. To confirm a decision arrived at, on the telephone.

The format of a memo is different from that of a normal letter. Since a memo moves from one department to another or one employee to another, there is no salutation. The writer's signature is put without a subscription or complimentary close. It is written in a direct style.

Inter-Office Memo

Date: ________

No. ________________

To

____

____

From

____

____

Subject: ____

1.
2.
3.

Signature

In government offices, the format of a memo is little different.

Memo

Office of the: ________

Date: ________

Subject: ____

1.
2.
3.

Signature

Designation

To,

XYZ

Department/Section

Sometimes, it becomes necessary to instruct subordinates to refrain from certain undesirable habits. The most common among them is 'not being punctual'. Such an attitude disturbs office discipline. Under such circumstances, an office chooses to issue a memo, which is a sort of mild warning, to such an employee which also sets an example before other subordinates.

MODERN ELECTRONICS PVT. LTD.
Near Vanaz Factory, Paud Phata,
PUNE 411 052

Date:__________

Reference: A note dt. ....
from the Time Dept.
and Production Dept.

From:

Ramesh Bhatia
General Manager

To:

Shri. V. N. Gupta
Jr. Supervisor

Subject: Frequent absence from duty

It has been reported by the Time Department and the Production Department that you are in the habit of remaining absent from duty quite often without any prior intimation. This adversely affects the smooth working of the department and also disturbs company discipline. This has been brought to your notice by the heads of the above department. However, you seem to have taken no notice of these instructions and continued the habit of remaining absent from duty without any pre-information and sanction.

You are, therefore, strictly instructed to be very careful in future and avoid this undesirable habit. For your information, repetition of such instances in future would badly reflect on your personal record, and may result in penal action against you. This is the last opportunity given to you for showing improvement in your attendance performance.

(Ramesh Bhatia)

Some more specimens of memos

A memo requesting quarterly reports:

PHEONIX TEXTILES LTD.
Head Office: Ahmedabad

MEMO

Date: 22nd January, 2012

No.: Progress Report 102
From: S.P. Joshi
Marketing Manager

To: All Branch Managers

Subject: Quarterly Report

The Board of Directors in its meeting held on Jan. 20th, 2012 has decided that the sales performance of the company be reviewed every three months. Therefore, you are requested to make sure that your quarterly report reaches the head office on 7th March, 7th June, 7th September and 7th December.

Thank you.

C.C. to:
R.P. Gupta, Mumbai Branch
S.K. Basu, Pune Branch
P.P. Mehra, Nasik Branch
S.S. Cheema, Solapur Branch
N.K. Menon, Kolhapur Branch

A memo regarding safety precautions:

NEW AGE STEEL MILLS
MEMORANDUM

Date: 5th March, 2012

No.: 1701/2012
From: The Managing Director

To: The Human Resource Manager

Subject: Safety Precautions

We have decided to launch a drive to improve our safety precautions throughout the factory due to the accident which occurred in the blast furnace recently. First of all, we want to eliminate the source of accidents by removing all possible danger points. Secondly, we want to make all our employees fully aware of dangers to themselves if they do not follow instructions:

1. Please prepare a report for me on what you consider the danger points to be and how to protect the staff against them.
2. Inform all employees of the risks to them if they do not follow all safety precautions.

(S. K. Kanade)

There is also a category of 'warning memos'. It is generally served to a subordinate whose work and/or conduct is unsatisfactory or against rules and regulations and does not show improvement inspite of oral instructions from time to time. A warning memo is not a threat of removal from service. However, it categorically states that better work and behaviour is expected from the individual concerned. Failure to show improvement may result in holding up of incentive, increment, promotion etc.

When such a warning memo is issued, care needs to be taken that vague statements about defaults should not be made. Such charges of irregularity, inefficiency should be supported by documentary evidences such as attendance register, leave register, work/performance, record/reports etc. Otherwise the management is likely to face trouble from the union.

A typical warning memo is given below:

Dear Mr. C. L. Manjarekar,

It has been brought to my notice that your attendance at the office has been very irregular. You have reported late by more than half an hour, six times during this month, in spite of warnings by your supervisor.

There have also been several mistakes in your work, indicating inattention and lack of application. These mistakes were pointed out to you by your supervisor on each occasion, yet you have not shown any concern to perform better.

You are expected to be regular and punctual in attendance and to be careful and attentive in your work. If you do not improve your conduct and your work, disciplinary action in the form of withholding increments, may be taken.

Yours sincerely,
Manager

(c) e-mail

e-mail stands for Electronic Mail. It is also known as email or e-mail. It is more than inter-personal messaging. At present, number of business organisations are increasing their business activities because of e-mail applications. It is used in personal as well as corporate correspondences. It is comparatively cheap and less expensive as well as it saves valuable time of business organisation. Electronic mail is a simple tool for receiving and sending messages between large number of users.

e-mail in business can be used for intra-company communications, marketing purposes and coordinating with business partners, suppliers and customers. Some common advantages of business email over traditional mail or telephone communications are cost reductions, lower overheads and speed. e-mail software is available for every size business, allowing for customized business and employee interactions.

An e-mail consists of two parts, one is address and second is message. The message is delivered to the receiver of the letter. This electronic letter looks just like any other simple document created using the word processor. This is one kind of file transfer from one computer to another computer, where a copy of the document gets copied from the computer which is sending message to receiver. An e-mail address works on the same principle as the telephone number.

Function

Although it cannot and should not replace all face-to-face communication and others forms of communication, internal e-mail usage can cover many areas within the business. Internal e-mails can function as an effective communication for sharing basic information, such as new product prices, paper use guidelines or security precautions, for example. Sending simple messages to an entire workforce with just the click of a mouse is fast, easy, convenient and can save the company money. If saved, the e-mail can function as proof of a message sent or received, and is easily accessible to remind the recipient of pertinent information. Many businesses use e-mail as part of its marketing efforts to share information with prospects, customers, vendors.

Format

- Business e-mails should be concise and to the point.
- One should use plain text and common fonts with a simple signature line.
- Fancy graphics, fonts, and backgrounds can take up unnecessary storage space in the recipient's inbox and may load slowly, or not at all.
- One should stick to one topic in a business e-mail and write only the things that are appropriate for anyone to read, as e-mail forwarding makes it possible for originally unintended parties to receive the e-mail.
- Proper grammar and spelling is very important in business e-mails because it reflects on the sender and his or her abilities.
- Attachments should be prepared in a format that any recipient can easily access/download.

Uses of e-mail:

1. e-mail reduces valuable time and cost.
2. e-mail can distribute information to company as well as to vendors, suppliers and customers.
3. It increases accuracy.
4. It is a very fast means of communication. An e-mail can reach its destination within seconds of pressing the ENTER Key on keyboard.
5. It can work as multi-media, as it can be used to send sound, video and computer software.
6. e-mail is cheaper than postal mail.
7. While sending e-mail it is necessary to be concise in transfer of matter.
8. In official letters precise words are used to save the valuable time.
9. e-mail is one of the most widely used of all internet applications.
10. e-mail fits in somewhere between telephone conversations and letters.
11. e-mail is not as immediate and informed as a telephone conversation.
12. e-mail is one of the most widely used of all internet applications.
13. It is method of exchanging digital messages.

Mail Etiquette

1. It is necessary to keep messages and its replies as brief as possible. An e-mail must be just, and to the exact point. No unnecessary details are to be added. Any reference must be make as short as possible and to the point.
2. Valuable internet time may be saved by making messages as brief as possible.
3. It is good idea to include details like name of the organisation at the end of the mail.
4. Whether sending an internal or external e-mail, the subject line should accurately represent the content of the e-mail.
5. One should use proper punctuation and capitalization, and use bold, italics and white space to make important information stand out from the rest of the content.
6. Use a salutation to begin the e-mail and only send to individuals that readily need or request to receive information from your business.
7. Business e-mails should be brief, positive and professional.

5.1.1 Circulars

- A **circular** is a form of business communication aimed at mass distribution to a large number of people at one time. It is a communication meant to notify or convey to all the customers, business friends, shareholders and employees, certain fundamental changes or important information. A circular is a letter which contains common information about the business which is printed or photocopied in sufficient number of copies and sent to all the concerned parties.
- They are used for giving relevant information to people within the organisation or for communication from the organisation to outside stakeholders. With the advent of computers and photocopying, they can be prepared only once, copies of which can be made for wider circulation. Each of these copies can look like the original. With modern technology, it would be possible to personalise them as well.
- The audience for a circular can be widespread and the number depends on whom the writer intends to contact. Occasions when circulars are issued can be many. Some examples are:
  (a) Establishment of a new business.
  (b) Starting a new department.
  (c) Business expansion, acquisition of new business, new collaboration etc.
  (d) Downsizing, break-up in partnerships/collaborations ec.
  (e) Change in business location.
  (f) Starting of new branches of business.
  (g) Admission of a new partner into the business.
  (h) Announcing price changes.
  (i) Clearance sale.
  (j) Sale of a business
- Circulars used for advertising and making public statements can be in the form of posters, leaflets or handouts that make people aware of a particular subject.

Example of a Circular

Giving Information about the New Price List to Franchisees

May 20, 2017

Sir/Madam,

We call your attention to the new price list of our product range, which we enclose. We are happy to announce that we have reduced the prices of some of our key products by 10% flat. This has been possible because of the strategic shift we have introduced in our inventory policy and also because of better logistic management. We have upgraded our capsule filling machine recently. This will help us scale up our production, thus ensuring seamless supply of these drugs in the coming years. This will help us service the market better.
We would also like to draw your attention to the technical specifications of items 4 to 7, where we have retained the old price, but there is vast improvement in the quality standards. We would welcome reactions from your customers regarding these products.

In view of the price reduction mentioned above, we have retained the credit period and payment terms as before.

You are welcome to visit our facility to understand firsthand the improvements we have brought in. That will give you an opportunity to interact with our marketing team

Please study our new price list and contact us if you have any questions. Your orders shall always receive best attention from us.

Yours faithfully,
ABXY Pharmaceuticals Ltd.

5.2 JOB APPLICATION WITH RESUME

- The first official communication related to your career is the job application. Since it is the beginning of your career, sufficient care has to be taken to prepare and draft your application for a job. Until recently, the conventional method of drafting a job application was simple and straightforward. It was in the form of a letter addressed to the employer explaining or giving personal details such as name, address, qualifications, experience (if any) and references. This information used to be general, accommodated in a single page. However, times have rapidly changed and so has the method of sending job applications.
- The job application is divided into two parts viz. Forwarding letter and Bio-data. The forwarding letter is in a simple form addressed to the prospective employer with one's bio-data in a table form covering different aspects of one's personal details. The forwarding letter simply contains one's name, date, address to whom it is to be sent, the job one is seeking, the source through which the vacancy is known and a sentence indicating that the bio-data is enclosed.
- The second part is the bio-data which is also called 'Resume' or Curriculum Vitae, abbreviated as C.V. This is the most important document in the process of job application. Hence, this has to be prepared with utmost care and should highlight the applicant's achievements. It should emphasize the details to show one's suitability for the job one is applying for. Bio-data is organised in a table form. Tabulation ensures a disciplined and orderly presentation and allows no omissions. In general, the information to be included in the bio-data is classified into the following five broad categories:
  1. Personal details.
  2. Educational qualifications.
  3. Extra curricular activities.
  4. Experiences of work.
  5. References.

1. Personal details:

This should generally cover the items given below:

(a) Name in full, ppreferably in Block Capitals in the order of the first name, middle name (father's name) and surname. If specifically demanded, the surname may be written first, followed by first name and middle name.

(b) Date of birth

(c) Height (in cms)

(d) Weight (in kg)

(e) Marital Status (married or unmarried)

(f) Mother tongue

(g) Other languages known

If applying for a job outside the country, include the following details:

(i) Nationality

(ii) Passport number

(iii) Period of validity of the passport

(iv) Foreign languages known

(v) Level of skill in the foreign language such as certificate level, graduation, etc.

2. Educational Qualifications:

(i) A complete record of academic qualifications with ranks must be provided, beginning with S.S.C., H.S.C, followed by further hierarchy such as graduation, post-graduation etc. Details such as name of institutions attended, universities from which diploma/degree obtained etc. should also be mentioned. Nowadays, a reverse order may also be followed, giving details of the last degree or diploma obtained first, and then of subsequent previously obtained qualifications, with school details coming in the end.

(ii) Additional courses, trainings.

(iii) Award or prizes, scholarships etc.

3. Extra Curricular Activities

Under this heading the co-curricular activities one has participated in are to be explained. They may be relating to sports, drama, study circles, workshops, seminars etc. A mention of prizes, awards, distinctions achieved should be made. These aspects throw light on one's personality traits such as teamwork skills, inter-personal skills, organisational skills etc.

4. Experience of work

Include in this, the information related to one's:

(i) Present employment indicating the employer.

(ii) Previous employment stating the name of the companies.

(iii) Job description of the present as well as the past employment.

(iv) Specific achievements, if any.

(v) Duration of employment with each employer

Fresh candidates who are beginning a career can mention here if they have performed any voluntary work for any social, cultural, charitable organisation, short period work during vacation etc.

5. References:

Generally, three references are expected. They should include teachers and/or guide, an employer or a respectable individual in the society. Names, titles, addresses, telephone number should be given indicating one's relationship with them. It is necessary to obtain permission from these people to include their names as references.

A specimen of an application with Resume/Bio-data/Application to Modern Automobile Factory, Pune 16, for the post of a Junior Engineer.

XYZ
19 Zilla Peth, Jalgaon
10th November, 2012

To,

The Manager,
Modern Automobile Factory,
PUNE 411016.

Subject: Application for the Post of Software Engineer.

Sir,

This has reference to your employment notification in the Indian Express of 5th November, 2012. I, the undersigned, wish to apply for the post of a Junior Engineer in your organisation. Enclosed is my bio-data showing my qualifications, experience, references and other details

I hope my qualifications and experience suit the job which you have advertised. If I am given an opportunity to serve in your organisation, I will leave no stone unturned to work satisfactorily. I am ready to accept the scale of pay as per your advertisement. As I come from Pune, the job would be advantageous for me and convenient to my aged parents.

I am enclosing a copy of my resume which would highlight my credentials.

Thanking you,

Yours faithfully,
R. K. Deshmukh

Encl.: Bio-data.

BIO-DATA / RESUME / C. V. (CURRICULUM VITAE)

1.	Name	: RAMCHANDRA KRISHNAJI DESHMUKH OR (DESHMUKH RAMCHANDRA KRISHNAJI)
2.	Address	: 19, Zilla Peth, Pune 10 Phone No.: ____________
3.	Date of Birth	: 10th April, 1988
4.	Height	: 159 cm
5.	Weight	: 50 kg
6.	Marital Status	: Unmarried
7.	Mother Tongue	: Marathi
8.	Other languages known	: Hindi, English
9.	Nationality	: Indian
10.	Passport No.	: EA.02493782
11.	Validity of Passport	: 10th April, 2020
12.	Foreign Languages known	: German and Japanese
13.	Level	: Certificate Course in German Certificate Course in Japanese

Educational Qualifications

Year	Exam.	School / College attended	University	Subjects	Marks	Class
2003	B.E. (Inf. Tech.)	College of Engineering, Pune	Pune	Marks list showing the marks obtained in different heads of subject is enclosed		I. Held University Scholarship
1999	H.S.S.C.	Modern High School	H.S.S.C. Board Pune	Physics, Chemistry, Mathematics	97%	I. with Distin. M.J. Memorandum Prize Holder
1997	S.S.C.	Modern High School	H.S.S.C. Board Pune	All compulsory subjects	85%	I. with Distin. National Scholarship Holder

14. Extra Curricular Activities
    (i) Represented Pune University in the Cricket Team during all four years.
    (ii) Participated and won prizes in Drama Competitions.
    (iii) Was a state Badminton Champion during the year 2000-2001.
    (iv) Was an active member of the University Youth Club. Held Presidentship in the year 2002.
    (v) Attended Workshops and seminars arranged by the College.
15. Work Experience:
    (i) Worked as a Software Engineer with Sunrise Infotech Pvt. Ltd. during the period 1.8.2003 to 31.12.2005.
    (ii) Working as a Software Engineer with International Information Systems Pvt. Ltd. since 1.1.2005.
16. References:
    1. Prof. S.S. Mohite,
       Head, Department of I.T.
       C.O.E.P. Pune
    2. Prof. M. N. Kamat,
       Professor of Computer Science
       C.O.E.P.
       Pune 020-25333431
    3. Shri. P.V. Majgaonkar,
       Chairman, Ph. 020-25454564
       Vasudeo Sahakari Sugar Mill,
       Indapur, Dist. Pune Ph. 020-25898235

Conventional Application in letter form

- Sometimes, applications for jobs are also sent in a conventional letter form. Generally, such traditional letters begin with reference to the advertisement, announcement or declaration of vacancies by the employers. It is also possible these days, that the agencies providing candidates for different jobs direct the candidates registered with them to submit applications to those who place their demand for recruiting staff. Under such circumstances, applications may be submitted in a general letter form. In such cases, the only difference is in the method of furnishing personal information. Instead of the table form it is written in a continuous text form. A specimen of such an application in letter form is given below.

From: XYZ
33, Ghorpade Peth
Pune 411 002

Pune 411 002
Dt. 5th November 2012

To,
The Manager (Administration)
Hindustan Steels Pvt. Ltd.
Pune 411 007

Subject: Application for the post of Assistant Manager

Sir,

With reference to your advertisement in the Times of India, dated 1st November, 2012, I, the undersigned, wish to submit my application for the post of Assistant Manager in your company; Furnished herewith are the details of my qualifications, experience and other aspects for consideration.

Education:
M. Com. I Class with Business Management as a special subject from BMCC, Pune 2001-
Diploma in Computer Application, 1999
Diploma in Industrial Electronics, 2004

Experience:
(i) Store Superintendent from _____ to _____ with M/s. Universal Hardwares, Ltd. Pune.
(ii) Personnel Supervisor in the Time Department of Majestic Steel Manufacturing Company Ltd., Pune from 1998 till this date.

Personal Data:
Age - 28
Married
Languages known: Marathi, Hindi, English.
Hobbies and Games: Cricket, Badminton, Travelling, Photography.

References:
(i) Shri. A.B. Surve
Manager, Universal Hardwares Ltd.
Pune
(ii) Shri. D. N. Madhok
Personnel Officer, Majestic Steel Manufacturing Company Ltd.,
Pune
(iii) Shri. V.B. Dhole Patil
MLA
Yerwada, Pune 411 009

Thanking you,

Yours faithfully,
XYZ

5.3 Business Correspondence: Enquiry Letter, Order Letter, Complaint Letter and Adjustment Letter

Introduction

- As was in the past, business correspondence functions as the breathing process of the modern corporate world which consists of various types of business concerns and industrial empires. In the vast expanse of the modern world, it is almost impossible to perform all business or production activities through personal contacts or through oral communication. Non-verbal communication which includes writing of letters is the best tool for exchanging information, furnishing evidence, maintaining records for future reference, building goodwill and achieving other allied objectives. It is not, however, that easy to achieve these objectives unless the business correspondence is effective.

Quality of Business Letters

- Business correspondence becomes effective only if it possesses the following qualities:
  - (i) Clarity
  - (ii) Simplicity
  - (iii) Conversational style
  - (iv) Correctness
  - (v) Courtesy
  - (vi) 'You' attitude
  - (vii) Persuasion
  - (viii) Sincerity
  - (ix) Completeness
  - (x) Public Image.
- Business correspondence is really an art in itself and requires constant and consistent practice. This can be achieved through alertness, careful observation, desire to learn and retentivity of what is learnt.

Layout of Business Letters

- A typical layout of business letters has been discussed earlier. It can briefly be stated as under for quick recollection. Business letters contain the following points:
  - (i) Heading
  - (ii) Date
  - (iii) Subject
  - (iv) Reference
  - (v) Salutation
  - (vi) Body of the letter
  - (vii) Complimentary close
  - (viii) Signature
  - (ix) Identification marks
  - (x) Enclosures
  - (xi) Copy distribution
  - (xii) Postscript

  Business letters are typed and sent on company letterheads.

Physical Appearance of a Business Letter

- As the content of a business letter is important, its physical appearance is also equally important. The physical appearance should be attractive, pleasant and project a good impression on the reader. Its appearance creates goodwill of the business house. The factors which are required to be kept in mind from this point of view are enlisted below:

(a) Paper of the letter	–	Should be of good quality.
(b) Quality of typing	–	Font, neatness in layout, error-free, punctuation, good margins.
(c) Appropriate paragraphs	–	The letter should be divided into appropriate paragraphs related to the matter
(d) Folding	–	Folding of the letter should be appropriate to the size of the envelope.
(e) Address	–	Should be uniform on the letter and the envelope in which it is to be sent.
(f) Envelope	–	The quality, size, colour etc. of the envelope should be attractive and appropriate to the size of the paper used for the letter.

Types of Business Letters

- As a matter of fact each and every letter or correspondence that a business house of an industrial concern undertakes, cannot be categorically classified. Every letter written on a new subject, issue, event, situation in itself becomes a type. However, for the sake of convenience and understanding, some of the important types of business letters can be classified into different categories such as:
  (a) Letters of enquiry.
  (b) Replies to enquiry letters.
  (c) Letters of quotation.
  (d) Letters for placing orders.
  (e) Complaint letters.
  (f) Adjustment letters.

(a) Agenda

- An organised meeting, whether regular or not, will be effective if the topics to be discussed are organised in a pre-determined manner, the overall time for the meeting and time allocated to each topic/speaker is indicated along with their topic of presentation. In short each and every business meeting needs a well written agenda. Thus, a meeting agenda is the list of items that participants hope to accomplish at a meeting. The agenda should be distributed to participants several days in advance of a meeting, so that they have the opportunity to prepare for the meeting.

Significance of the Agenda

- An agenda is the statement of the business, including supporting documentation, for consideration/ noting at the meeting. The procedure for preparing the agenda for a meeting varies according to the level at which it is held. If it is a review meeting or a regular departmental meeting, the concerned manager drafts the agenda with the help of the support staff. At the board level, however, the secretary of the organisation drafts the agenda and the chairperson either suggests amendments or approves it. Once the agenda is finalised, it is circulated to the members and other interested parties.
- Without a written agenda, business meetings tend to be chaotic, run longer and deviate from the main objective. On the other hand, preparing an agenda will help you structure the meeting, focus it on the most important problems, and identify its desired outcomes. Besides, all participants of the meeting will be equally aware of the issues that will be taken up for discussion.

- The creation and timely circulation of an agenda gives the participants an opportunity to prepare for the meeting. It also enables them to give the Chairman feedback about the subject and topics of the meeting and to suggest changes to the agenda if necessary. Deviation from the agenda often makes it difficult for the moderator to manage the meeting. It results in digression and loss of focus. The outcome will not be satisfactory, making follow-up activity difficult. It will be difficult to match the agreed objectives to the outcomes and so hinder the subsequent monitoring of the actions according to the decisions taken. Such a chaotic meeting is a waste of time and expenditure.
- Hence, having no written agenda can never be an option for organising a professional meeting. The manager may consider himself to be very sharp, organised and has excellent memory. Yet it is necessary to write down the meeting plan for two reasons: (a) Important items may be forgotten, and (b) The participants need prior information so that they can organise their own material and time.

Contents of an Agenda

The following items should appear on a typical meeting agenda:

- Purpose or subject matter. Basically, an objective has an active component, and chalks out what needs to be done or decided during the meeting.
- Date and start and end time of the meeting.
- Venue.
- Names of the participants, and their position.
- The items for discussion and action to be taken to accomplish the purpose of the meeting.
- Anticipated for each item, and time allotted for breaks, questions and answers session etc.
- Preparation required for the meeting, such as review of any documentation circulated, data, minutes of the previous meetings etc.
- Facilities available at the venue such as flip charts, overhead projector, voice recording, video recording, arrangements for recording the minutes etc.
- Any special requests such as keeping cell phones on silent mode etc.
- Any refreshments that will be served and timing.

Tips for Preparing a Meeting Agenda

- Meetings without firm direction are unlikely to be constructive, and do not lead to tangible or effective results. In addition, meeting participants may become disengaged if a meeting moves away from the relevant topics, setting a poor precedent for future meetings and possibly discouraging future attendance. This will affect the credibility of the manager as a leader. If your job description requires that you have to organise meetings frequently, prepare an agenda template, in to which you can feed the details of each meeting.

> "Begin with the end in mind." Stephen Covey in the book *Seven Habits of Highly Effective People*.
>
> The purpose of the meeting will determine the meeting focus, the meeting agenda, and the meeting participants.

- Determine the goals of the meeting and mention it clearly in the agenda. Discuss this with a few key participants and finalise.
- Determine the date, time and venue for the meeting. Make sure the venue is available and that all necessary facilities are available there.
- Prepare a cover sheet giving the main idea, date, time, venue, the names and contact details of the organisers and the date before which confirmation of participation should be received. Where necessary, give details of the registration procedure, directions to reach the venue, transportation and other arrangements.
- Determine the list of participants or subject specialists you would like to have at the meeting and send letters of invitation to them with details of the meeting.

- Inviting the right participants will enhance the chances of successful deliberations. Do this by asking yourself some useful questions:
  - Who needs to know the information you are distributing?
  - Who has the experience or expertise to share with the group and provide the right information to guide decision making and contribute to the success of the meeting?
  - Who must provide permission or resources to accomplish the meeting outcome?
  - Who must support the meeting outcome and also the implementation of any solutions or tasks that emerge from the meeting?
  - Who might oppose the implementation of any outcome, solutions or direction? Should you invite them? Or can you plan a strategy to deal with them?
- Identify agenda items for the meeting. Make sure the schedule is full of activities. This is more so in workshops, consultations and conferences running into a number of days.
- Within the overall objectives of the meeting, either determine the topics and sub-topics in consultants with subject specialists invited for the meeting or ask them to suggest. This will help you prepare a list of topics or agenda items to discuss. Ask them to indicate the amount of time to be allotted to each segment. All this will require a lot of time and continuous interaction with them. Follow-up.
- Determine if you need help from other staff in planning the meeting. Identify them and have a meeting to discuss the details.
- Once these topics are identified and priorities fixed, give numbers to each of them in the order in which they will be taken up. Usually, this exercise is done about a month before the meeting so that everyone is able to prepare the issues they would like to discuss during the meeting.
- Organise the agenda considering the priority items or those requiring lengthy presentation and discussions. Put them at the top. Exclude those agenda items that aren't aimed to achieve the stated objective of the meeting.
- When allocating time on each of the agenda items, remember that time is needed not only to present a certain issue, but also to discuss it. Therefore, take the discussion time into account when planning your duration expectations.
- Assign presenters for each agenda item. Not all of the agenda items should be presented solely by you. It is better to involve more people in active participation. Assign the persons who will be responsible for taking the minutes.
- Make sure you include people who can contribute meaningfully. Do not include anyone who is not associated with the subject unless it is desired by the top management. Sometimes, the top management may want an under-study to observe and learn.
- Prepare all attachments or reading material that you would like to enclose with the agenda. It may be minutes of the previous meeting to be confirmed or any information pertaining to the subject to be discussed.
- Give advance directions to the participants**.** Prior to despatching the agenda, request the speakers to hand in any paper or documents with appropriate references that they would like to attach to the agenda, so that they can be distributed in time.
- Check the agenda for mistakes, as well as for points you may have missed.
- Distribute the agenda well in advance, at least two weeks before the meeting, so that the participants can schedule them.
- Place agenda copies at the table of each speaker. Conform to the agenda at the meeting.
- At the end of the meeting, the moderator should sum up the discussions and the decisions taken according to the agenda points. This becomes a reference for future meetings on the subject.

- Writing an agenda for a programme, especially for a multi-day programme, is sometimes challenging as it has to give all necessary details but in a brief manner. A manager has to take all care to plan the events, draft the agenda, consult with others where necessary and finalise it in time, such that it is possible to circulate it to the participants and invitees and get their response ahead of making all the ground level arrangements at the venue.

Example of a Business Agenda

Date: 17 July 2013

Time: 11 am to 1.30 pm

Location: Conference Room of Marketing Department, Head Office

Attendees: Mr. Anil Gupta, Ms. S. Joshi, ..., ..., ..., ...

Objective: Develop strategy to introduce the new cell phone application developed by the company. Participants are expected to come prepared to discuss the scope for the product in their regions and the support they require. A brief description of the product has already been sent to them.

Arrangements: PC and LCD projector.

Schedule:

11.00 to 11.25: Mr. Gupta, Marketing Director to elaborate on the new application and its application areas.

11.25 to 11.40: West Zone

11.40 to 12.00: South Zone

12.00 to 12.20: North Zone

12.20 to 12.40: East Zone

12.40 to 01.00: Questions and answers

01.00 to 01.30: Response from marketing team, Summing up of major recommendations and actions with specific dates.

Responsibilities

Recording the Minutes: ....

Note taking: ....

Moderator: ....

Contact Person: ....

Contact Details:

The meeting will be followed by lunch served at the same venue.

Memorandum

- Compared to a circular or a notice, a memorandum is usually less public, has a smaller audience and is usually more exclusive. They are often used as a way of reminding someone of something that needs to be done, or to pass on a proposal or idea of some kind. They are most commonly typed or even hand written.
- The literal meaning of the word memorandum is 'a note to assist the memory'. The term 'Memo' is derived from the Latin word 'memorate' meaning 'to mention' or 'to tell'. It is a short form of the word memorandum and its plural form is 'memoranda'. A memo is generally used for internal communication. Whenever an officer or a company executive wishes to issue certain directives or instructions to his staff, he writes a memo and sends it to the concerned persons.
- A memo has to be very brief and to the point. It is written directly in a less formal style.
- **Contents:** Since it moves from one department or employee to another, it must contain the following:
  - Words like 'from' and 'To', to indicate where it has come and to whom it addressed.
  - Heading,
  - Body and
  - Signature.

- There is no need for a salutation (Dear ...) nor a complimentary close (Yours ...).
- **Significance:** It is the most convenient and inexpensive device for communication within the organisation. A memo can be used for the following purposes:
  1. To seek or give suggestions.
  2. To communicate policy changes.
  3. To issue instructions to the staff.
  4. To request for help or information.
  5. To confirm a decision that was arrived at over the telephone.
- Sometimes, it becomes necessary to instruct subordinates to refrain from certain undesirable habits. The most common among them is being late, spitting in the office premises etc. Such an attitude disturbs office discipline. Under such circumstances, an office chooses to issue a memo, which is a sort of mild warning, which also sets an example for other subordinates.
- There is also a category of memo, known specifically as 'warning memo'. It is generally served to subordinates whose work is unsatisfactory or against the rules and regulations of the company. In spite of several oral instructions, they have not shown improvement in their attitude or behaviour. A warning memo is not a threat of removal from service. However, it categorically states that better behaviour is expected from the concerned employee. Failure to show improvement may result in holding up of incentive, increment, promotion etc.
- While issuing a warning memo, the manager has to make sure that there is enough documentary evidence to show the errant behaviour of the recipient. The evidence could be by way of attendance register, leave register, work/performance record/reports etc. Issuing warning memos based on vague charges may lead to conflict, which the unions can escalate.

Email Etiquette OR Netiquette

- **Netiquette,** short for "Network Etiquette" or Internet etiquette", is a subject for study as part of business writing, because the Net has taken the place of sending paper copies by post, fax etc. It is faster and very convenient for organisations as today the Internet is ubiquitous. Netiquette is the etiquette of cyberspace. Etiquette means "the forms required by good breeding or prescribed by authority to be required in social or official life." In other words, Netiquette is a set of rules for behaving properly online. Netiquette has a dual purpose, to help net newbies minimise their mistakes, and to help experienced cyberspace travelers guide the newbies. It realises that most people would rather make friends than enemies, and if a few basic rules are followed, the Net users can make many more friends than enemies. With social networking becoming more and more popular, etiquettes are being compiled for other portals. 'Twitter-quette' is one example.
- Computer networks bring people together who would otherwise never meet. However, this medium is impersonal and so, the correspondence becomes less personal. It is said that humans exchanging email are often like those behind the wheel of a car. They curse the other drivers, make obscene gestures, and generally behave rudely. Such behaviour is hardly seen at home or at work, where people meet and talk. Since there is a machine in between, they think they can be at their worst. The purpose of Netiquette is to convey to users that such behaviour is not acceptable. The network should be used in a positive and constructive manner to express oneself freely, access a world of information, enrich themselves and explore strange new worlds without stepping on the toes of others.
- The e-mail has replaced many functions of the business letter. Here are some tips on writing effective e-mail. Email has made a number of major changes in the way business functions. One of the biggest advantages in business is the convenience it provides. It is easy and costs nothing to send mail to many in the company. Even in education, cyberspace has the potential to be a tremendous resource for students and teachers at all levels.

- **Netiquette** is described as a set of social conventions that facilitate interaction over networks, ranging from Usenet and mailing lists to blogs and forums. These rules were described in IETF RFC 1855 document. (The Internet Engineering Task Force (IETF), which develops and promotes Internet Standards.)
- The USENET netiquette highlights using simple electronic signatures, and avoiding multi-posting, cross-posting, off-topic posting, hijacking a discussion thread, and other techniques used to minimise the effort required to read a post or a thread. Similarly, some Usenet guidelines call for use of unabbreviated English, unlike the instant messaging protocols like SMS.
- Companies like IBM post their own netiquette guidelines for their employees highlighting the basic need for professionalism, amiable work environment, and protecting the company's intellectual property. Some companies limit the use of the Net in their offices, as accessing the Net and communicating on the social networking sites becomes addictive.
- Like many Internet Phenomena, the concept of netiquette and its application keep changing, and vary from community to community. However, the basic Do's and Don'ts should be understood and followed.
- It is often possible to contact powerful people directly, because many executives read their own email, rather than having it screened like their paper mail and phone calls.

> *"I am the only person who reads my email so no one has to worry about embarrassing themselves or going around people when they send a message."* **... Bill Gates**

- There are some managers who do not know how to cope with email facility and are perpetually "email overloaded", while others may be so accustomed to it that they become "snail-mail ignorers", i.e., they forget or do not bother to read their paper mail.

Tips for email etiquette

- In addition to the netiquette for exchange of mails among family and friends, there are a few additional conventions for email at work, which are distinct. *Please note that all the tips given above apply here as well.* Here are some tips:
  1. **Do not get too informal:** The tendency when it comes to e-mail is to write less formally. Remember that you are still writing a business letter, except that the mode of conveying it is technology based.
  2. Though professional email can still be informal, a lot of the humour and wit that is appropriate in a discussion group is out of place at work.
  3. Do not discuss personal problems or company politics or criticise a colleague or boss on email or social networking groups. You never know where it lands and create problems for you.
  4. Email etiquette forbids spreading of misinformation, personal or confidential information, or any statement that is needlessly damaging to the organisation and to others.
  5. When you introduce yourself via email, you are not only making a first impression, but also leaving a written record. So use caution, especially when dealing with powerful people. Make sure that they do not mind receiving an email.
  6. In order to ensure that people know who you are, include a line or two at the end of your message with contact information. You can save a ".sig" or a "signature" file and add it to the end of your messages. It takes the place of your business card. It should not exceed 4 lines. Generally signature files are not used in work email within the organisation. It is used only when the email is sent to external recipients.

     Keep the mail messages short! Do not exceed 100 words per mail, and 65 characters per line. But if the message is a lengthy one, prepare the reader by including the word "Long" in the subject header. Lengthier files can be sent as attachments. A good rule of thumb would be not to send a file larger than 50 Kilobytes.

Remember that many people pay for connectivity by the minute, and the longer your message is, the more they pay. They would not appreciate lengthy mails.

Be careful when addressing mail. Know to whom you are sending the mail. Unsolicited mail costs the receiver as well.

7. **Format:** Use the format provided by the software properly. The 'To' is for the individuals the letter is directed to. 'cc' segment is for those you want to have a copy and 'bcc' refers to those you want to send a copy but you do not want people to know that they are in. Fill in the subject line. The recipient will know the subject and accordingly open it immediately or keep it for later. It will also help him recall and bunch all mails pertaining to a subject. Begin the e-mail with a normal salutation.
8. **Contents:** As with the business letter, keep it brief. All the business letter essentials apply to the e-mail as well.
9. **Links:** If you are referring to a web site, provide the link.
10. **Follow the ABCs of communication:** Accuracy, Brevity, Clarity and Courtesy.
11. Remember, mails on the 'Net can be archived. It is easier to revert to electronic mails than to hard copies stored in files and stacked away out of reach. In short, take care what you send on the 'Net!
12. While forwarding or re-posting an official message, do not change the wording. You may shorten the message and quote only relevant parts, but be sure you give proper reference.
13. **Be tactful:** "Flaming" is what people do when they express a strongly held opinion with full flow of emotion. It gets immediate attention. It is a long-standing network tradition and can be a lot of fun for the writer and the reader. Netiquette does not forbid it, but does not appreciate its perpetuation, like a series of angry letters, most of them from two or three people directed toward each other, that can dominate the tone and destroy the camaraderie of a discussion group. They monopolise the bandwidth. It is unfair to the other members of the group.
14. Common rules for e-mail and USENET such as avoiding flame wars and spam are constant across most mediums and communities. (**Flaming**, also known as **bashing**, is hostile and insulting interaction between Internet users)
15. **Be conservative** in what you send and liberal in what you receive. You should not send 'heated' messages even if you are provoked, not even in response to 'flames'. Wait overnight to send emotional responses to messages. If you still feel strongly, use FLAME ON/FLAME OFF to indicate your mood.
16. **Try to avoid sarcasm:** It may be misunderstood and remembered. Be careful with slang or local acronyms.
17. **Use mixed case:** ALL UPPER CASE LOOKS AS IF YOU ARE SHOUTING. Instead use symbols like * for emphasis and underscores for underlining.
18. If you send encoded messages make sure the recipient can decode them.
19. Unless you are using a protective firewall, you should assume that mail on the Internet is not secure. Your email address can be subject to hacking. Never put in a mail message anything you would not put on a postcard.
20. Do not open unsolicited mail or mail from unknown persons or respond to request for personal information. Do not respond to unsolicited mail asking for information. It is unsafe.
21. Check all your *mail subjects* before responding to a message and respond to only those, which are directed to you. Make sure you respond to mails where you are the primary recipient and not cc'ed.
22. Watch 'cc's when replying. If it is meant as a 2-way communication, do not include others.
23. **Do not use emoticons:** There are two reasons for avoiding these. (1) While many net-savvy users are familiar with it, there are still a lot of people who do not understand the abbreviations and symbols used in emails. (2) You are writing a business letter and

emoticons have no place in it. However, if the situation warrants,use smileys to indicate your tone of voice, but use them sparingly, and make sure the recipient knows what the character position stands for.

24. **Do not use Acronyms**: The acronyms may be misunderstood and result in a war of words.

25. It should be remembered that one's posts or can easily be made public. It can have disastrous consequences. Users generally take the privacy of their posts for granted. Even one-on-one communications, such as private messages on chat forums and direct SMSs, may be breached.

26. Emails can be forwarded easily and messages can spread faster than one can imagine, resulting in grave situations.

> This is a true story of a company CEO who misjudged the situation. He observed for several days that the car parking lot was fairly empty at the beginning of the work day and even before the closing hours. The indignant CEO sent a group e-mail to all managers to enforce discipline among their subordinates or they would be sacked. The recipients forwarded this to hundreds of other employees with a warning. This leaked to the media and the public. This resulted in a catastrophic drop in the company's stock price by the end of the day.

27. Emails can be stored and retrieved easily. When you communicate through cyberspace via email or on discussion groups, your words are written and stored in such a way that you have no control over them. In other words, if your mail has not been worded carefully and has hurt someone's feelings, there is a good chance they can create problems for you.

28. Beyond matters of basic courtesy and privacy, e-mail syntax allows for two different types of recipients, such as the main recipient and the CCs. The primary recipient defined can be expected to respond. If you are a 'CC' recipient, you are not expected to respond. You have received the mail for information only

29. It is advisable to use the traditional mailing list than CCs, as it can be misused. The "reply to all" facility quickly expands the number of responses manifold. In extreme cases, if the number runs into millions, it can bring down the mail server. In cases like this, rules of netiquette require efficient sharing of resources.

30. Another rule for online behaviour is to adhere to the same standards of behaviour online that you follow in real life. In real life, most people are fairly law-abiding, either by disposition or because they are afraid of getting into trouble. In cyberspace, the chances of getting caught may seem slim and so people may think that a lower standard of ethics or personal behavior is acceptable in cyberspace. This is a wrong notion.

31. The writer should realise that the receiver of an email or a message is also a human being with his/her own sensitivities. When you communicate electronically, all you see is a computer screen. You do not have the opportunity to use facial expressions, gestures, and tone of voice to communicate your meaning. Whether it is an email exchange or a response to a discussion group posting, it is easy for the respondents to misinterpret your meaning. All the writer has is words, and they should be used judiciously and caringly. Every time they write a mail, the users should ask themselves: "*Would I say this to the person's face?*" Be pleasant and polite. Do not use offensive language or 'flame'.

32. *Breaking the law is bad Netiquette.* If something is illegal in real life, chances are it is also bad Netiquette.

33. While writing mails, mistakes like spelling errors or even flame may creep in. The writer may have asked an irritating or even a stupid question. Netiquette requires that the recipient overlook the mistakes and not criticise the writer. If it is a minor error, it should be ignored. Even if you feel strongly about it, think twice before reacting. If you need to, then point out the mistake politely, in a private email, and not in public online fora. No one has the license to correct everyone else. Give people the benefit of the doubt. Arrogance and self-righteousness will make the dispute go out of hand. To avoid conflicts arising out of mistakes, companies can develop their own codified internal manuals of style to set an acceptable limit for user behavior.

34. Respect other people's time and bandwidth. When you send email or post to a discussion group, you are taking up their time. Make sure that the time they spend reading your posting is not wasted. Bandwidth is the information-carrying capacity of the wires and channels that connect everyone in cyberspace. There is a limit to its data carrying capacity at any given moment. People pay for this.

35. People access the Net at their convenience. Do not expect instant responses to all your questions or your arguments. It should be worth the trouble.

36. People as a matter habit, CC mails to people whether it is necessary or not. This is rude, as they will be wasting their time reading unwanted mail. Ask yourself whether they really need to know. If the answer is no, do not waste their time.

37. Most discussion group readers are already spending too much time sitting at the computer. They will have other matters to attend to.

38. Make yourself look good online. You will be judged by the quality of your writing. Spelling and grammar do count. Pay attention to the content of your writing. Check your facts. Bad information spreads like wildfire on the net and may come back to you in a totally distorted manner. You are responsible for what you post and not for what anyone else does with it.

39. Make sure your notes are clear and logical. It is perfectly possible to write a paragraph that contains no errors in grammar or spelling, but still makes no sense whatsoever if unwanted irrelevant words are used. Keep it simple.

40. Whether your mail is for giving information or in response to a question, try to include all the necessary information without digressing or being incoherent. It is a good practice to edit your email and make it perfect before sending it. It makes you look good and the reader will also appreciate it. It is not a good idea to send supplementary emails for the same purpose to indicate corrections.

41. Swear words are not nice. If you feel that cursing in some form is required, it is preferable to use amusing, tactful substitutes for harsh words or the classic asterisk filler, like s***. The meanings of these are known to all. This playfulness is somehow appropriate to the net, and you avoid offending the reader needlessly.

42. Check your mail 'inbox' preferably three or four times a day, but at least twice a day. You should always check your mail in the morning when you come in to the office and in mid-afternoon, or an hour or two before you leave.

43. It is bad manners to constantly watch the computer or cell phone screen while talking to someone. It is disrespectful and non-productive.

44. If you are really too busy to check your mail that often, or if you are going to be out for more than a day, consider delegating the task by putting a temporary password. Some systems also allow you to set up a limited-access password for the person screening your mail. Another way to handle the situation is installing an "answermail" facility. The message tells your correspondent how long you will be gone and whom to contact if the information is urgent.

45. If you think the importance of a message justifies it, immediately reply briefly to an e-mail message to let the sender know you have received it, even if you will send a longer reply later.

46. If you come across someone who claims to be too busy to answer emails, the option is either not to send emails to that person, or have the return receipt feature on the system. If it is important, follow-up on phone. In extreme cases, send a follow-up email and copy it to your superior even if it is going to make you unpopular with that person. But then your boss may view it as your incapability to work with that person, and form a negative impression about you. So, this step should be adopted with caution.
47. If people with whom you communicate are located across the globe, it takes time for the message to reach the recipient and for him/her to respond. Do not send reminders unnecessarily.
48. One way of using the Net is to enrich our knowledge. When in doubt about anything, one can ask questions online. Many persons may give intelligent answers. When you post a question to a discussion group that you do not visit often, it is customary to request replies by email instead of to the group. When you get all those responses, post the summary to the discussion group. That way, everyone benefits from the experts who took the time to write to you. Similarly, if you have researched a topic that you think would be of interest to others, post it. Sharing your knowledge is fun. It makes the world a better place.
49. Respect other people's privacy. Do not read other's emails. Even the system administrator has that power.
50. If there is a request for someone else's email address, take that person's permission and give it, along with work phone number if it is available.
51. Do not send an email notification that you have sent a document by post or any other medium. This is rude as it assumes that the intended recipient does not check his/her snail mail regularly, unless you know for sure that the recipient ignores paper work or the document being sent by post or courier. If the document is urgent, market it so and send it.
52. Many companies impose restrictions on the usage pattern of the Internet in the same way as they limit personal use of office telephone.
53. Some companies do not make it easy for their employees to send mail to everyone in the organisation. Employees may be designated for specific periods to screen every message that is sent out to all employees. Other companies have mail areas called "Junk Mail" or "Fourth Class Mail" that employees can use for classified ads, requests for general information etc. Such mail is separated from normal person-to-person mail, so that busy people are not bothered by it. If your company offers this benefit, use it, but follow the organisation's rules and check everything twice before you send it out.
54. Delete old and unwanted mail regularly. In most companies, mails are retained on the employee's computer only for a limited period. The main server of the organisation retains them. If there are important emails, save them in hard copy or on your own workstation. Make a list of situations where problems have arisen due to lost emails, and request the management to allow for additional storage facility.

The list is long. Net users should remember these tips and develop appropriate cyber world habits.

Significance of Internet access in India

- We in India have taken to the Net in a big way, and the number of converts to this is increasing by the day. Obviously, it is cheaper and faster than postal communication. For some, it is a status symbol, to indicate that they are keeping pace with this communication marvel! For some, it is an addiction, especially posting, tweeting, playing games etc. For many, it is convenient to do all bank transactions, pay bills, buy products, stay connected with their friends and relatives through email and social networking portals and also find some entertainment. For many business managers, it is an essential facility, especially in these days of quick correspondence, telecommuting, and groupware, to share computer screens and work together. Whatever it is, the Net is here to stay and expand into all aspects of our personal and business world.

- According to a survey conducted by JuxtConsult, an online research solutions consultancy, urban areas of India are home to over 30 million Internet users. Of these Web users, about 20 million users access the Web daily, and this is a significant number. Again, home usage accounts for 59 percent! Broadband reaches 77 percent of home Internet in urban India, and 74 percent of office-based Internet users. With the development of mobile communication technology of 3G (third Generation level), it is now possible to access the Net even on cell phones, as the Developed World moves to the 4G era, which helps in faster data transfer, both audio and visual communication. With cloud computing, optimisation of the networking facility round the world is being attempted.
- Statistics like these generate a lot of interest. One way of looking at it would be to remark that the country has arrived on the map of 'Net users. However, it should be remembered that the initial users of the Net were technically minded, and understood the nature of the medium and the protocols. Today, there are many who are new to the environment and are unaware of the etiquette. They should understand to behave properly on the Net and learn how one should 'Post' or 'Send'.

5.3.1 Need and Functions of Business Correspondence

- The success of any business depends upon sound relationships within the organisation and with other stakeholders and the public. Business Correspondence means exchange of letters or written communication, which is the best and cheapest means of communication. Correspondence constitutes an indispensable organ of business world. In modern days, trade and industries have expanded to national and international borders. Business activities are diversified. The businessman has to maintain contacts with the suppliers, the customers and the government in carrying out daily transactions. In other words, written communication is at the root of a successful commercial and industrial activity.
- Speech came before writing. Humans like to see and hear the people they communicate with. However, verbal communication is not always the most effective method of communication. Many times the message is not clearly understood because of other factors such as environmental distractions, body language, tone of voice and emotions. The message is lost because the audience is distracted by the other factors. Even in staff meetings or training sessions the way in which it is delivered can cause misunderstanding or loss of interest in the message and so written communication is used as a back-up.
- Writing is more unique and formal than speech. Written communication involved in business communication has great significance in today's business world. Effective writing involves careful planning of the ideas to be communicated, choice of right words, their organisation in correct order, in cohesive sentences, and in paragraphs. While speech is spontaneous and fast, writing is more valid and reliable than speech but it can cause delay and take time as feedback is not immediate.

Advantages of Written Communication

1. Written communication can be well thought out and organised to convey the message in a more precise and explicit manner. The writer can take time to formulate the document appropriately.
2. It can be erased, revised, rewritten and edited before it is dispatched.
3. It is effective as a standalone medium.
4. It can be targeted to reach specific individuals or parties.
5. Written communication helps in laying down apparent principles, policies and rules for the running of an organisation.
6. It can be stored and reproduced.

7. It is a permanent record of communication. Thus, it is useful where record maintenance is required. It is accepted as evidence in the courts of law.
8. It assists in proper delegation of responsibilities. While in the case of oral communication, it is impossible to fix and delegate responsibilities on the grounds of speech as it can be taken back by the speaker or he may refuse to acknowledge.
9. Effective written communication develops and enhances an organisation's image.
10. It can be cost effective, as it can be reproduced using appropriate technology and distributed to a large number of parties. With the Internet, it is faster and cheaper to communicate with a large number of parties and get feedback.

Disadvantages of Written Communication

1. Written communication adds to the time and costs in terms of stationery and the manpower employed in writing/typing and delivering letters.
2. Without the help of body language and vocal intonation to guide the reader, written communication is open to his/her interpretation, which can lead to huge misunderstandings. There is lack of personalisation in letters, emails, text messages and instant messenger tools. This lack of personalisation means there is greater need to think carefully about how these messages will be perceived. Poor written or verbal communication can lead not only to a distortion of the facts, but also of the writer's thoughts and intentions.
3. Perhaps the biggest problem with written communication is that once it is in the hands of the reader, it does not provide a second chance to make your point in a different way. Once the reader is lost, it is difficult to get them back. This is why you need to pick and choose your words carefully, and present your points in a style, manner and sequence that best suit the message you are sending.
4. If the receivers of the written message are separated by distance, and if they need to clear their doubts, they will have to wait.
5. Feedback is not immediate. The encoding and sending of message takes time.
6. Effective written communication requires great skills and competencies in language and vocabulary use. Poor writing skills and quality have a negative impact on an organisation's reputation and can create confusion and even disputes.
7. Too much paper work and e-mails creates a burden.

Types of Written Communication / Business Correspondence

- The different forms of written communication and business correspondence and their methods of delivery are described below. Each form has a use and, in some cases, a phraseology of its own. Only black or blue ink will be used in communications. A small margin of about one inch will be left on all sides (left, right, top and bottom) of each page of communications to ensure filing and better preservation of records as at times the paper gets torn from the edges, making reading of the documents difficult.
  1. **Letter:** This form is used for corresponding with individuals, various authorities, organisations and the public. A letter begins with an appropriate salutation. A letter can be informal, formal or in government parlance, semi formal or 'demi-official.'
  2. **Office Memorandum:** This form is generally used for corresponding with other departments or in calling for information from or conveying information to its employees. It is written in the third person and bears no salutation or supersession except the name and designation of the officer signing it.
  3. **Circular:** This form is used when an important and urgent piece of information or decision taken internally has to be circulated within a department for information and compliance by a large number of employees.
  4. **Inter-departmental note:** This form is generally employed for obtaining the advice, views, concurrence or comments of other departments on a proposal or in seeking clarification of the existing rules, instructions etc. It may also be used by a department when consulting its attached and subordinate offices and *vice versa*.

5. **Office order:** This form is normally used for issuing instructions meant for internal administration, e.g., grant of regular leave, distribution of work among officers and sections, instructions, financial sanctions, disciplinary actions appointments and transfers etc.
6. **Notification:** This form is mostly used in notifying the promulgation or declaration of statutory rules and orders. The government notifications of certain types are contained in publications in the Gazette of India.
7. **Report:** This is a detailed document describing an event, the current status of a project, marketing information etc. This is a time consuming activity and requires a lot of planning, so as to include all necessary information in a logical manner. For instance, a marketing survey report will contain information about market expectations, market size and shares of competitors, qualitative information regarding the activities, strategy and future plans of the competitors, government policy implications and the future outlook for the client party. The report will also give specific recommendations based on the study.
8. **Resolution:** These are policy decisions taken at company level meetings and are recorded as 'resolutions'. These are communicated to the shareholders and the executing authorities base their actions on these resolutions. In the functioning of the government, important matters of policy, e.g., the industrial licensing policy, appointment of committees or commissions of enquiry etc., are passed as resolutions by the concerned legislature or at cabinet meetings. These are published in the Gazette of India.
9. **Press Communiqué/Note:** The purpose of this is to give wide publicity to any matter of significance to the issuing organisation and the public. A press communiqué is more formal in nature than a press note and is expected to be reproduced in exact words by the press. A press note, on the other hand, is intended to serve as a hand-out to the press. It contains information which may be edited, compressed or enlarged by the media according to the importance that is attached to the matter.
10. **Endorsement or expression of support:** This form is used when a paper has to be returned in original to the sender, or the paper in original or its copy is sent to another department or office, along with approval, for information or for action. It is also used when a copy of a communication is forwarded to other parties.
11. **Advertisement:** This form is used for communicating with the general public to create awareness and may take the form of audio-visual or written communication.

Media for Despatch of Written Communication / Business Correspondence

- Written communication or Business Correspondence can be sent by the following media:

1. **Telegram:** This form is used for communicating with out-station parties in matters demanding prompt attention. The text of the telegram should be as brief as possible. With the increasing popularity of email, telegrams are not common. This form of written communication is soon closing down.
2. **Fax facility:** In urgent and important matters (including legal and financial messages), managers may use fax facilities to send messages, wherever available.
3. **Registered Post, Speed Post:** This method of delivery is used in sending written communication when receipt of despatch is required. With the advent of IT enabled services, movement of the document till it reaches the addressee can be tracked on line.
4. **E-mail:** This is a paperless form of communication to be used by managers having computer facilities supported by internet or intranet connectivity and can be widely used for subjects where legal or financial implications are not involved.
5. **Various media of mass communication**, such as press, hoardings, audio-visual communication, Internet based publicity.

Functions of Written Communication / Business Correspondence

- The need for written communication and business correspondence arises because of the need to achieve or "do" something, such as to:
    - Inform
    - Respond
    - Request
    - Report
    - Develop goodwill
- **Significance of written communication / business correspondence in a manager's work life:** People move through the workforce in stages. Each stage or position of employment has its own set of characteristics and requirements. This includes good written communication skills. A manager's communication skills indicate his/her level of professionalism and leadership qualities. In a professional environment, the clearer the manager is in written and verbal communication, the more proficient he/she will seem. If a manager's letter is badly worded or his/her presentation is badly planned and disjoint, it might appear that the person has not taken the matter seriously, or lacks the skills required for his/her position. Such a manager is termed as unprofessional. The effectiveness of a manager is judged by his/her ability to get across his/her points or views clearly in as few words as possible. Such a manager will demonstrate the capabilities of a good communicator who knows his/her subject well.
- Written communication usually requires more thought and effort. Writing must have a goal. It must be concise, informative and easy to read as both an informative and instructional tool. The importance of written communication in business is evident by the plethora of forms, manuals and materials that companies publish each day.
- It should be noted, however, that the reader's reaction to a written communication can be coloured by his/her knowledge, biases, values etc. A written communication will fail to convince if the reader has had an earlier bad experience with that person or organisation or has developed a prejudice through hearsay. Just like oral communication, written communication can also encounter barriers. The manager has to be aware of this and develop appropriate skills.

5.3.2 Enquiry Letter

- Letters of enquiry are broadly categorised as:
  (a) General enquiry letters and
  (b) Specific enquiry letters.
- General enquiry letters are written when the information required is of a general nature whereas specific enquiry letters are written with an intention to know about particular or specific information about a specific product, or a specific service provided by the organisation, etc. In the business world, these enquiry letters are mostly received from the buyers/customers and are concerned with goods and/or services. A seller's replies are generally concerned with giving information, quoting prices and conditions of supplies.
- In order to get a positive and quick response, enquiry letters should begin courteously, should be short, without much formalities and straight to the point. The phrases like *'could you please ...'*, *'would appreciate ...'* or *'please ... '* would make a good impact on the receiver and may make him send his reply quickly and wholeheartedly. The letter should be ended with friendly words like *'An early reply would be highly appreciated ...'*. Such an ending helps to create and establish a friendly relationship. In short, an enquiry letter should include the following:
  (i) courteous request
  (ii) what is required
  (iii) purpose of asking for the information
  (iv) conclusion with friendly words

- A few examples of enquiry letters are given below:

Enquiry Letter No. 1:

Date: 10th November, 2012

Dear Sir,

During my visit to your stall in the 'EXPO' exhibition recently held at the College of Engineering grounds, Pune I had seen some samples of your newly developed tiles. I plan to purchase them for my newly constructed house in Aundh area. Will you please let me know:

(i) whether you can send your salesman to inspect my premises and advise me.

(ii) in how many colours the tiles are available?

(iii) what the period of guarantee is and the durability of the tiles?

(iv) whether you undertake the job of fixing the tiles.

(v) whether you offer any discount.

I shall really feel obliged if you will respond to my enquiry as early as possible.

Thanking you,

Yours faithfully,

XYZ

Enquiry Letter No. 2:

Date: 20th November, 2012

Dear Sir,

We are conducting a general survey about the customers' demand for electrical and electronic goods and its trend during the past ten years. This survey would be of immense use to the producers and sellers of these goods for their future planning of production and sale. It will also help us to exploit an opportunity to introduce new products in the market or to improve the existing ones. Your comments on this aspect will also be of great value. Kindly, therefore, send the information about the sale of the following goods during the year 2002 to 2012.

1. Ceiling fans
2. Pedestal fans
3. Table fans
4. Tube lights (normal)
5. Tube lights (slim)
6. Television sets (Sony, BPL, LG, ONIDA, AKAI, Samsung, Philips, Videocon)
7. Powerhouse (Sony, BPL, LG, ONIDA, AKAI, Samsung, Philips, Videocon)
8. Microwave Oven (Samsung, LG)
9. Home Theatre (Sony, BPL, LG, Philips)
10. Computers

If you have any suggestions about these products or ideas about introducing new ones, they will be valued very much. We will certainly take due cognizance of your useful views.

We hope that you will co-operate by sending the information and help to serve industries and their developments which will ultimately contribute to the individual development of our country.

Thanking you and waiting for your early reply.

Yours faithfully,

XYZ

Enquiry Letter No. 3:

UDYAM AUTOMOBILE COMPANY, PVT. LTD.
12, Laxmipuri, Kolhapur

Date: 25th November, 2012

Fax:
Telegram
To

The Manager
Bharat Automobiles
Mumbai 400001

Subject: Price of automobile parts.

Dear Sir,

We have been in the automobile spare parts business for the last thirty years. Our firm is considered as one of the leading business concerns not only in the State of Maharashtra but also in neighbouring states like Karnataka, Andhra, Gujarat. We have gone through your advertisement in the Times of India dated 23rd November, 2012. We are very much interested in having an agency of your products for Maharashtra and other states.

Kindly, therefore, arrange to send us a complete catalogue and price list of your products applicable to the sole selling agents. It will also be appreciated if you would send a brochure of your products and a booklet indicating terms and conditions of the business. Our firm is enjoying a high status in the field of automobile spare parts dealers. Thus, we would also like to know if you could extend any special privileges to us.

Waiting for a prompt and favourable reply from your end,

Yours faithfully,

Partner
Udyam Automobile Company Pvt. Ltd.

Enquiry Letter No. 4

- Write a letter to an Electric Company asking for an estimate of the cost of installing electric lights in your house.

XYZ
349, Sindh Housing Colony,
Pune 411007
5th Nov. 2012

To,

The Manager
Electric Installation Company
PUNE - 411 001

Subject: Estimate of installing electric fittings.

Sir,

I have recently constructed a building in Sindh Housing Colony, Pune. The Maharashtra Electricity Board has agreed to give me a connection and hence I wish to get the internal wiring of my building completed within 15 days, so as to obtain the Completion Certificate from Pune Municipal Corporation at the earliest.

It is, therefore, requested that you will kindly visit the site for inspection and prepare a detailed estimate of the cost of the installation of lights and other fittings including labour. The material required will be purchased from your company on cash payment.

Looking at our urgency, you are requested to kindly give the complete estimated figure within three days, indicating the probable date by which the work will be completed after giving the final order, after which work of installation can be started as early as possible.

Thanking you,

Yours faithfully,

XYZ

Enquiry Letter No. 5

- Write a letter to a bank regarding provision of negotiations of bills on approved drawees facilities for the company.

THE ABC COMPANY LIMITED
Manufacturers of Automobile Spare Parts,
KARAD

Date: 25th October, 2012

To,

The Manager,
State Bank of India,
Karad Branch,
KARAD.

Subjects: Negotiations of bills on approved drawees.

Dear Sir,

Due to tremendous increase in the business of our company, it is felt desirable to have an account with your bank.

We wish to know whether the facilities for negotiation of bills on approved drawees could be made available to our company. Our yearly transactions shall be to the extent of ₹ 2 crores approximately. I enclose herewith a copy of the company's latest balance sheet which indicates the company's latest balance sheet which in turn indicates the company's sound financial position.

I shall be glad if you will be kind enough to supply the required information together with your terms and conditions for undertaking the business. I hope you will send us the necessary forms for opening the account and making formal applications for the negotiation facilities.

Yours faithfully,

Managers

Enquiry Letter No. 6

Date: 21st December, 2012

Dear Mr. Mujumdar,

I have learnt from a friend of mine that you have evolved a system to successfully handle the problem of indiscipline and absenteeism among your workers. As my company is similar to yours, the system evolved by you would be with a little modifications useful and suitable to my establishment.

Will you please, therefore, make it convenient to communicate to me the details of your system? A team of personnel management experts from the Indian Institute of Management, Ahmedabad had visited my company last month. A copy of their report is sent with this letter which may be of help to you in understanding my problem.

Waiting hopefully for your co-operation and useful advice,

Thanking you.

Yours Sincerely,

XYZ

Enquiry Letter No. 7

- A letter to the Employment Exchange Officer, requesting him to sponsor a few candidates for the post of Junior Engineer.

M/s. ABC & COMPANY
(Manufacturers of Automobile Spare Parts)
29, Bandra Industrial Estate, Bandra Mumbai – 400 050

Date: 10th November, 2012

To,

The Employment Exchange Officer,
Technical Division
Pimpri-Chinchwad.
Pune

Subject: Post of Junior Engineer in our organisation.

Sir,

We have planned expansion of our organisation in the near future. For this purpose we would like to employ five junior engineers in the production division of our firm.

We hope that you will recommend suitable candidates who would be capable of taking up challenging jobs. The candidates who fulfil the following requirements may be recommended for consideration:

1. The candidate should not be of more than 30 years of age.
2. He should be a holder of a Diploma in Mechanical Engineering. A candidate with a Bachelors Degree in Automobile Engineering will be preferred.
3. He should possess at least three years experience in a production section.
4. He should be capable of supervising the work of at least ten workers of his section.
5. The scale offered is ________ . In the beginning, the total emoluments would be about ₹ ________ per month, inclusive of all allowances.
6. The candidate should possess sound health.
7. He should be able to speak, read and write Marathi, English and Hindi.
8. Preference will be given to the candidate who is a good sportsman.

Please, therefore, arrange to send a list of about fifty candidates. Before sending the list, the concerned candidates can be informed that they will be required to attend a written test as well as a personal interview on the data suitable to us. If successful in these tests, they will be required to undergo a medical examination. No expenses for attending the interview shall be reimbursed i.e. they will have to attend the interview at their own cost.

The selected candidate shall be on probation for a period of one year. After completing the probation period successfully, they will be placed in a suitable time scale which will certainly carry more pay than that during probation. The candidate will also require to furnish two references of reputable persons. Once getting regular appointment, they shall also be required to execute a necessary bond for serving the company for atleast two years.

We hope to receive a list of candidates from your end at the earliest.

Thanking you,

Yours faithfully,

Personnel Manager
ABC and Company, Bandra
Mumbai

Replies to Enquiry Letters

- Replies to enquiry letters should be considered as tools to build up clientele and enhance market goodwill by a business house of any sort such as commerce, industry, banking, etc. Considering this aspect, it is imperative that utmost care is taken while replying enquiry letters. Replies to enquiry letters should be drafted in such a way as to favourably influence the reader, because these readers are the potential customers who would enhance any company's business. The points to be kept in mind while replying enquiry letters are briefly stated below:
  - **(i) Promptness:** Replies to enquiry letters should be given promptly. The market today is extremely competitive. As such, any delay in giving a reply may result in the prospective customer to be attracted by other dealers and thereby losing a business opportunity.
  - **(ii) Courtesy:** The reply should be very courteous. The reader should be impressed with the friendly attitude reflected in the letter. Even if the reply is a little negative, the style of writing adverse matters should also be attractive and should not hurt the reader's feelings. If you want to supply a substitute for what is asked, you must be skillful enough to persuade the reader to accept the alternative product you recommend.
  - **(iii) Completeness:** An enquiry letter should be carefully read and understood. All the queries or points included in the enquiry letter should be properly replied. Whenever necessary, brochures, catalogues, price-lists should be forwarded. All the details such as sizes, colours, prices and other relevant details about the product should be mentioned. The information about the terms and conditions of business, credit facilities, extent of commission should be very clearly mentioned to avoid complications in future. If there are certain expenses other than prices quoted in the catalogue, such as packing, transportation, insurance, forwarding charges etc. a clear mention of such expenses should invariably be made. This helps to avoid disputes.
  - **(iv) Validity of prices:** The period, for which the prices quoted in the price-list are valid, should be mentioned in the letter in view of the fluctuating situations in the market. As far as possible, the period within which an order should be placed, also should be mentioned in the reply letter.
  - **(v) Persuasion:** Do not stop merely by sending a catalogue or a price list. You must specify the advantages involved in purchasing your products. You must make a convincing appeal to the concerned buyer/consumer to go in for your goods and commodities, services or products. Gift schemes, trade discounts, longer credit facilities would certainly attract prospective customers. Attractive and colourful brochures create a good impression on the reader i.e. customer.
- Some illustrative replies to enquiry letters are given below:

Reply to Enquiry Letter No. 1

Dear Sir,

Thank you very much for your enquiry letter dated 10th November, 2012, regarding our newly developed tiles. We enclose herewith a copy of our detailed catalogue about our products which shows the designs and colours in which the tiles are available.

Messers Naik and Naik Co., Market Yard, Pune are our sole selling agents and deal with all our products in Pune and the surrounding areas. We have separately advised them to approach you and clarify all your points of doubt and show some outstanding samples.

Our newly produced tiles are very promising and hard wearing. They are durable and moisture resistant if they are laid according to the instructions provided by our representative firm. We are confident that our products will provide you with long lasting satisfaction.

Hoping to receive your order and the opportunity to add you to our esteemed group of customers,

Thanking you,

Yours faithfully,

XYZ

Reply to Enquiry Letter No. 2

Date 5th Dec. 2012

Sir,

Received your letter dated the 20th Nov. 2012 requesting us to provide data regarding sale of electric and electronic goods - mostly consumer durables - during the period of ten years from 2001 to 2012.

We are very sorry to state that the data for the period from 2001 to 2012 are not available with us. Had it been with us, it would have been a great pleasure for us to make it available to you. We would however, suggest that the information could be available from the Chamber of Commerce where they collect and consolidate the information for all goods. You may choose the relevant one from it.

Thank you for approaching us in this respect.

Yours faithfully,

XYZ

Reply to Enquiry Letter No. 3

BHARAT AUTOMOBILES LTD.
22, Dalal Street
Mumbai 400001

Phone: **Date 5th Dec., 2012**
Fax:
Telegram:

Subject: Prices of Automobile Spare Parts.

Dear Sir,

We express our thanks for your letter dated the 25th Nov. 2012.

We have great pleasure in sending you our catalogue and price list separately within a day or two, as our multi-colour brochure is under print.

The automobile spare parts manufactured by our company are of excellent quality and are claimed by the industry as the most durable and economical.

We normally offer a trade discount of 20%. If, however, your turnover exceeds ten lakhs, an additional 5% incentive discount shall be given. Payment of bills within fifteen days is our prerequisite condition. Forwarding and insurance charges shall be extra.

Thanking you,

Yours faithfully,
for Bharat Automobiles

Letters of Quotation

- Whenever an organisation needs to purchase material from the market on a large scale, it is a general practice to call for quotations from the prospective suppliers either by sending a letter to each individual prospective supplier, or by publishing a notice in the newspapers. This enables the firm to select the most suitable deal. Prices quoted by different suppliers can be compared so that the lowest can be selected to save funds. It is also possible to compare the terms and conditions of suppliers offered by different sellers. This ensures the best service and benefits. For getting all details of the suppliers, it is essential that the letter or notice for quotations should contain:
  (i) Details of the material required
  (ii) Quantity and quality of the material required
  (iii) Time limit for the supplies
  (iv) Place of delivery
  (v) Packing, forwarding and insurance
  (vi) Discount/Concessions
  (vii) Penalty for default.

- Following is an example of a letter written for calling a quotation from a manufacturing company.

POONA BICYCLE SPARE PARTS SUPPLIERS
36, Shaniwar Peth, Pune 411 002

Outward No. 20 **Date: 6th November, 2012**

To,

The Manager

Hero Bicycles Ltd. Fort,

Mumbai 400 001

Subject: Quotation for Supply of Bicycle Spare Parts

Sir,

We have been a leading wholesale suppliers of bicycle spare parts to Pune market, catering to the needs of retailers of this city for the last twenty years. We own in the city three well equipped show-rooms with spacious storage capacity. We wish to place an order for supply of the following items. Kindly send us your quotation letter at the earliest.

Description	Quantity	Specifications
1. Frames (for ladies bicycles)	500 pieces	20" black colour
2. Frames	1000 pieces	20" black colour
3. Chain covers	2000 pieces	Half and Full with different colours
4. Chain pieces	1000 Nos.	It should be strong enough
5. Handles	1000 Nos.	Chromium plated
6. Rims	2000 Nos.	Should be useful for various types of bicycles
7. Tyres and Tubes	5000 Nos. each	Dunlop

Terms and Conditions of supply

1. The above mentioned items should be securely packed before dispatch to avoid any damage during transit.
2. The consignment should be insured against all risks.
3. The payment of the material will be made only on receipt of the material in proper condition.
4. The rates mentioned in your quotation letter must be adhered to while preparing a bill for the supplies.
5. In case the material is not received within fifteen days, we are free to cancel the order.
6. The responsibility of damage during transit will be yours.
7. Packing, forwarding, insurance, taxes, etc. will have to be borne by you.

We expect kind co-operation from your end and wish to establish a good business relationship with you.

Thanking you,

Yours faithfully,
XYZ
Pune Bicycles Spare Parts Suppliers

- A reply to an enquiry letter for quotation is very important from the supplier's point of view. The success of quotation results in a confirmed order for supplies. Utmost care should be taken while sending a quotation. Such a letter should invariably begin with words of thanks. It helps to create a good impression on the mind of the purchaser. A tabulated statement indicating prices, quality etc. of goods should be prepared and sent with the letter. When the market conditions are normal, usual prices should be quoted. If, however, there is acute competition, the lowest affordable prices should be quoted as far as possible. Do not fail to mention the advantages in buying goods from you. The conditions for supply should be mentioned in a polite and courteous manner so that the prospective buyer's feelings are not hurt. Also the quotation should reach in time.

- After sending the quotation, pursue your buyer either through a reminder or a phone call. This will make the buyer understand your urge for the business.
- If you are sending a consolidated price list or catalogue as a quotation, do not forget to indicate the relevant page number and serial number of the articles the customer desires to purchase.
- A reply to the letter calling a quotation is given below:

A reply to the letter asking for a quotation.

HERO BICYCLES COMPANY LTD.
Fort, Mumbai - 400001

Date: 10th Nov. 2012

Fax:
Telegram:
Telephone:

To Poona Bicycle Spare Parts Suppliers,
36, Shaniwar Peth,
Pune 411 002

Subject: Quotation for supply of bicycle spare parts.

Dear Sir,

We thank you for your letter No. 20 dated 6th November, 2012 asking us to send a quotation for the supply of bicycle spare parts.

We enclose herewith our price-list for your consideration. The items mentioned by you in your letter appear in the price-list as serial numbers and pages given below:

(i) Cycle frames (ladies bicylces) – Sr. No. 23 Page 4
(ii) Cycle frames (gents bicylces) – Sr. No. 22 Page 4
(iii) Chain covers – Sr. No. 123 Page 9
(iv) Chains – Sr. No. 30 Page 5
(v) Handles – Sr. No. 89 Page 7
(vi) Rims – Sr. No. 39 Page 5
(vii) Tyres – Sr. No. 160 Page 12
(viii) Tubes – Sr. No. 161 Page 12

Terms and Conditions:

1. The material will be properly packed.
2. The consignment will be insured against all risks.
3. The rates quoted are valid for thirty days only. Thereafter these prices will not hold good.
4. The damage in transit will be equally shared.
5. We will bear the packing, forwarding and insurance charges.
6. Since the order is in bulk we will allow a special discount of 15% on the total price.

We are eagerly waiting for your confirmed order at the earliest.

We specifically wish to mention that the goods manufactured by us are of the best quality and our prices are competitive. We will give you our best services, if you give us an opportunity.

Hoping to establish continued patronage,

Thanking you,

Yours faithfully,

Sales Manager
Hero Bicycles Co. Ltd.
Mumbai

5.3.3 Order Letter

- Letters relating to placing of orders are the most important ones. Almost all business letters are an outcome of order letters. No business can be carried out if goods and/or services are not ordered. Hence great care is required to be taken while finalising, drafting and dispatching order letters. Compliance of order letters creates goodwill and establishes business relations with existing customers and attracts new customers. Before placing an order, the relevant information should be kept ready.

  This information includes details such as:

  (i) Product required.

  (ii) Specifications of the product required i.e. size, colour and other details.

  (iii) Quantity required.

  (iv) Quotation/offers if any, duly approved by the appropriate authority.

  (v) Completed order form (if available).

  (vi) Nature of order – First or trial order or confirmed order or repeat order or routine order.

- An order letter can be drafted once these details are finalised. A reference should invariably be given about the previous communication such as an earlier letter or a quotation received from the supplier. Apart from the details about the goods required, a mention should invariably be made about the terms and conditions of the order, such as packing, forwarding, insurance, transport, damages during transit, credit facilities, discount/rebate/incentive, settlement of accounts, right to reject order, after sales service, and other factors which are thought important for communicating to the supplier. A mention of these details in the order letter helps to avoid disputes and complications in future. A specimen of a routine order is given below:

Letter of Order

GOVERNMENT POLYTECHNIC STUDENTS' CO-OP. STORES
Govt. Polytechnic Compound
Ganeshkhind Road, Pune 411 016

Date: 10th Dec. 2012

To,

The Manager,
A.B.C. Company
(Manufacturers and Suppliers of Drawing Material)
Budhwar Peth,
Pune 411 002

Subject: Supply of Drawing Material.

Reference: Your quotation No. 244 dated 25th Nov. 2012

Sir,

Your above mentioned quotation for the supply of drawing material is approved by the managing committee of the stores. We are, therefore, pleased to place an order for the supply of the following articles of drawing material:

Sr. No.	Name of the Article	No. required
1.	Drawing Boards (Standard size)	200
2.	Instrument Boxes	200
3.	Workshop tools	100 sets
4.	Drawing sets (Imperial size)	2000 students
5.	Mini-drafters	100
6.	Wires files	500
7.	Box files	200

As already communicated and agreed by you through your quotation stated above, the supplies should be made on the following terms and conditions:

Terms and Conditions of Supply

1. The material should be securely packed;
2. The material should be dispatched by road transport;
3. Damages, if any, during transit should be compensated;
4. Prices of the articles as quoted should be adhered to;
5. No taxes should be charged since the material is required for educational purpose;
6. The bill should be in the name of the Chairman of our Society;
7. The payment shall be effected only after the material is received as per specification and in good condition;
8. Since the material is required by students, a special discount of 15% be allowed.

We expect to receive the material at the earliest but not later than 15 days from the date of order.

Thanking you,

Yours faithfully,

Secretary

Reply to order or compliance of the order placed

- Any reply letter to any order should necessarily begin with an expression of gratitude or thanks. It should ensure the compliance of the order within the stipulated time and under the approved terms and conditions. Assurance must be given about the quality of goods supplied and satisfactory services thereafter. A specimen reply letter can be in the following form:

Date: 12th Dec. 2012

Dear Sir,

Thank you for your order dated 10th Dec. 2012, for different items of drawing material.

We shall deliver the goods well within the time specified. The bill will be prepared and sent with goods which may please be settled as early as possible preferably within a fortnight from the date of receipt of the material. As desired, a special discount of 15% shall be allowed.

The material supplied will be of the best quality. You will not have any reason for complaint. We would also expect your continued patronage.

Thanking you,

Yours faithfully,

Manager

5.3.4 Complaint Letter

- Once an order containing clear and specific terms and conditions is placed for supply of goods and/or services with the supplier, it is expected that the requirements are properly, promptly and carefully fulfilled. However, due to one reason or the other (such as a mistake or negligence or sometimes reasons beyond anybody's control) the expectations of the purchaser may not be fulfilled. This may result in monetary loss, inconvenience, setback to goodwill, loss of time in correcting the matters etc. The reasons for complaint may be one or more of the following types:

  (i) Delay in fulfilling the order.

  (ii) The quality of goods not as per samples.

  (iii) The quantity supplied is not as per the order.

  (iv) Goods received are in damaged condition.

  (v) Improper packing.

(vi) Delivery of goods at a wrong place.

(vii) Machines failing to function.

(viii) The actual quantity of goods supplied may not tally with the one shown in the bill.

If the order is concerned with services, the complaints may be about:

(i) Work not executed as per design or plan.

(ii) Work not executed in time.

(iii) Inferior quality of material used.

(iv) Staff engaged not courteous.

(v) Work not properly supervised.

(vi) Prices/rates shown in the bill may not tally with those in the quotation/tender.

(vii) Unnecessary reminders issued for payments.

(viii) Advances paid not adjusted in the bill.

- These are only a few examples of reasons for complaints. They may vary according to different situations. The obvious result of these shortcomings, defaults, mistakes or negligence is that, it can cause anxiety to the purchaser. The only alternative before the purchaser is to bring these factors to the notice of the supplier promptly without the loss of temper or balance of mind. While writing a complaint letter, besides maintaining courtesy, the following factors should be made known to the supplier for prompt compliance:

(i) The complaint letter should be written and sent without any loss of time.

(ii) The order number and date on which the order was placed should be mentioned.

(iii) The fault or mistake in clear and precise terms should be expressed.

(iv) The shortcomings and if possible the ways to correct them should be stated.

(v) The inconvenience or loss caused should be stated.

(vi) A request to investigate the reasons and precautions for the future should be taken.

(vii) A request for proper adjustments or negotiations should also be made.

- A few illustrations of complaint letters are given below:

Complaint Letter No. 1

A letter complaining about the poor performance of a Diesel Engine.

ABC & COMPANY LTD.
(Farm Management & Supply of Agricultural Implements)
PUNE 16
Date: 25th Nov. 2012

To,

The Manager,
Machine Supply Corporation, PUNE.
Your Ref: No. REF/86/215 of 20th November, 2012

Dear Sir,

We have purchased from you one diesel engine vide receipt No. MNP-Sale - 1234 dated 15.10.12. A guarantee card of the said machine bears Sr. No. 352 of 15th October, 2012.

We regret to inform you that the diesel engine supplied by you is not satisfactorily working. We have invited your local agent twice in order to repair it. Inspite of his efforts, it is still not properly working. Further, it is found that the machine is not at all economical in so far as consumption of fuel and oil is concerned. As a result, our work is getting hampered.

It has given a very poor performance so far. We feel that there is some major manufacturing defect in the engine. In the opinion of the mechanic, it is beyond repair(s). It serves no purpose to get it repaired time and again. Since the guarantee period of one year is not yet over, we request you to replace the engine supplying a new one in its place.

Expecting you to acknowledge the receipt of this complaint promptly and to do the needful immediately,

Thanking you,

Yours faithfully,
ABC & COMPANY LTD
PUNE 4

Complaint Letter No. 2

- A letter to the General Manager, Central Railway, claiming compensation for goods damaged in transit.

PUNE ELECTRICAL GOODS SUPPLY AGENCY
14, Chhatrapati Shahu Bus Stand, Swargate,
PUNE 411 030

Date: 27th November, 2012

To,

The General Manager (Claims)
Central Railway, Chhatrapati Shivaji Terminus.
Mumbai 400 001

Subject: Compensation for damaged electrical goods.

Sirs,

Ours is a leading electrical goods supply company in Pune region, catering to the needs of consumers. We had placed an order for two hundred table fans with the Mumbai Company. The company dispatched by railway parcel the consignment duly packed in wooden boxes, on 10th November, 2012.

When we received the consignment, it was found that the two boxes were badly damaged in transit, probably because no proper care was taken while loading and unloading those precious goods. At the time of handling those boxes at Mumbai, they were intact and were in proper condition. As the two boxes from our consignment were badly damaged when the goods were received at Pune, our agent refused to take delivery of these damaged boxes. It Is the duty of the railway authorities to hand over the goods in a proper condition. Panchanama of the damage has been done by the agent in the presence of the railway authorities and railway police authorities on the spot.

We request the authority to make full compensation of the damaged ten fans of those two boxes as has been guaranteed in the railway receipt. Application in the prescribed form with a copy of the 'Police Panchanama, in original is enclosed.

Early action in this respect is expected. We hope to receive a prompt reply from you. We are enclosing the necessary papers with this letter.

Thanking you,

Yours faithfully,
Pune Electrical Goods Supply Agency,
Pune

Complaint Letter No. 3:

POONA BICYCLES SPARE PARTS SUPPLIERS
27, Chhatrapati Shahu Bus Stand, Swargate,
Pune 411 030

No. Order/169 **Date: 25th November, 2012**

To,

The Manager

Hero Bicycles Ltd.

Mumbai.

Subject: Delay in supply of bicycles spare parts.

Ref.: Our order letter No. 20 dated 16.10.2012

Dear Sir,

This has a reference to our purchase order No. 20 dated 6th October, 2012. We have already placed with you an order for the supply of bicycle spare parts in the first week of the month. In our order, we had specifically insisted on the delivery of the spare parts within a period of 15 days. In spite of specifying our conditions, we have neither received the goods so far nor have we received any communication from you about the extension of time limit for effecting the supply of the ordered material. However, we are giving you an extension of time limit for one more week to effect the supplies. In case your dispatch intimation is not received within a week from the date of receipt of this letter, we will be compelled to cancel the order and make alternative arrangements to procure the material from any other company. In good business relations, such an action is not desirable as it affects market prestige of a firm.

This delay has certainly caused us great inconvenience and has also resulted in many awkward explanations to our retail suppliers.

We trust in your promptness and hope that the material will be supplied by the end of this month at the latest i.e. within the extended time limit now communicated.

Thanking you,

Yours faithfully,

XYZ

Poona Bicycle Spare Parts Suppliers

Complaint Letter No. 4

MAHARASHTRA INDUSTRIAL TRAINING INSTITUTE

No. Store/460 **Kothrud, Pune 411038** **Date: 7th Dec. 2012**

Fax:

Telegram:

E-Mail ID:

To

M/s Godrej and Company
37, LBS Road, Pune 411 030

Subject: Supply of Furniture

Reference: Our order No. 1234 dt. 5th Nov. and your invoice No. 786 dt. 5th Dec. 2012.

As per the order referred to above we had ordered 5 Godrej Storewell Cupboards, 7 filing cabinets, 3 office tables with drawers, 7 computer tables, 20 folding chairs, 7 stools for our office.

The transport contractor you engaged i.e. M/s Anisha Transport Co., Bhavani Peth, Pune effected the delivery of the packed consignments along with your invoice-cum-bill for the suppliers today morning.

We, however, regret to state that the item of seven stools has not been received, although included in the invoice. This has been brought to the notice of the above transport contractor who promised to convey this to your office. The invoice in question is therefore not signed and has been returned with him.

We, therefore, request you to do the needful at the earliest in order to settle the matter and effect payment of the bill.

Yours truly,
Store Manager
for Maharashtra Industrial Trg. Institute
Pune

Encl.:

C.C. to M/s Anisha Transports Company, 27, Bhavani Peth, Pune 411030 for information and confirmation with M/s. Godrej and Company, Pune 411001.

5.3.5 Adjustment Letters

- In the modern business-world of acute competition, a customer is the monarch of the market. His satisfaction should be taken as a fundamental duty by the suppliers of goods and/or services. An adjustment letter should therefore, be taken or considered as an opportunity to correct mistakes and faults if any, in order to please customers. This becomes the most essential factor particularly when there is a growing awareness among customers and there are legislations to protect customer's rights. A prompt action on the complaints from customers becomes imperative for the suppliers of goods and services. A quick reply from a seller certainly pacifies the dissatisfied customer and half the battle to regain the customer's confidence is won. This enhances the goodwill and reputation of a firm in the market.
- Adjustment letters are classified into the following different categories:
  (i) Letters granting adjustment
  (ii) Letters refusing adjustment
  (iii) Letters offering a compromise
  (iv) Letters regretting the mistake/fault/error/complaints.
- Sometimes, in order to get time for proper investigation and to ascertain the genuineness of the grievance or complaint before finally accepting or rejecting an adjustment, an interim reply is sent to the customer informing him that an investigation is being carried out. However, an interim reply like this should invariably begin with words of regret for the inconvenience and annoyance caused.
- While drafting an adjustment letter, great care has to be taken for handling the delicate situation, and retaining or regaining a customer's confidence. The letter should therefore, invariably begin with an expression of regret for the inconvenience caused. If the seller is convinced about the mistake or fault, he should straight-away volunteer the adjustment without undue arguments or discussion. When the supplier feels that there is no fault or mistake on his part but it is the error on the part of the customer himself, the supplier has to refuse the adjustment; however the refusal of adjustment should be expressed in genuinely polite and courteous words without hurting the customer's feelings. He must be convinced that the fault is his and not that of the suppliers.

- Sometimes, it becomes necessary to offer a compromise, when there might be a difference of opinion among the supplier and customer, or the cause of complaint may be due to reasons beyond the control of either party, so as to maintain a cordial relationship.
- Even a partial adjustment helps to clear the situation. In brief, an adjustment letter should contain:
  (a) Regret for the inconvenience caused.
  (b) The result of investigation into the complaint.
  (c) If an error on the part of supplier is detected, accept it and openly offer an adjustment. Also the steps taken to avoid recurrence of mistakes should be stated.
  (d) Repair/replacement at own cost.
  (e) If an error is on the part of the customer, an adjustment should be politely and courteously refused.
  (f) An apology if there are irreparable losses.
  (g) The customer should be persuaded for continuing patronage.
- A few examples of replies to complaint letters are given below:

Reply to Complaint Letter No. 1

MACHINE SUPPLY CORPORATION LTD.
PUNE 411032

Fax: .... **Date: 27th Dec. 2012**
Telegram: ....

To

The Manager,
ABC Company
Pune 411004

Subject: Working of the Diesel Engine.

Dear Sir,

This has a reference to your letter dated 25th November, 2012.

We are extremely sorry to know that the diesel engine we have supplied to you is not working satisfactorily causing you great inconvenience and telling adversely on your expenses. Our local agent has also informed us accordingly just yesterday. In our opinion, the only alternative is to replace the machine and return it to the manufacturer since there is certainly a major manufacturing defect.

When the machine was supplied to you, it was not expected that it will cause any trouble. However, in view of this situation, we have informed our agent to supply a new machine to you at the earliest and send the old unserviceable machine to the manufacturing company under intimation to us. We hope this will solve your problem and your work will no more suffer.

We express our thanks for bringing the matter to our notice. This has given us an opportunity to approach the manufacturer for removal of the drawbacks in the machine to improve the performance.

We always assure to provide prompt services in future and expect continued patronage and co-operation from you.

Thanking you,

Yours faithfully,
Sales Manager
Machine Supply Corporation Ltd.

Reply to the Complaint Letter No. 2

CENTRAL RAILWAY

No. Accounts/Claims/3962 **Dt. 10th Dec. 2012**

To

The Manager,
Pune Electrical Goods Supply Agency,
14, Budhwar Peth, Pune 411002

Subject: Compensation for damaged electrical goods.

Sir,

We are in due receipt of your letter dated 27th November, 2012 on the above subject. It is regretted that two boxes from the consignment meant for you, were damaged during transit. The investigation report and the 'Panchnama' signed by the Railway Police, Pune have been just received in our office. In order to workout and sanction compensation for the loss, it is necessary to complete certain official documents. The various blank forms to be completed and signed by you and countersigned by the station superintendent and the Railway Police, Pune, are enclosed with this letter. Kindly complete them in all respects and return to us at your earliest.

Kindly inform us the name in which the cheque for the compensation should be issued.

The inconvenience caused is very much regretted.

Yours faithfully,

Encl.: Blank forms

Accounts officer (claims)
Central Railways

Reply to Complaint Letter No. 3

HERO BICYCLES
121, Fort, Mumbai 400001

No. Supply - 239 **Dt. 5th Dec., 2012**

To

The Manager,
Poona Bicycles Spare Parts Suppliers,
36, Budhwar Peth,
Pune 411 002

Sir,

This has reference to your Letter No. Order/169, dated 25th November 2012, regarding delay in the supply of bicycle spare parts.

At the outset, we regret the inconvenience caused to you due to the delay in supply of the ordered spare parts and apologise for it.

Never in the past has any delay occurred in effecting a supplier order. But this time, it has happened due to some unforeseen cause. Recently, the trade union of our workers abruptly declared an indefinite strike. Attempts were sincerely made to negotiate with the leader to arrive at an acceptable compromise, and it was expected the strike would be called off within a day or two. Unfortunately it did not happen and the strike continued for a longer period. This has resulted in non compliance of your order in time. We hope you will kindly understand our helplessness.

The strike has been called off on the 2nd only. 3rd December being a Sunday the workers have resumed their duties on Monday i.e. only the day before yesterday. The company's work will soon resume. We, therefore, assure you that the spare parts ordered will be dispatched within a week's time.

We once again regret the inconvenience caused and appreciate your co-operation in extending the time limit for effecting the supplies voluntarily without waiting for any explanation from us.

We sincerely hope that your whole-hearted co-operation will continue to encourage us to give better and satisfactory services in future.

Thanking you,

Yours faithfully,
Sales Manager
Hero Cycles

Reply to Complaint Letter No. 4

M/s. GODREJ AND COMPANY
37, L.B.S. Road, Pune 411030

No. Sale/235 Dt. 10th December, 2012

To

The Principal,
Maharashtra Industrial Training Institute,
Kothrud, Pune 411038

Subject: Supply of furniture.

Sir,

This is with reference to your Letter No. Store/460 dated 7th December, 2012 sent by your Stores Manager.

We sincerely regret the inconvenience caused to you due to non-receipt of office stools.

On investigation it is found that the item of seven stools was kept ready in the delivery department. As a co-incidence there was another order of seven stools. Inadvertently the lot of the seven stools meant for your institute was dispatched to the other firm. This was sheer negligence on the part of the person concerned.

We have today itself separately dispatched the ordered stools to your stores section. I hope the delivery might have been effected by now.

Kindly arrange to acknowledge the receipt. We hope you will kindly continue your co-operation in future also.

Thanking you,

Your faithfully,
Manager
Godrej and Company

5.4 REPORT WRITING: ACCIDENT REPORT, FALL IN PRODUCTION, PROGRESS REPORT AND INVESTIGATION REPORT

- The dictionary meaning of the word 'report' is *to give people information about something that you have heard, seen or done* etc. In the business world, it falls in the category of upward communication. It is a logical presentation of facts. A report is a useful tool in the hands of executives for evaluating progress, for reviewing the processes or policies, for planning future actions, for taking decisions, for assessing performances, for keeping a check on activities, for reviewing market conditions, for knowing consumer trends and for many other purposes. Depending upon the purpose, reports are classified as:
  (i) Routine reports or
  (ii) Special reports.
- Different types of reports fall in these two categories. Classification of reports can be given as follows:

Reports

Routine Reports	**Special Reports**
(i) Progress report	(i) First Hand Information report
(ii) Inspection report	(ii) Survey report (or feasibility report)
(iii) Periodical report	(iii) Investigation report
(iv) Appraisal report	(iv) Project report
(v) Performance report	

- A report is written either by an individual or a committee. If an individual report is very small, it can be written in a letter form. A committee report is never in the letter form. It is in the impersonal style, therefore, the use of the word 'committee', instead of either 'I' or 'We' is made. A report has its own form or structure and is divided into the following sections:
  1. **Terms of reference:** It shows the details of assignment and purpose and scope of the investigation or research.
  2. **Procedure:** It indicates the method or methods used for collecting the information.
  3. **Findings:** This is the main part of the report. Whatever information is collected is systematically organised in a presentable form and properly analysed. It generally ends in a brief summary of the findings.
  4. **Recommendations:** Based on the analysis of the collected information, proposals for taking suitable actions or following proper procedures are made.
  5. **Signature, place and date:** Dated signatures of the committee members are obtained. At the end, the place where the report is finally written and the date are mentioned.
- An engineer is expected to write several reports during his career. The purpose of a report, as earlier mentioned, is to give information to a person or body of persons who do not possess full knowledge of all the facts of the subject of the report. A writer of the report must therefore, be equipped with sufficient knowledge of the subject matter. An engineer either in a private organisation or in the government department is often called upon to write reports on different subjects. Since the subjects on which the reports are required to be written are innumerable, it is not possible to formulate exact rules for the collection of information and the adoption of a specific course in presenting, analysing and organising the material, whatever the material is. However, the following general principles can be made use of for writing reports:
  1. Reports written by an individual (expert) to his client and those by an employee to his employer are usually written in a letter form.
  2. Individual reports are generally written in first person. (by using 'I' and 'We' as the subject).
  3. A report should be brief, clear and convincing.
  4. A report should be concluded with the considered opinion and summarised recommendations of the writer.
  5. A report should be informative.
  6. A report should contain the relevant information required by the client.
- Due to increased professional skills and business developments, the need for report writing has increased. An engineer must acquire the skill of writing good reports.

Essential Characteristics of a Good Report:

1. A good report must be objective. There is no place for expression of feelings of the writer in a report. It must be based on observations and inferences.
2. A report must be factual. It should be limited to general observations and must aim at clarity, directness and conciseness.
3. It must help the reader to get what he requires quickly. The use of an index, well labelled tables, charts, graphs and good summaries is necessary. They save the reader's time in understanding the report.
4. In report writing, simple language should be used.
5. Unnecessary and lengthy sentences should be avoided.
6. In report writing, it is necessary to use specific words rather than general ones. For example, 'the metal is melted' is the right way to say; if we write 'the metal is heated until it turns to a liquid' it becomes unnecessary.
7. Standard abbreviations should be used in writing technical reports.

5.4.1 Accident Report

- An accident report must state:
  - (i) The circumstances under which the accident took place;
  - (ii) The type of injury caused;
  - (iii) Action taken for relief and treatment;
  - (iv) Recommendations or measures to be adopted for avoiding such accidents in future.
- The following is a report about an accident which took place in a workshop reported by the engineer supervisor in charge of the Assembly Department.

Accident Report - 1

ASSEMBLY DEPARTMENT SECTION 'C'
ABC AND CO. LTD.
20, Hadapsar Industrial Estate,
Pune 411 028

Date: _______

To,

The Manager
ABC & Co. Ltd.
Pune

Sir,

I regret to inform you that a worker Shri. Patil C. G. was involved in an accident in the assembly section of our factory at 8.00 a.m. on the 2nd Nov. 2012. In the capacity of a supervisor in charge of the shift on the day, I am submitting my report about the accident for further action.

At eight o'clock in the morning, Shri Patil was lifting a heavy iron bar. He could not keep up his balance properly and as a result he slipped on the floor. The heavy rod he was lifting fell on his foot and caused injury. He was immediately removed to the Company hospital by our Service Department staff and then taken to a hospital in our ambulance car. He was given proper medical help by our Company doctors without delay.

His condition is improving. All arrangements to give him the required medical assistance have been made. As per the opinion of the doctor, it is probable that he will be discharged from the hospital within fifteen days. After the accident, his family members were also informed about the incident. They are now with him.

Reasons for the Accident: The reasons for the accident were as follows:

1. He was unable to judge the weight of the bar.
2. The iron bar was quite heavy and it was beyond his capacity to lift it.
3. He tried to remove the bar from the passage as instructed by the operator.
4. He should have taken help of some other workers to complete this work.
5. It was only because of his negligence and over confidence that this accident (caused) occurred.
6. The floor was also a little slippery due to oil, grease, etc.

Suggestions to avoid such accidents in future:

1. No worker should be allowed to lift such heavy things without the help of other colleagues.
2. Handy cranes and carriers be purchased for lifting and carrying such heavy bars from one place to another.
3. Workers should be given training for using the carriers. They must personally ascertain the weight of the material to be lifted or moved from one place to another.
4. No such shifting of material should be allowed without the help of appropriate conveyors.
5. A supervisor should be present whenever material from the store section is to be carried to various sections of the Production Department.

If these suggestions are scrupulously followed, we shall be able to avoid such accidents in future.

Thanking you,

Yours faithfully,
Supervisor, 'C' Section
Assembly Department
ABC & Co. Ltd. PUNE

- The following is an example of another accident report written by a class representative in the class for submission to the Workshop Superintendent.

Accident Report 2

FROM A. A. PATIL
Class Representative 'C' Division
(Name of the Institute)
Date: 27th Nov. 2012

To,

The Workshop Superintendent,

____

____

____

Sir,

As a Class Representative, I have been asked to submit a report about the accident which took place in our work-shop at 3 p.m. yesterday, when the complete class had workshop practicals. My friend Shri. Zambre, A. V. Roll No. 16 was working on the Lathe Machine preparing for his job. As per the norms of the workshop, all the students are expected to wear aprons while working on machines. He forgot to bring his apron and did not demand one from any other student. He was doing a job that required operation of taper turning and threading. He was in a loose dress. The instructor objected to his working in that loose dress, but in vain. As he bent over the machine, the thread caught hold of the front portion of his dress and started winding it. He cried for help. Many of us rushed to save him from the danger and someone switched off the motor. As the machine slowed down, we managed to overcome the disaster with much difficulty. His shoulders received minor bruises. He was immediately removed to Civil Hospital and was given medical treatment. His parents have already arrived and are attending to him. He has been advised two days rest as he is in shock.

The following suggestions are made to avoid such accidents in future:

1. No student should be allowed to enter the workshop without an apron. Use of aprons should be made compulsory.
2. There should be a separate instructor to look after the work of first semester students.
3. Instructors must give a clear understanding about the dangers before students handle any such machines.

If these suggestions are made compulsory for all the students, it will be possible to avoid such accidents in future.

Thanking you,

Yours faithfully,
Class Representative,
'C' Division

Accident Report 3

Accident in the workshop

WORKSHOP SUPERINTENDENT
ABC & Co. Ltd.
15, Jalgaon Ind. Estate
Jalgaon
Date: ______

To,

The Company Secretary
ABC Company Ltd.
JALGOAN

Sir,

As conveyed through a telephonic message on 9.11.2012 you are already aware of the fire accident that took place in the Production Section of our Jalgaon Plant at 9. p.m., on 9.11.2012. I submit herewith a detailed report about the facts of the accident and the approximate damage caused to our machinery, material and building on account of the fire.

Facts about the Accident: The foreman of the Production Section 'C' of Jalgaon Plant handed over charge of his section to the Foreman of the third shift on the 9.11.2012. After about an hour or so, he first smelt something unusual. While searching for the cause of the smell he looked at the corner where broken empty wooden boxes in which the newly purchased machines were kept. Those wooden boxes were piled up in a heap. Some electric lines pass from that area. Electric wiring of that part of the shed was uncovered and found to be defective. This fact was reported to the electrical maintenance department nearly fifteen days back to get the defective wiring system changed from the contractor who had executed the work of electric wiring.

There was a short circuit in the electric unit. The piled up wooden boxes ultimately caught fire.

On noticing this, the Foreman immediately contacted the fire-brigade. Within about ten minutes all the available fire-fighting equipment in the factory were put into operation. The warning given to all the workers helped in extinguishing the fire. Their timely help certainly saved the adjoining section where highly inflammable chemicals and waste cotton was stored. Had the fire spread to this adjoining section, the loss would have been very high.

Due to timely action, the damage was restricted only to some machinery in the 'C' Section of the Plant.

Observations:

1. The fire took place in the 'C' section of the Production Department on 9.11.2012.
2. The main cause of the fire was the faulty wiring system put in by the contractor. He had left some of the electric points in the corner of the 'C' Section of the Production Department uncovered.
3. The fire eventually burnt out some portion of the Production Section which is involved in the production of some important parts of machines.

Arrangements:

The following arrangements have been made during the period required for rebuilding the damaged part of the shed:

1. Directives have been given that required repairs must be completed within two months.
2. The Accounts Department has also been directed to make available funds to the extent of ₹ 80,000/- for reconstruction and for purchasing machinery and equipment for replacing the damaged ones.

3. Meanwhile the Jobs which the 'C' Section has been involved in have been given to other firms on a contract basis purely as a temporary measure.

Changes required to be made in the present working system:

1. All the old worn-out wiring which cannot sustain the work load, particularly in the Production Section, must be replaced/repaired without any further lapse of time to avoid any loss in future.
2. Every month, our electrical maintenance department must check up all the defective points of the electrical supply system and the wiring condition and remove/repair faults whenever necessary.

Conclusions:

The damage caused is approximately to the tune of ₹ 90,000. No worker was injured in the accident. The main reason for the damage was the defective wiring system and negligence on the part of the Maintenance Department as well as Stores Department which did not remove the empty wooden boxes immediately after the machines were taken out. Hereafter, no such boxes will be allowed to be stored in the Production Department. If these suggestions are properly implemented, it is likely that such accidents can be avoided in future.

Workshop Superintendent
ABC Company Ltd.
JALGAON

Damages and losses due to fire
Comments:
Report Regarding Damages:

- A detailed report on the damage caused due to fire and the arrangements made for carrying out the production programme during the period of rebuilding / rehabilitation:
- It is presumed that the Work-shop Superintendent in a Company is writing about the damage caused in one of the sections of the Production Department for information of the Company Secretary of ABC Company Jalgaon.
- The information or the details regarding the damages caused due to fire or any other means should be clear, simple and to the point. It should include the following points in particular:
  1. The actual incident.
  2. Causes for the accident.
  3. The extent of the damage.
  4. The Details about the harm/damage caused.
  5. Extent of repairs to machinery and/or building.
  6. Arrangements proposed for the repairs.
  7. The period required for repairs and rebuilding/rehabilitation.
  8. Probable expenses for repairs and rebuilding/rehabilitation.
  9. Arrangement for production during the period of rebuilding /rehabilitation.

5.4.2 Fall in Production

- Usually a report addressed to the General Manager on the fall in production is written either by the Production Manager or by the Departmental Head, to the Production Manager. Such reports contain a detailed account of the fall in the quantity of production in the factory unit. Following are the important points which are worth noting for preparing such reports:
  1. The report must clearly mention the decrease in production in relation to the target allotted by the Board of Directors or the Top Management.
  2. It should mention the period during which the production has gone down.
  3. It must include the causes of the fall or decline in the production.
  4. It must include suggestions to overcome the situation and restore the production level.
  5. The report must be factual, brief and to the point.

6. It must include concrete statistical data as far as possible.
7. It should be expressed in a simple style.

ARCHANA ELECTRICAL WORKS
56, Bhosari Industrial Estate
Pune
Date: 28th November, 2012

To,

The Production Manager
Archana Electrical Works
Bhosari, PUNE

Subject: Report about the fall in production.

Sir,

I regret to report that during the last ten months we have not been able to maintain the level of production of Electrical goods in our Company. It has registered a sharp fall of almost 25 percent as compared with the last year production level, in January.

In my opinion, the following are the main causes of decrease in the production level:

1. There has been a shortage of electricity supply. The plant could not, therefore work to its full capacity.
2. Some of the workers adopted 'go-slow' tactics. They are demanding higher wages. The company's agreement with the Trade Union regarding wages and other allowances expired almost six months back. The Union has placed a fresh Charter of Demands. As a result of this, 'go-slow' method and 'work to rule' tactics have been adopted by the workers. Hence, there has been a sudden fall in production.
3. Sufficient raw material was not available during this period because of transport difficulties.
4. We have not been able to exercise strict control over the quality of our product. Some of our electrical goods have shown a tendency to fail within the guarantee period offered by us to our customers.
5. There is a keen competition in the market and our products do not excel in quality as compared to that of the products of other companies.

The following measures would be helpful to restore and maintain the right level of production:

1. It would be helpful to install our own power generators. Though it may be costly in the initial stages, it will certainly pay good returns in the long run.
2. We must keep sufficient raw material in stock to meet emergency requirements.
3. An immediate dialogue should be started with the union leaders for settling the worker's demands, while giving a reasonable wage rise. Production norms should also be fixed.
4. There should be strict supervision for quality control.
5. Sufficient advertisement campaigns need to be launched to compete with other companies in the market.
6. Our sales representatives should be given better incentives if they sell our products over and above the targets given to them.

I hope that within about a month, the optimum production level can be achieved if the above mentioned suggestions are implemented.

Thanking you,

Yours faithfully,
XYZ
I/C Production Section
Electrical Goods.

5.4.3 Progress Report

- Whenever new work is undertaken, it becomes necessary to have information about how far the task has progressed and how much remains to be done. Such reports equip the executives to plan future actions so as to complete the job in time. Timely completion of any project depends upon the facts reported in the progress report. It becomes possible to remove any bottlenecks in the work which may be regarding labour, material, finance and other factors. Here is an example of a progress report about the construction work of a school building. The points to be covered are work to be done, work so far completed, remarks regarding difficulties encountered, other assistance needed etc.

Progress Report of the Construction Work

From XYZ,
Site Engineer, Village Sangvi,
Tal. Haveli, Dist. Pune 411 027
Date: ________________

To

The Chief Engineer,
Public Works Department
Pune Division, Pune 411 001.

Subject: Brief report regarding the progress of construction of a School building at Sangvi village.

Sir,

I am submitting a brief report about the progress of constructing a school building for making further financial provision.

The actual construction work of this School commenced in the month of .... It is scheduled to be completed in the month of .... There are about .... months at our disposal before the handing over of the completed building to the Pimpri-Chinchwad Corporation authorities.

During the last three months, nearly two third of the construction work has been completed by the contractor. As the school has only one storey, there is no difficulty in completing the work within the prescribed time limit. During the first month, construction up to the plinth level was completed as proposed.

If as per recent instructions the building is to be completed earlier, i.e. 15 days before the scheduled time, the following difficulties shall have to be overcome:

1. Labour supply shall have to be increased by 25%.
2. Increased cement supply must be effected at once. Nearly one thousand cement bags should be allotted from the Central supply. The contractor is pressing on this point. I have promised him the supply in the interest of work without waiting for your approval. He has also been instructed to pay the advance towards this additional supply immediately.
3. The contractor who is supplying bricks for the project is in difficulty, because his workers are on strike. It is presumed that within three or four days the strike will end and the brick supply will become normal.
4. The contractor has already placed orders for the supply of cement sheets required for the roof. The delivery of those cement sheets from Hyderabad is expected next week.
5. Manufacturing of mosaic tiles is going on according to schedule.

Remarks: The project-work of the school building can be completed fifteen days before the scheduled time as the school authorities are pressing hard for it. The following provisions are:

1. Financial clearance for making payments of the pending bills.
2. Supply of cement.

I feel, if the above provisions are made immediately, there need not be any cause for delay in completing this prestigious project as desired.

Yours faithfully,

Site Engineer

5.4.4 Investigation Report

- A friend's machine is out of order. He requests you to find out what is wrong with it. Draft a letter about your investigations and ways and means for restoring it to working condition.
- **Comments:** Following are the important points required to be considered while writing a letter on damages:

  Investigation:

  1. The letter should be clear, simple and to the point.
  2. Ambiguous language should be avoided.
  3. It should be factual and authentic.
  4. The exact damages caused to the machine should be pinpointed.
  5. Whether the damage has been caused due to an accident or due to wrong operations should be stated.
  6. What your suggestions are for the repairs to the machine, etc.
  7. The extent of repairs needed.
  8. Harm that is likely to be caused if the immediate repairs are not undertaken.
  9. Probable time required for completing the repairs.
  10. What are the probable expenses required for the repairs/replacement of damaged parts, etc.

From XYZ
15, Shivajinagar, JALGAON
Date: 27th Nov. 2012

To

Shri Sakharam Patil,
Zilla Peth, JALGAON.

Dear Sir,

Thank you for inviting me to examine the lathe machine of your workshop for its repairs. Necessary details about the causes of damage and expenses for repairs to your lathe machine shall be furnished separately.

On careful examination of the machine, I could make out that the newly purchased lathe machine seems to have been very badly handled by workers, most probably by the workers who are not trained for this type of work. Your workers do not possess sufficient knowledge about operating and handling the machine.

The job on the machine was not properly adjusted. Without making sure of its proper positions, the worker might have started operating it. This mistake has caused damage to the whole lathe machine, and thus the machine has gone out of order.

In fact, such costly machines need to be operated by workers who are trained to handle such machines. The cost of repairs to the lathe machine would be nearly ₹ 3,000/-. The parts which have been completely damaged due to wrong handling are to be replaced.

Regular oiling and cleaning to the machinery is quite necessary. However, it seems that the operator has not taken care to oil the various parts of the lathe machine properly before operating it. It is thus sheer negligence on the part of the operator. The main moving part of the lathe has gone out of order and requires heavy expenses to replace it.

If new parts are to be purchased, nearly ₹ 5,000/- would be required.

I suggest that it would be better to replace the whole part which has gone out of order instead of going into repairs. Even if repaired, the machine will not work for a long period. After some period, the lathe will again start giving trouble. Although replacement may require ₹ 5,000/- it is worth it in the interest of longer duration of the working of the machine.

General Suggestions:

1. Such costly machines, must be operated by trained operators alone.
2. The machines must be properly maintained.
3. Without the permission of the supervisor, no one should be allowed to handle such precious machines.
4. As far as possible, new parts should be purchased for replacing the old instead of repairing them.

If these suggestions are followed, there would not be any difficulty in operating the machine properly. I request you to please let me know about your decision immediately so that the work of putting the machine in order again can be taken up.

After replacing some of the suggested parts, I can give a guarantee of one year for the machine.

Thanking you,

Yours faithfully,
XYZ
(Maintenance Engineer)

- Some more examples of different types of reports are also given below for the guidance of students.

Feasibility Report

1. Write a feasibility report about a Small Scale Industrial Unit you wish to start.

Comment:

1. When one wishes to start a particular industrial unit, one must have a mental picture of the clarity and objectives of the project.
2. There are a number of products which can be manufactured in a small unit.
3. One has to prepare a feasibility report about the project.
4. Project feasibility follows:
   (A) Identification
   (B) Pre-selection
   (C) Analysis
   (D) Evaluation.

5. Feasibility report must include the following points:

Feasibility Report:

1. Name of the manufacturing organisation XYZ
2. Introductory information ....
3. Detailed discussion about the product ....

Marketing:

4. Who are the main customers?
5. Arrangements for selling the product
6. Other possible competitors.

Analytical Report

- Write an analytical report in a letter form to the manager on the suitability of certain premises for the office of the organisation.

Presume necessary details:

- **Comments:** Large organisations are required to establish new branches, when their business expands and the area of their office proves to be inadequate. New spacious premises have to be searched out. In certain cases, industrial organisations have to identify new sites for their expansion and diversification of manufacturing activities. Under such circumstances, they depute their executive engineers and project managers to locate suitable sites and to submit reports regarding suitability of those sites. For this purpose, the following points require special consideration in a report:
  1. Refer to the resolution or order authorising the report writer to prepare the report.
  2. Brief references are to be made to the growing requirements of the organisation.
  3. Suitability of proposed sites or premises (whatever the case may be).
  4. Suggest the most suitable one referring to its advantages.
  5. Mention the drawbacks of the site, if any, and suggest the means to overcome them.
  6. Clearly mention the terms and conditions on which the site (or premises) can be acquired.
  7. Budgetary requirements i.e. cost involved; recommendations in clear terms.
  8. The report must be factual and it should not be biased. (disinterested.)

REPORT

NAME OF THE ORGANISATION

Date:

To

The Manager,

____

____

Pune,

Subjects: Report on suitability of site (premises).

Dear Sir,

In accordance with your instructions, I have observed the present area of our plant. In my opinion, it is very difficult to continue further expansion and diversification of our product manufacturing on the existing premises. There is a definite shortage of space.

Our organisation started in Pune in the year 1985. The activities around our site were then considerably limited. The population of the area has now gone up and our activities have also shown definite growth. Consequently, the presently occupied area is highly inadequate. Every inch of the available space has been pushed into service. The whole atmosphere of our present plant therefore appears to be very congested and overcrowded by men and material. Such an atmosphere can hardly be conducive for further diversifications of our products. If we continue to function here, I fear that further progress is likely to be jeopardised.

As was instructed, I concentrated my efforts on searching for an area suitable for accommodating our growing requirements in this very locality. The area which I wish to suggest has close proximity to our present site. It is only of five minutes walking distance from us. If we occupy that area, not only will our employees not be put to inconvenience, but it would be most convenient for our business and production activities. It is a levelled plot measuring one acre. The only difficulty is that it does not have an approach road at present. We would be required to construct an approach road, which would also be useful for the people of this locality. We also have to bear the cost of constructing a small culvert over the canal that flows between the main road and the proposed site.

I have already had preliminary negotiations with the owner of the area and he is willing to sell the area to us at the market rate (which is about ₹ 100 per sq. ft.) Legal charges would be extra to be equally shared by him and our company.

Recommendations: The price of the land which the owner expects is a bit high but in view of its strategic location, I have no hesitation in recommending that this area may be acquired immediately.

The detailed estimates of land cost, development costs, overhead costs and a preliminary consent letter are enclosed. The cost is well within our expected budget. Since there are many other customers intending to purchase this land, a quick processing and final decision would be highly appreciated.

Yours faithfully,
Executive Engineer

Survey Report

- **Survey Report: A brief report on the suitability of a plot of land surveyed for setting up a small-scale industry on the outskirts of your city.**
- While writing the report on survey for establishing a small scale industrial unit the following points should be considered:

Essentials of the Reports:

1. Purpose of the report must be clearly mentioned.
2. The report must be able to present full facts of the subject.
3. The writer of the report must possess full knowledge of the matter to be reported.
4. Such reports give clear information to the reader and help him in the process of decision-making.
5. An engineer must possess the skill of writing such reports. Success in the engineering profession depends upon the ability of the presentation of an idea in a convincing manner.
6. A good report must be objective. Stress should be on the actual observations and not on personal opinions.

7. A technical report must be factual. It must be limited to general observations and aim at clarity and directness.
8. A report should help the reader to get what he wants quickly.
9. A report must give tables, graphs, good summaries wherever necessary.
10. It should be devised to save time.
11. Use of simple language impresses the reader.
12. Use of specific words rather than vague and general terms makes a report graceful.

Outline of the Report:

A site selection report should include the following points:

1. Introduction
2. Description of various sites.
3. Requirements of the Industry.
4. Facilities available at the selected site, i.e. availability of raw material, fuel power, labour, transport and market should be clearly indicated.
5. Conclusion: It must include future possibilities of expansion etc.

A Report on Suitability of the Site

- **Introduction:** The present report is prepared as a contribution to the development of the brick and tiles industry at Pimpri. It is hoped that it might be of service to any industrialist who needs information regarding the suitability of a site. Nearness of raw material, fuel, transport facilities and markets on the outskirts of Pimpri is conducive for the establishment of a profitable tile industry.
- **Description of Site:** After surveying a number of plots on the outskirts of Pimpri area the area known as the MIDC area near the river bed, 4 kilometers from Pimpri railway station is found to be suitable for the industry.
- **Availability of material required for the brick and tiles industry at the selected site:** At the selected site there are extensive and almost inexhaustible supplies of clay. On testing of the clay it is found that it satisfies all the requirements of brick clay. There are no soluble salts in the clay. It possesses sufficient hardness, when burnt at a moderate temperature, it gets a uniform buff colour on burning. Water of the river Pavna can be freely used for wetting.
- Wet bricks and tiles can be dried on the banks of the river and burnt in kilns. For this purpose, coal can be obtained from Pimpri station where it comes from Chandrapur district.
- Local artisans can make wet bricks on a large scale with the help of dies or machines. Labour is comparatively cheaper and readily available. It can be used for various types of manual labour such as drying the bricks, setting them in and taking them out of the kiln, arranging them for transport etc. As the site is just 4 kms off the railway, bricks can be easily transported to different parts of the state both by road and rail very economically.
- **Present demand and future possibilities:** Pimpri is a rapidly growing industrial town with ever increasing building activities. Bricks and tiles produced will be just sufficient to cater to the local demand. If, however, the production cost increases, bricks and tiles can be supplied in wagons or carried to Mumbai which is not far from Pimpri.

Report No. 4.

A newly purchased lathe machine in your work-shop is out of order since last week. Submit a report to the work-shop Superintendent giving suggestions on its testing.

ASSTT. WORK-SUPDT.
ABC Co., Ltd., Pimpri, Pune
Date: _______

To

The Superintendent of Workshop,
ABC Company Ltd.,
Pimpri, Pune

Subject: Report on the newly purchased lathe.

Sir,

A newly purchased lathe machine was installed in the Production Department of this Company just fifteen days back. I want, to report however that the lathe is not operating well since last week or so.

The lathe in question was purchased by our Company Representative one month back after making due inspection at Mumbai showroom and was transported from there to our factory site. During this transit, it seems that some defects have occurred due to a faulty transport and handling system.

The defect was noticed only at the time of its installation. Such a major defect in a newly purchased machine was never expected. I personally contacted the transport company and enquired about the transportation of the lathe. The Company has accepted that there were some defects in the crane used for lifting the lathe. As a result, the complete consignment was dropped on the ground, while lifting and placing it on the vehicle. In my opinion, we must claim compensation for the damage either from the transport company or the manufacturer as per the terms of the contract. I do not know whether the consignment was insured before dispatch. In that case, we can claim compensation from the concerned Insurance Company.

As it is an important lathe in our work-shop, we cannot afford to keep it out of work for a long time. The jobs which were expected to be completed on the lathe have been given to another company on contract basis. Such an arrangement cannot be continued for a long time.

In the interest of keeping the production schedule undisturbed, it is suggested that the company which has supplied this lathe machine may be contacted for its immediate replacement, without any delay.

There is no point in getting it repaired as it may develop some other defects in future due to the damage done. As per our agreement, guarantee period of the machine is one year.

We have to sustain the expenditure which we have made on its installation. Meanwhile, the jobs are to be given on contract. There is no other alternative with us, as we have to keep up and maintain the finance schedule of the supplies of our products to our clients.

It is hoped that action is expedited in this direction to avoid further losses.

Yours faithfully,
Asstt. Workshop Supdt.
ABC Company Ltd.,
PIMPRI

Project Report (General)

A project report for starting a small scale unit for manufacturing envelopes, files, file-boards, registers, exercise books, letter pads, etc.

Report No. 5

A PROJECT REPORT

Object: It is proposed to set up a small scale industrial unit for manufacturing envelopes, files, file-boards, registers, note-books, pads etc.

Location of the Unit: A suitable plot has been acquired in the M.I.D.C. area near Pimpri, Pune. The plot is quite suitable for a small industrial unit and it is not very far from Mumbai-Pune Road.

Introduction: There is always enough demand for various items of office stationery, both in Pune and Mumbai. The proposed items of stationery materials are constantly required in huge quantities in Government offices, schools, colleges, banks and in factory offices. A market survey was conducted in and around both the cities. The survey revealed that there are good prospects for such a unit. It is hoped that a manufacturing unit will prosper beyond doubts within a very short time.

Manufacturing Process: The raw material such as paper and board is proposed to be purchased from M/s. Paper and Pulp Conversion Ltd. Khopoli. Raw paper will be purchased on a contract basis. Detailed terms of contract have been finalised with that company. The raw material will be cut into pieces of required size and shape and printing as well as folding will be done in our unit. Sufficient and up-to-date machinery has already been ordered, and it is expected to arrive at our factory site very shortly. The said machinery is entirely manufactured in India. There is no difficulty in getting the spare parts. Thus, the purchase of foreign machinery has been completely avoided.

Technical Aspects: To begin with, the following three items of stationery articles will be manufactured:

1. Envelopes.
2. File boards.
3. Note books.

Paper and boards will be first cut into pieces of required size and shape with the help of a paper cutting machine. The paper cutting section will be kept separately. Experienced and skilled labour is required in this section. No wastage will be allowed. One diploma holder has already been engaged for upkeep and maintenance of the machinery and for controlling and supervising the work in this section.

Before the paper and boards are used for manufacturing envelopes, file boards, note books, etc., they are required to be sent to the Printing Section. Two printing units are proposed for the job. One will be engaged in printing work while the second one will be used for the printing of lines and borders required for exercise books and file boards. The printing machinery is semi-automatic. Ten workers will be engaged in this section. Also one supervisor has already been appointed. He will be in charge of the printing section.

The third section will look after the jobs of folding, binding, finishing and packing. In this unit, the basic material will be collected and finished goods will be dispatched to the Stores Section for disposal/sale.

Financial Implications: A statement of the expenditure involved in the project is given below:

Particulars	**Amt. (₹)**
1. Non-Recurring Expenditure	
(A) Land	3,00,000
(B) Construction of Factory Shed	6,00,000
(C) Office and Store Building	2,00,000
2. Machinery and Equipments	2,00,000
Recurring Expenditure for one year	13,00,000
3. Raw materials:	
(A) Product A	50,000
(B) Product B	50,000
(C) Product C	50,000
	1,50,000
4. Staff (Indirect Labour)	1,00,000
5. Direct Labour (Skilled, Semi-skilled and Unskilled)	1,00,000
	2,00,000
Total expenditure for one year:	
Direct and Indirect Labour	50,000
6. Consumer Stores	
7. Other items of expenditure (Electricity, Water, Packing, Transportation, Miscellaneous etc.)	50,000
	1,00,000
Total Recurring Expenditure Per Year (3 + 4 + 5 + 6 + 7)	4,50,000
8. Total investment required (Non recurring + recurring expenditure for one year)	17,50,000 6,00,000
9. Total amount expected to be received from Sales	4,50,000
10. Cost of production	50,000
11. Depreciation	1,00,000

A commercial bank has promised to provide finance.

This project report would clearly indicate that this small scale industrial unit is certainly a profitable business. This report may also be useful for those who are also interested in starting such ventures.

Project Engineer

Write a detailed report on the testing of a new lathe machine in the work-shop.

- **Comments:** Following are the important points which should be kept in mind while writing a testing report:
  1. The report should be clear, simple and to the point.
  2. The report must be based on actual experience/operations.
  3. Ambiguous language should be avoided.
  4. The various tests required to be carried out should be stated.
  5. It must mention the method and procedure of operation.
  6. It must state what precautions are required to be taken in order to keep the lathe always in good working condition.
  7. It must mention the period after which servicing/overhauling should be carried out.

Testing Report

ASST. WORK-SHOP SUPDT.

Date: 27th Nov. 2012

To

The Superintendent, Workshop

Sir,

I have been instructed to submit a detailed report about the various tests undertaken on the newly purchased lathe machine.

1. The various tests connected to the working of the newly purchased machine have been carried out.
2. It suits our requirements.
3. As we have to give this machine to our Final Year students utmost care needs to be taken while teaching them the various operations of this machine.
4. The lathe machine has a special feature. If the job to be done on the machine is not properly fitted, the lathe does not operate.
5. I have tested the various parts and the revolutions of the machine and in my opinion they are according to our specifications.

After conducting various tests on this lathe machine, it was found that all the parts were in good condition. There was no difficulty in putting the machine in operation.

Care to be taken while operating:

1. No student should be allowed to operate the machine unless the operator is present.
2. The operator must give sufficient instructions to the students before operating the machine.
3. Careless handling of the machine should be avoided.
4. The lathe machine should be oiled every week and servicing should be done periodically after every two months.

This being a very important machine in our workshop care is required to be taken while handling it. All the different types of operations have been performed by me personally and I did not find any defect in this lathe.

Thanking you,

Yours faithfully,

Astt. Workshop Suptdt.

Report on an Educational Tour

1. **Title of the Report:** Educational tour to Pune, Pimpri-Chinchwad Industrial belt undertaken by the Mechanical Engineering students of final year.
2. **Objective:** To study the working conditions and transport facilities available to the Industrial workers of Pimpri-Chinchwad area of Pune.
3. **Purpose of the Report:** To give a brief description about the educational tour to the junior students and enlighten them in advance, about the things expected from engineering students during the course of their study and what they should observe and study when such technical tours are arranged by the institute in future.

REPORT

Introduction: During the first terminal vacation in 2012, we, the Final Year Mechanical Engineering Students, visited the Pimpri-Chinchwad Industrial belt of Pune city with a view to study the general working conditions of the industrial workers and particularly the transport system provided to them.

A brief report of the visit is given below:

Layout of the Pimpri–Chinchwad Area:

It is a rapidly developing area adjacent to Pune City on the Mumbai-Bangalore highway. When our bus was moving towards that area, our teacher informed us about the exact area which we were expected to study. We could see a chain of industrial units on both the sides of the highway.

Our first stop was near Atlas Copco factory. We kept our note books and writing material ready with us. Our teacher contacted the Public Relations Officer of the Company, who came out and escorted us to the factory. He showed us the various departments and general working of the factory. We asked several questions of the workers who willingly answered them. We had already written down those questions in our note books and replies given by workers were immediately noted down by all of us.

From there, we visited several medium and small scale industrial units. Our questions were mainly directed towards their working conditions and the transport arrangements provided by their organisations.

Summary of the Discussion with Workers:

After discussing various matters with the workers, we arrived at the conclusion that there were a lot of difficulties and drawbacks, in their transportation system. It was seen that the workers were required to spend on an average two to three hours to reach their working site from their residence which is certainly a laborious task. An efficient transport system will ultimately increase their productive efficiency. In all the industrial cities of India, such problems exist with only a difference of degree.

(i) Report on Education Tour

Introduction: During the first term vacation (holidays) in the month of January 2012, the First Year Civil Engineering Class Students' batch visited Mumbai with a view to study the planning of the city in general and the working of some of the educational institutes engaged in imparting technical education.

This report would be useful for those who wish to arrange such study tours in future.

The lay-out of the City: Mumbai is a well planned city. There are broad roads. Two parallel central roads are made in such a way that they connect all the important buildings, on both the sides. The chief buildings of the city include the Secretariate and Assembly Hall, in the Fort area. The commercial centre is suitably located. It is crowded with shops, commercial offices, hostels, lodges, theaters, stadiums etc.

Noteworthy Models of Modern Architecture

Innumerable grand structures in and around Mumbai are Chhatrapati Shivaji Terminus, Chhatrapati Shivaji Airport, Aarey Milk Colony, Jijamata National Park, G.P.O., Essel World, Hanging Garden, Taraporewala Aquarium, Sea Beaches, many new constructed commercial buildings, flyovers, etc. They are important and worth visiting. In the Fort Bandra-Kurla Complex and Nariman Point area foreign trade office buildings reflect the grandeur and richness of modern architecture. They are airconditioned and their glass panels with multi-coloured curtains add to the charm. All these magnificent buildings make this well planned city a centre of tourist attraction.

Conclusion: Such visits are definitely useful for technical students. It actually gives some insight into their future profession.

Inspection Report

A Specimen proforma for an inspection report.

Inspection Report on Taperecorder

Number Date	Mark Model Type Serial No.
Checked by .... **Note:** Please put a tick ✓ against the relevant item.	
(a) Visual Inspection	Case broken/Scratched/normal
(b) Mechanical Movement	Stopped/Obstructed run/intermittent/noisy/smooth
(c) Amplifier	Playback: dead/poor/distorted/humming /normal
(d) Recording	1. Direct proper/poor /distorted normal.
(e) Frequency Response	2. Other sources recording: dead/ noisy/poor/normal.
(f) Suggested Repairs and Replacement	3. Microphone: dead/noisy/ poor/ normal high frequency missing/low frequency missing /speaker
(g) General Remarks.	
Signature	

Requirements of a Manufacturing Process:

1. Production capacity of the proposed unit.
2. Capital Expenditure (Fixed assets).
3. Expenditure on working capital.
4. Probable profit per year.
5. Requirements of Raw Material.
6. Expenditure on water, electricity, fuel etc.
7. Manpower requirements.
8. Expenditure on stationery, taxes, postage etc.
9. Total yearly expenses, sales, net profit.
10. Percentage of profit:
    (A) Percentage of Profit to total capital invested.
    (B) Percentage of Profit to total sales.

Book Review

- It has almost become a common feature of newspapers, magazines, periodicals etc. to devote some columns in their publications for book reviews. The main concern of any book review is the general assessment of a published book and its readability for the readers at large. It is normally written by a person who is supposed to possess thorough knowledge about the subject concerned. In writing a book review a critic essentially takes into consideration the following broad aspects about the book under consideration:
  1. What is the theme of the book?
  2. Author's view about the theme.
  3. Treatment given by the author to the subject.
  4. Whether the view of the author enhances or retards the popular notions.
  5. How does the present book stand in comparison with other books on the same subject published earlier?
  6. How far does the author's treatment of the subject significantly differ from that of the others?
  7. What new light does the author want to throw on the subject?
  8. Style and language used.
  9. How far does the style and language suit the subject?
  10. Are there any lacunae in the book in respect to theme, treatment, or views of the author?
  11. What additions, alterations or eliminations would make the book popular among the readers?
  12. Comments about the general assessment of the book.

  The columns on book review are being widely read these days because of the fact that people hardly get time to read books in totality. Book reviews help people to choose only such books as would appeal to them.

5.5 DEFINING, DESCRIBING OBJECTS AND GIVING INSTRUCTIONS

Framing Definitions

- The word definition comes from the Latin original *'definre'* ('de' means down and 'finre' means to end which comes from 'finis'). The dictionary meaning of this word as given in the Concise Oxford Dictionary is 'exact description of the nature, scope or meaning of something'. When we use the words 'exact meaning' that itself means that in any definition there is no scope for elaborating the thing but to state its meaning very precisely. Inclusion of unnecessary words tends to make a definition loose or at times meaningless. Such group of words may misguide the reader about the correct meaning of the word being defined.

- You read many text books in your curriculum. Whatever concepts or theories you study in your curriculum, are first and foremost defined in the text book, to make further studies easy and clear. Most studies generally begin with a definition of that concept. Take for example, the word communication itself. You will find in Chapter 1 of this book as many as eleven definitions of the word communication. These definitions are framed by different scholars on the basis of how they conceive the meaning of the concerned concept, theory or term. Different people think differently about the same thing. However, one thing is common in their definitions and that is they are precise, exact and to the point. They are not elaborate and do not contain unnecessary words. The main aspects of any definition are generally as follows:
  (i) Conciseness.
  (ii) Precision.
  (iii) Exactness.
  (iv) No ambiguity or circumlocution.
  (v) Conveying exact meaning.
  (vi) Limiting the scope.
- While framing any definition, care has to be taken to see that it is framed, keeping the above aspects in mind.
- No definition becomes final in one stroke. Scholars frame them as they conceive. When others read it, they subject the definition to their thoughts and thoroughly study and critcise it to bring it to exactness. Definition of anything is never final but subjected to a long process of study. Each new definition is further studied in detail by other scholars to eliminate ambiguity, redundancy, vagueness, circumlocutions in the definition, if any, etc.
- Thus, framing definitions is a very important task in any study. It requires careful thinking and proper attention to every aspect of the concept being defined. Following are some examples of definitions regarding some theories or concepts.
- **Economics:** 'Economics is a study of the allocation of scarce resources among competing ends and uses'.
- **Entrepreneurship:** 'Entrepreneurship is the ability to create and build something from practically nothing'.
- **Organisation:** Organisation is defined as 'the process of identifying and grouping the work to be performed, defining and delegating responsibility and authority and establishing relationships for the purpose of enabling people to work most effectively together in accomplishing objectives'.
- **Bank:** Banks are institutions whose debt, usually referred as 'bank deposits', are commonly accepted in the final settlement of other people's debts'.
- **E-mail:** This stands for 'Electronic Mail'. It is the name given to the 'messages transmitted and received through electronic devices like computers via satellites'.
- **Engineer:** An engineer is 'a person who thinks and achieves proficiency in his respective field of engineering'.
- **Calculator:** It is 'a device based on an electronic circuit and works faster than the human brain'.
- **Management:** Management is 'the task of planning, coordinating, motivating and controlling the efforts of others towards specific objectives'.
- **Floppy disk:** A floppy disk is 'a data interchange or distribution media device used to store data'.
- **Data:** Data is 'unstructured raw facts in isolation to be used as raw material of the information system'.

5.5.1 Defining Objects – Appearance, Its Use

- There are several occasions when a person has to describe certain objects including living and non-living ones. Description of anything is a piece of writing or speech that says what somebody or something is like. It can also mean the act of writing or saying in words what somebody or something is like.
- Non-verbal skills play an important role in written communication. Similarly, visual signs also play an important role in understanding objects. Describing an object can also be done with the help of graphs, tables, pictures, images, paintings, photographs, etc. Pre-literate man might have widely experienced visual expressions. Following are the important visual forms of communication today:
  1. Tables
  2. Charts and graphs
  3. Pictograms
  4. Maps and plans
  5. Slides
  6. Film strips.
- In visual communication, very elementary and simple ideas, orders, warnings are used. It can be effectively used only in combination with other mediums. Diagrammatic representation combines the written with the visual. Pie charts and pictograms also combine the two because there is always a written indication of what is shown in them. Such forms are very useful in the presentation of statistical data.

Advantages of Describing Objects

1. Charts and graphs can be used for effective communication.
2. It makes people remember things for a longer period. In this case pictures and graphs are more effective.
3. Description of objects helps in making comparisons easy.
4. Description of objects helps in making publicity more effective.
5. It educates people.
6. Sometimes it is worth a thousand words.
7. It is useful for scientific and technical subjects.

Limits of Describing of objects

1. Descriptions cannot be used for conveying all kinds of information.
2. They are not effective for discussion.
3. Descriptions are not more effective in explaining ideas in detail.
4. Sometimes it may lead to misunderstandings.

Processes and Methodologies

- In today's world of advanced technology and science, effective communication plays a significant role. Particularly in science laboratories, engineering workshops, technical institutes, practicals are carried out on various subjects, processes and mechanical functions. In order to make the students understand these aspects thoroughly, merely conducting or carrying out these experiments in the laboratory, workshop or shop-floor is not enough. Although students may understand the processes, experiments and functions of machines and equipment by seeing them personally, it is essential that they are written down to be used for reading time and again or for verification or assessment and remembrance. It is for this purpose that educational institutes prescribe writing of journals for their students. These journals are also forms of non-verbal communication.

- Students are made aware about the journals which incorporate experiments carried out in their science laboratories instead of reporting of experiments made according to the will and wish of students. They are to be reported in a prescribed and uniform manner. For many technical and science subjects, specific journals containing the methods and manners in which the processes of these practicals are to be written, are prescribed or printed. Take any science journal and you will find that the experiments are written in a particular chronological order. For example, an experiment is to be completed and its process is reported in the following particular order or chronology:
  1. Aim of the experiment.
  2. Equipment required.
  3. Functions of each equipment.
  4. Arrangement of equipment in the proper order of an experiment.
  5. Description of the process of experiment with appropriate figures.
  6. Findings of the experiment.
  7. Recording the findings.
  8. Required graphs and charts.
  9. Results and conclusion.

Example:

1. A typical example of a laboratory experiment is reproduced below for proper understanding of the students.
   1. Aim – Propagation of sound through liquid
   2. Equipments required – Glass Trough
Metallic Disk
Water
Tuning Fork
   3. Function of each equipment –
      - Glass trough for storing water
      - Metallic disk for creating sound
      - Water for studying the propagation of sound through it
      - Tuning fork to assess the propagation of sound.

Arrangement of equipment

- Place a glass trough on a table. Fill it with water. Place the metallic disk on its surface. Fix a tuning fork with stem on the metallic disk.

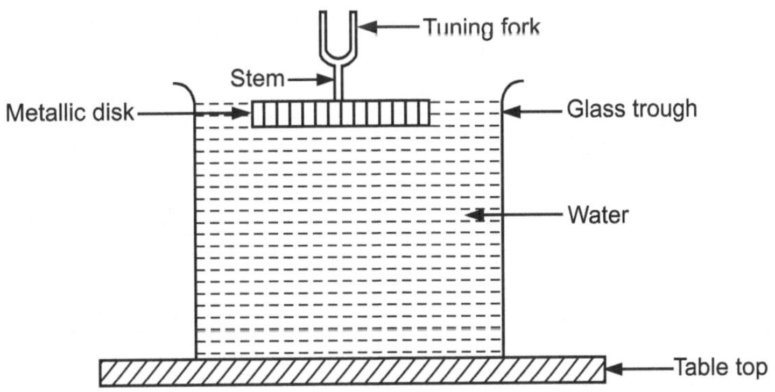

Fig. 5.1

- **Process:** The tuning fork with a stem is attached to a metallic disk, and is set into vibration and held with the surface of water in a glass trough kept on the table top. The vibration travels through water to the bottom of the glass trough. The water also begins to vibrate and then emits a loud sound.

Findings or result of the experiment:

- The vibration travels through water to the bottom of the trough. Water begins to vibrate and emits a loud sound.
- There are many other occasions when different processes are required to be reported. It is not the scientific or technical processes only that are required to be written. We constantly come across many situations wherein processes are required to be brought to the notice of the people concerned. Almost every action and reaction in human life follows a certain process. Take the example of printing a book. The task of printing a book involves different processes taken together.
- When a book is to be printed, an appropriate author is to be approached. He is given the subject and its outline for writing the matter. After the author writes the manuscript, it is given to the typist or D.T.P. operator for typing. Rough proofs are taken out. The proofs are then corrected by proof readers. The final proof is given to the author for his permission to print the book and the matter is printed. Figures, graphs, pictures are added at appropriate places. The cover page and other front pages are also finalised. When the author says 'O.K.', the book is printed. Thus, printing of a book involves certain processes.
- Every development in human life or incidences around us takes place following certain processes which people report from time to time.

5.5.2 Giving Instructions: Precise, Directive, Imaganistic Statements of a Futuristic Stance

- There are two types of people in any work force. One group consists of those who issue instructions for carrying out the desired work, while the other consists of those who actually carry out the tasks as per the instructions given to them by their superiors i.e. Supervisors, Engineers, Managers, Officers etc. Instructions may be regarding what to do and how to do it. To ensure that the required tasks are performed by concerned people requires proper thinking, proficiency, ability to give proper guidance, and a proper attitude. Proper instructions enable subordinates to understand the objectives of the supervisor and manager in giving those instructions. Instructions should lead to achievement of the management objectives. Giving instructions involves the handling of human beings. It is one of the most important parts of management functions and means directing the subordinates into action. Subordinates do not work unless they get clear cut instructions from their superiors. Instructions form a part of directing the organisation towards certain desired goals. Proper technique has to be followed while issuing instructions.

Instructions

- Giving instructions is a continuous task of making decisions. It forms a major daily function of effective leadership. It also includes co-ordinating day to day functions in order to ensure that those efforts are properly directed to achieve desired results. Instructions, whether in oral form or written form, should be given to a small group or to all the employees at a time. Written instructions require certain standard operating procedures.
- Instructions should be intended to arouse the work force into action that ultimately leads to achieving company's desired objectivities. In this respect, supervisors play an important role in securing the required level of standard in team work. Although giving instructions and inducting team work are the necessary activities of supervising staff, they must also exercise direct personal supervision and control if maximum and positive results are to be achieved. Supervisory staff is concerned with enforcing orders given to them by the management. It also deals with giving appropriate training to operators or workers. Trained and experienced subordinates require less instruction and direction. Hence a greater number of trained workers can function under one supervisor.

Need of issuing instructions

- A supervisor or a foreman is a person who is primarily in-charge of a section or a group of employees in an industry. He is responsible for the successful completion of the assigned task or job of his section. He is also responsible both for the quality and quantity of production. This ensures efficient performance of the equipments under him. He is also responsible for the efficiency, boosting of morale and training of his subordinates. He is expected to issue instructions to get the desired results. A supervisor can play a significant role in increasing productivity if the operators are provided with correct and timely instructions in respect of production schedules. Timely instructions can save wastage of material, machine hours and labour. He must see that the instructions given by him are properly understood by the operators. He should devise a system of feedback so as to ascertain whether the instructions given to the workers are correctly followed by them or otherwise. On the basis of such feedback, he should be prepared to modify, improve or change the instructions if necessary. This may create an atmosphere of understanding, co-operation and co-ordination among the workers in his group or in his unit of operations.

Important characteristics of instructions

- Before issuing instructions the instructor or supervisor should first assess the feasibility of those instructions himself. He should also think of his workers' choices and carefully consider whether or not the allotted job is within the capacity of the workers and whether it can be completed efficiently according to the specifications laid down. An important function of a supervisor is to get results according to the specifications and targets. While specifications refer to the quality of work, targets refer to the physical output. Every supervisor has, therefore, to see that he gives results according to the objectives of the organisation and ensure fulfilment of those targets according to the specified schedule.
- Sometimes it happens that the given instructions lack clarity. As a result, the subordinates often get confused while following such orders. Many times, timely supply of raw materials is not certain. Adequate services to the subordinates are also not provided. This makes the operator's job difficult.
- Following are the important points in respect of clarity, completeness and feasibility of giving instructions:
    1. While assigning a job to an employee, a supervisor must give clear instructions.
    2. The supervisor should see whether the person whom he is assigning the job is capable of completing it in time as well as according to the specifications.
    3. The feasibility of completing the given instructions must be examined before assigning a job to a particular worker.
    4. The supervisor must take into consideration the capability of the worker, his knowledge and his previous practical experience in performing such tasks.
    5. The instructions given must have enough clarity in them to enable proper understanding for the workers.
    6. The supervisor should have perfect knowledge in terms of Job description, Job satisfaction and Job analysis.
    7. Knowledge or instructions or job specifications give him a clear idea about the personal qualities required to be possessed by the workmen to do the job according to the set standards.
    8. The supervisor should be conversant with induction procedures; good induction means ensuring a good start for the employee.
    9. While giving instructions to subordinates, the supervisor should see that the worker is not subjected to fatigue or stress but is at ease.
    10. While performing the job, the supervisor must provide essential information wherever and whenever needed.
    11. The supervisor should ask his workers to confirm clarity in understanding of the instructions with respect to the assigned job.

12. He should develop the learning interest of a new worker when he allots the job to a new worker.
13. He must equip workers with know-how about the use of materials, tools, jigs, etc. required for the job.
14. To get greater clarity in performing the job, the supervisor should show the worker a finished job, so that he can get a clear idea about its form, quality and the efforts he is required to put in for a similar work or job.
15. Sufficient instructions should be given to see that the worker pays attention to all safety measures.
16. To get quality in performance, the supervisor should tell, show and illustrate different stages of the work gradually taking one stage at a time.
17. The supervisor should help the workers in correcting the errors that occur while performing the job.
18. The supervisor must prepare carefully and clearly all the instructions, part by part, and provide them to the workers for their guidance and speedy execution of the job.

- Following are the points which a supervisor should try to avoid while issuing instructions:
  1. Giving instructions in an off-hand manner.
  2. Assume the worker's capability of understanding.
  3. Giving too many instructions at a time.
  4. Issuing conflicting instructions.
  5. Entrusting a job which the supervisor himself is unable to do.

Types of Instructions

- In an organisation instructions are passed from one person to another engaged in certain activities. There are different types of instructions depending upon different situations. Broadly, instructions fall into the following categories:
  1. **Precaution:** This type of instruction is given to the user of an object so that no mistake, mishandling or damages result. They are generally like guidelines to be followed while performing a job or handling certain tools, machines or equipment.
  2. **Caution:** Caution is an instruction to the user to avoid mishandling and carelessness. It is more serious in nature and needs to be followed very scrupulously.
  3. **Warning:** Warning alerts the person from danger in regard to certain consequences of actions or handling of tools, machines or equipment.
  4. **Danger:** It is a notice given to the user for being extremely careful in a particular situation or circumstances.
  5. **Notice:** Notice is given to a person to attract attention towards a certain situation.

Important Points

Common types of business letters

1. Circulars
2. Notices
3. Memos
4. Job applications
5. Enquiry letters
6. Orders
7. Complaints
8. Adjustments
9. Reports
   1. Circular letters are used when certain matter is to be made known to a large number of people.
   2. Notices: This form of correspondence is used in the corporate world for the people within the organisation and not for outsiders.

3. Memo: It is the short form of 'memorandum' and is used for internal communication.
4. Job application:
    1. Is the most important because with this we start our career.
    2. It requires careful wording.
    3. States personal details.
    4. Complete Bio-data/Resume/C.V. is provided (either on a separate sheet or incorporated in the body of the application).

Letter of Enquiry (and replies thereto)

It may be either for a general purpose or specific purpose.

1. Used when specific information about products/goods or services etc. is required.
2. Such letters must be courteous and short.
3. Particularly pleasing language is required.
4. Letter must end with friendly words.

Replies to an enquiry letter:

1. Must be so worded as to enhance market goodwill.
2. It must influence the readers.
3. Promptness in reply is a must.
4. Reply must be courteous.
5. Validity of the information supplied must be mentioned.
6. It should contain enough persuasion.

Letter of Quotation:

Quotations are asked:

1. When purchases are required to be effected by an organisation.
2. Conditions of supplies are mentioned in the Quotation letters.
3. Mostly, information in tabulated forms is supplied.
4. Quotations are always time-bound.
5. Conditions of supplies are mentioned as a matter of precaution.

Letter for placing orders

1. Such letters are issued for placing orders of goods or services.
2. Specifications of the material or goods and services required are mentioned in these types of letters.

Reply to order letters

1. Begins with thanks.
2. Quick compliance of the order is expected.
3. Quick responses build company's reputation in the market.

Complaint letters:

1. When certain requirements are not fulfilled, complaint letters are written.
2. Even though there is a complaint to be made, we should not lose the balance of mind.

Adjustment letters:

1. An adjustment letter contains opportunity to correct mistakes.
2. Various topics with respect to adjustment are included in such letters.
3. Such letters might be for accepting or rejecting adjustments.
4. The rejection should also be in a polite language.
5. This type of letter indicates the relationships between a supplier and customer.
6. Should not hurt customer's feelings and confidence.

Report writing:

1. Reports give information.
2. It is upward communication.
3. Facts are logically mentioned.

4. It is useful for planning future action.
5. There are a variety of reports such as special reports, routine reports etc.
6. Reports may be individual or committee report.
7. Report contains findings, recommendations etc.

Important qualities of a good report

1. It must be objective.
2. It must be factual.
3. Reader must get the required information.
4. It must be in simple language.
5. Specific words must be used whenever required.

Structure of an investigation report

1. Terms of references
2. Procedure adopted
3. Findings of the investigation
4. Recommendations
5. Signatures (with place, date) of the reporting officer/officers.

Types of Reports

(A) Accident Report

It should contain:

1. Circumstances under which the accident took place.
2. Type of injury caused.
3. Action taken for relief and treatment.
4. Recommendations to avoid repetition.

Report regarding damages/losses

It should contain:

1. Actual incident.
2. Causes for damages/losses.
3. Extent of damages/losses.
4. Extent of repairs.
5. Arrangements for repairs.
6. Probable expenses involved.
7. Arrangements during transition.

(B) Report on the Fall in Production

It should contain:

1. Extent of fall in production.
2. Its assessment in relation to the target.
3. Period during which production has fallen.
4. Causes for the fall in production.
5. Suggestions to overcome the situation.
6. Findings must be supported by statistical data.

(C) Progress Report

It should contain:

1. Work to be done.
2. Work so far completed.
3. Difficulties encountered.
4. Ways and means to overcome difficulties.
5. Probable period required for completing the task.

Framing Definition

Definition states the exact meaning of something.

Main aspects of definition

Shortness

Precision

Exactness

Unambiguity

Exact meaning

Limits the scope.

Classification and Description of Objects

Form of non-verbal communication

Description may be in the form of

Tables

Charts and Graphs

Maps and Plans

Slides

Film strips

Advantages

Overcomes language difficulties.

Communication becomes effective.

Comparison becomes easy.

Educates people.

Useful for scientific and technical subjects.

Limits

Not useful for conveying everything.

Not effective for explaining ideas in detail.

Sometimes leads to misunderstandings.

Explanation of process

Most useful in explaining experiments.

A useful tool for education.

Useful for proper understanding of things.

Writing Instructions

One of the major functions of management/leadership.

Need for Instruction

Essential for achieving desired objectives/aims.

Important Characteristics

Clarity

Completeness

Feasibility

Confidence

Practice Questions

1. Give in brief the main features that are required in processing message writing.
2. Write notes on:
   (A) Designing of message
   (B) Conciseness of message
   (C) Importance of clarity in writing a message.
3. What is meant by the 'You' attitude?
4. What are the qualities necessary for good writing?
5. Describe the factors necessary for planning a message.
6. What are the qualities required for organising a message?
7. Describe the method of writing complaint letters.
8. Describe the features of Technical writing.
9. Give a typical format of a general business letter.

BUSINESS LETTERS

10. Ms. ABC and Company has not paid you ₹ 20,000/- for steel furniture supplied to them in April 2012. Write a letter of complaint about non-payment of dues.
11. T. Ram and Company of Nagpur, has placed with you an order for measurement Instruments. Please write a letter asking them to supply you with two references because this will be your first transaction with T. Ram and Company.
12. Answer the following advertisement:

    Wanted a Production Engineer, young, hard working and with some experience of a production line. Apply enclosing attested copies of testimonials to Box. No. 526. 'The Times of India, Bombay'.
13. Write a letter to M/s. ABC and Co. asking them to supply machinery along with its spare parts required for a Kolhapur based company for the manufacture of ceiling fans.
14. Write a letter to the Bank Manager bringing to his notice certain errors in the credit and debit figures in the Company account and requesting him to carry out necessary corrections.
15. Write a circular letter to the company workers directing them to maintain discipline, cleanliness and a cordial social atmosphere for the betterment of the company.
16. Write a letter to the supplier of spare parts and informing him about the possible losses the company is likely to incur due to the delay caused by them in the supply of the parts for which an order had been placed long ago. (Assume your own data).
17. Write a circular letter to client firms informing them about the discontinuance of Mr. 'X' as an agent of the company and directing them not to undertake any business transactions with him.
18. Write a letter to an Employment Exchange, requisitioning them to supply a list of suitable candidates for the position of an engineer in the mechanical division of the factory. Mention details about your requirements such as qualifications, experience, salary they expect etc.
19. Write a circular letter to shareholders of the company communicating to them the progress the company has made and informing them about the proposed development and expansion in the company and also calling for their co-operation in further financing the future of the company.
20. Write short notes on:
    (a) Message writing
    (b) Essentials of good letters
    (c) Hints for writing letters
    (d) Mechanics of writing
    (e) Types of business letters
    (f) Circular letters
    (g) Memo
    (h) Job Application
    (i) Letter of enquiry
    (j) Active vocabulary
    (h) Passive vocabulary
21. What is definition? What part does it play in any study?
22. State how classification and description of an object forms a part of non-verbal communication?
23. State the different forms of verbal communication? Explain them in brief.
24. What are the advantages of describing objects?
25. What are the limitations of describing objects?
26. What is the role of explaining processes in science and engineering?
27. State the different stages in explaining a scientific process/experiment.
28. Explain any simple process you experience in your daily life.
29. Write down any process of an experiment in your laboratory/workshop.

■■■

Chapter **6...**

Oral Skills

Contents

6.1 PHONETICS AND PHONOLOGY

6.1.1 Introduction

- In this chapter, we sketch the pronunciation system of English. We begin with phonetics, a system for describing and recording the sounds of language objectively. Phonetics provides a valuable way of opening our ears to facets of language that we tend to understand by reference to their written rather than their actual spoken forms. Phonology concerns itself with the ways in which languages make use of sounds to distinguish words from each other.
- Teachers should be knowledgeable about the phonetics and phonology of English because:
  1. the sound system is primary and the basis for the spelling system;
  2. they may have to teach English pronunciation to students who are not native speakers of English;
  3. they may have to teach poetry, which requires that they teach about rhyme, alliteration, assonance, and other poetic devices that manipulate sound;
  4. it is important to understand accents and language variation and to react appropriately to them and to teach appropriate language attitudes about them to students;
  5. we are so literate that we tend to "hear" the sounds of our language through its spelling system, and phonetics/phonology provides a corrective to that; and
  6. phonetics and phonology provide systematic and well-founded understandings of the sound patterns of English.

Articulatory phonetics

- First, we introduce you to the ways in which the sounds of English are produced. Second, we develop a system for classifying speech sounds on the basis of how they are produced. Simultaneously we introduce an alphabet approximating that developed by the International Phonetics Association (IPA), which will allow us to refer to sounds quite precisely. When we want to indicate that letters are to be interpreted as phonetic symbols, we enclose them in square brackets, [ ], and when we want to indicate that letters are to be interpreted as letters from an ordinary spelling system, we enclose them in angled brackets, < >.
- The phonetic alphabet uses many of the letters of the English alphabet, but their pronunciations are very restricted and are not always the ones you might expect. In this system, there are no "silent" letters—every phonetic symbol represents an actual sound. Every letter always has the same pronunciation regardless of its context, no letter has more than one pronunciation, and no sounds are represented by more than one letter. To make fine distinctions, phoneticians add special symbols, called diacritics, to the basic letters. For some English sounds and for languages other than English, symbols not from the English alphabet have been devised.
- In the sections to follow, we describe the sounds represented by these symbols and how these sounds are made. As we go through these sections, pay attention to the ways in which individual sounds are ordinarily spelled in English, as well as to the phonetic spellings.
- To produce speech, air must flow from the lungs through the vocal tract, which includes the vocal folds (popularly called the vocal cords, though they are more like thick elastic bands than strings), the nose or nasal cavity, and the mouth or oral cavity .
- The vocal folds vibrate for some sounds but not for others. Air flows through the nose for certain sounds but not others. But the main creator of speech sounds is the mouth.

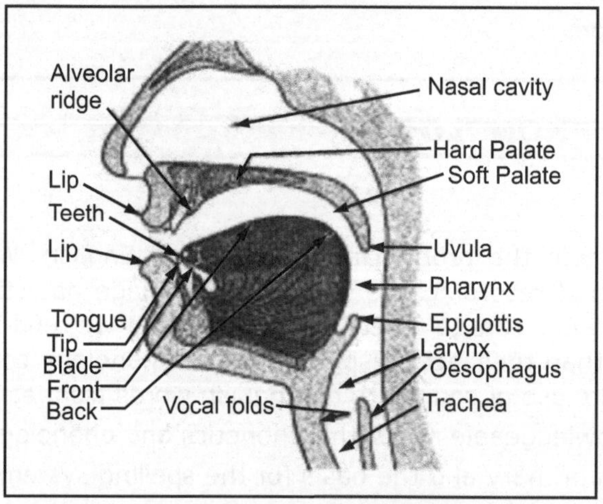

Fig. 6.1: Vocal Apparatus

- Languages can basically be thought of as systems—highly complicated ones—which enable us to express our thoughts by means of "vocal noises", and to extract meaning from the "noises" (speech sounds from now on) that are made by other people. *Linguistics is the study of the nature and properties of these systems and its various branches focus on different aspects of the communication process.* ***Phonetics*** *is the branch concerned with human speech sounds.* It can be classified into three different aspects:
  - **Articulatory Phonetics:** Describes how vowels and consonants are produced or "articulated" in various parts of the mouth and throat.

- **Acoustic Phonetics:** Studies the sound waves that transmit the vowels and consonants through the air from the speaker to the hearer.
- **Auditory Phonetics:** Looks at the way in which the hearer's brain decodes the sound waves back into the vowels and consonants originally intended by the speaker.

- ***Phonolog*** is another branch of linguists closely associated to phonetics. It pays attention to the way languages use differences between sounds so as to convey differences of meaning between words, each language having its own unique sound pattern. Phonology is really the link between phonetics and the rest of linguistics.

6.1.2 Phonetic Symbols

- **Definition**

  A written character used in phonetic transcription of represent a particular speech sound.

  Type of: character, grapheme, graphic symbol a written symbol that is used to represent speech.

Phonetic Symbols

Consonants

1	/p/	as in	**pen** /pen/
2	/b/	as in	**big** /bɪg/
3	/t/	as in	**tea** /ti:/
4	/d/	as in	**do** /du:/
5	/k/	as in	**cat** /kæt/
6	/g/	as in	**go** /gəʊ/
7	/f/	as in	**four** /fɔ:/
8	/v/	as in	**very** /ˈveri/
9	/s/	as in	**son** /sʌn/
10	/z/	as in	**zoo** /zu:/
11	/l/	as in	**live** /lɪv/
12	/m/	as in	**my** /maɪ/
13	/n/	as in	**near** /nɪə/
14	/h/	as in	**happy** /ˈhæpi/
15	/r/	as in	**red** /red/
16	/j/	as in	**yes** /jes/
17	/w/	as in	**want** /wɒnt/
18	/θ/	as in	**thanks** /θæŋks/
19	/ð/	as in	**the** /ðə/
20	/ʃ/	as in	**she** /ʃi:/
21	/ʒ/	as in	**television** /ˈtelɪvɪʒn/
22	/tʃ/	as in	**child** /tʃaɪld/
23	/dʒ/	as in	**German** /ˈdʒɜ:mən/
24	/ŋ/	as in	**English** /ˈɪŋglɪʃ/

Vowels

25	/i:/	as in	**see** /si:/
26	/ɪ/	as in	**his** /hɪz/
27	/i/	as in	**twenty** /ˈtwenti/
28	/e/	as in	**ten** /ten/
29	/æ/	as in	**stamp** /stæmp/
30	/ɑ:/	as in	**father** /ˈfɑ:ðə/
31	/ɒ/	as in	**hot** /hɒt/
32	/ɔ:/	as in	**morning** /ˈmɔ:nɪŋ/
33	/ʊ/	as in	**football** /ˈfʊtbɔ:l/
34	/u:/	as in	**you** /ju:/
35	/ʌ/	as in	**sun** /sʌn/
36	/ɜ:/	as in	**learn** /lɜ:n/
37	/ə/	as in	**letter** /ˈletə/

Diphthongs (two vowels together)

38	/eɪ/	as in	**name** /neɪm/
39	/əʊ/	as in	**no** /nəʊ/
40	/aɪ/	as in	**my** /maɪ/
41	/aʊ/	as in	**how** /haʊ/
42	/ɔɪ/	as in	**boy** /bɔɪ/
43	/ɪə/	as in	**hear** /hɪə/
44	/eə/	as in	**where** /weə/
45	/ʊə/	as in	**tour** /tʊə/

6.1.3 Consonants, Vowels and Dipthongs

(A) Vowels

- The letters *a, e, i, o, u and their sounds are known as vowels*. Consonants are produced by constricting the airstream to various degrees as it flows through the oral tract.
- **Vowels** are produced with a smooth, unobstructed airflow through the oral tract.
- Characteristic vowel qualities are determined by
  (a) the height of the tongue in the mouth;
  (b) the part of the tongue raised (front, middle, or back);
  (c) the configuration of the lips; and
  (d) the tension of the muscles of the oral tract.

Tongue Height

- Pronounce the words *eat* and *at*. Now pronounce just the vowels of these two words. Notice that as you go from the vowel of *eat* to the vowel of *at*, your mouth opens. If this is not obvious to you just by playing with these two vowels, look in a mirror as you produce them. Alternate the words, and then just the two vowels. Once you've become accustomed to the different degrees of openness of these two vowels, pronounce *ate* between *eat* and *at*. The degree of openness of its vowel falls between those of *eat* and *at*, so there is a continuous increase in mouth openness as you go from one vowel to another.
- These degrees distinguish **high, mid,** and **low** vowels. We will use the following symbols for this sequence of vowels:

    eat [i] High

    ate [e] Mid

    at [ae] Low

Front and Back Vowels

- Now compare the vowel of *beat* with that of *boot*. Alternate the words, and then just the vowels. It will be more difficult this time to monitor the activities of your tongue as you shift from one of these to the other, but try anyway.
- You produce the [i] of *beat* with the front (blade) of your tongue raised toward your palate. If you draw in your breath as you make this vowel, you will feel the cold air against your palate. As you shift from [i] to [u], the vowel of *boot*, you will find yourself raising the back of your tongue. (You will also find yourself pursing [**rounding**] your lips, but disregard this for the moment.)
- Because of the relative positions at which these vowels are made in the mouth, phoneticians call [i] and the other vowels in (1) **front vowels**, and [u] a **back vowel**. The back vowels, like the front ones, descend from high, through mid, to low, in a continuous sequence. You can observe this by pronouncing the words *coot*, *coat*, and *cot*, and then just their vowels. As you produce this series of vowels you'll find your mouth opening (monitor your lower jaw) as you go from *coot* to *coat* to *cot*. We use the following symbols for these back vowels:

    c**oo**t [u] High

    c**oa**t [o] Mid

    c**o**t [a] Low

Lip Rounding

- As you compared [i] and [u] you probably noticed that your lips changed shape as you shifted from the front vowel to the back one. Your lips were **rounded** as you produced [u]. They were **unrounded** (**spread** or **neutral**) as you produced [i]. As you moved through the series of back vowels you may also have noticed that lip rounding decreased as you moved from high to low. In fact the lips are unrounded during the pronunciation of [A]. In English, the only rounded vowels are back, though many languages, such as French and German, have rounded front vowels.

Vowels in multi-syllabic words

- Pronounce the words ***a**bove*, *sod**a***, *sof**a***, *comm**a***, *aren**a***, *patin**a***, *phot**o**graph*, paying particular attention to the vowel represented by the bold letters. Then pronounce this vowel in isolation. This vowel is called **schwa** and written [a]. Schwa is made at approximately the same place as [A] that is, farther forward than the back vowels and farther back than the front ones. Hence, it is central. In addition, [a] is mid, lax, and unrounded. It is heard primarily in unstressed syllables, as in the words above. It is the vowel we produce if we vocalise as we prepare to speak—*uh*. The tongue is said to be in its neutral position as we pronounce this vowel.

Diphthongs

- We have approached vowels as if they were articulated by a specific configuration of the tongue, lips, and oral cavity, which is held constant throughout their pronunciation. Vowels made like this are called **monophthongs**; others, called **diphthongs**, involve a change in the configuration of the mouth.
- The vowel sounds in the words *boy*, *by*, and *how* involves a change in the shape of the mouth as the vowel is being produced. The vowel of *boy* begins with approximately the mid back vowel [C] and finishes with approximately the high front lax vowel [I] (or the palatal glide [j]).

Classification of Vowels:

- In phonetics, a vowel is a sound in spoken language that is characterised by an open configuration of the vocal tract so that there is no build-up of air pressure above the glottis. This contrasts with consonants, which are characterised by a constriction or closure at one or more points along the vocal tract. A vowel is also understood to be syllabic: an equivalent open but non-syllabic sound is called a semivowel.
- In all languages, vowels form the nucleus or peak of syllables, whereas consonants form the onset and (in languages which have them) coda. However, some languages also allow other sounds to form the nucleus of a syllable, such as the syllabic *l* in the English word *table* /teIbl/ (the stroke under the *l* indicates that it is syllabic; the dot separates syllables).
- The word *vowel* comes from the Latin word *vocalis*, meaning "speaking", because in most languages words and thus speech are not possible without vowels.

Vowel – Articulation:

- The articulatory features that distinguish different vowels in a language are said to determine the vowel's *quality*. Daniel Jones developed the cardinal vowel system to describe vowels in terms of the common features *height* (vertical dimension), *backness* (horizontal dimension) and *roundedness* (lip position). These three parameters are indicated in the schematic IPA vowel diagram shown below.

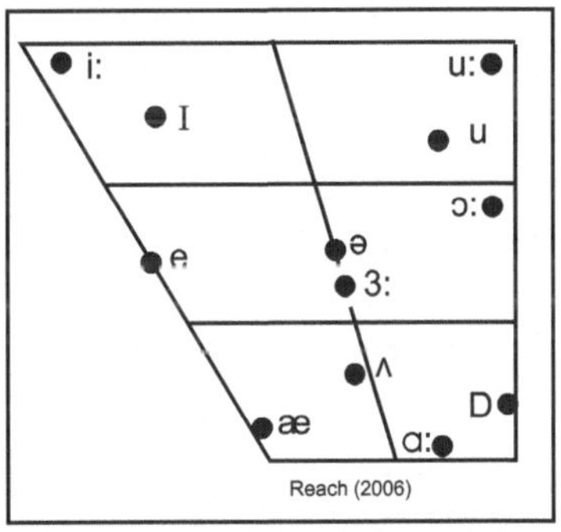

Fig. 6.2: Vowel - Height

- Height refers to either the vertical position of the tongue relative to the roof of the mouth or the aperture of the jaw. In high vowels, such as [i] and [u], the tongue is positioned high in the mouth, whereas in low vowels, such as [a], the tongue is positioned low in the mouth. Sometimes the terms *open* and *close* are used as synonyms for *low* and *high* for describing vowels. The International Phonetic Alphabet identifies seven different vowel heights, although no known language distinguishes all seven:

  (a) close vowel (high vowel)

  (b) near-close vowel

  (c) close-mid vowel

  (d) mid-vowel

(e) open-mid vowel

(f) near-open vowel

(g) open vowel (low vowel)

- The parameter of vowel height appears to be the most primary feature of vowels cross-linguistically in that all languages use height contrastively. The other possible parameters, such as backness and roundedness (explained below), are not used in all languages.

Vowel – Backness:

- Backness refers to the horizontal tongue position during the articulation of a vowel relative to the back of the mouth. In front vowels, such as [i], the tongue is positioned forward in the mouth, whereas in back vowels, such as [u], the tongue is positioned towards the back of the mouth. The International Phonetic Alphabet identifies five different degrees of vowel backness, although no known language distinguishes all five:

  (a) front vowel

  (b) near-front vowel

  (c) central vowel

  (d) near-back vowel

  (e) back vowel

  The highest number of constrastive degrees of backness is 3.

Vowel – Roundedness:

- Roundedness refers to whether the lips are rounded or not. In most languages, roundedness is a reinforcing feature of mid to high back vowels, and not distinctive. Usually the higher a back vowel, the more intense the rounding. However, some languages treat roundedness and backness separately, such as French and German (with front rounded vowels).

Nonetheless, there is usually some Correlation between Rounding and Backness:

- Front rounded vowels tend to be less front than front unrounded vowels, and back unrounded vowels tend to be less back than back rounded vowels. That is, the placement of unrounded vowels to the left of rounded vowels on the IPA vowel chart is reflective of their typical position.
- Different kinds of labialisation are also possible. The Japanese /u/, for example, is not rounded like English /u/, where the lips are *protruded* (or *pursed*), but neither are the lips spread to the sides as they are for unrounded vowels. Rather, they are *compressed* in both directions, leaving a slot between the lips for the air to escape.

Vowel – Nasalisation:

- Nasalisation refers to whether some of the air escapes through the nose. In nasal vowels, the velum is lowered, and some air travels through the nasal cavity as well as the mouth. An oral vowel is a vowel in which all air escapes through the mouth. French, Polish and Portuguese contrast nasal and oral vowels.

Vowel – Phonation:

- Voicing describes whether the vocal cords are vibrating during the articulation of a vowel. Most languages only have voiced vowels, but several Native American languages, such as Cheyenne and Totonac, contrast voiced and devoiced vowels. Vowels are devoiced in whispered speech. As in Japanese and Quebec French, vowels that are between voiceless consonants are often devoiced.
- Modal voice, creaky voice, and breathy voice (murmured vowels) are phonation types that are used contrastively in some languages. Often, these co-occur with tone or stress distinctions; in the Mon language, vowels pronounced in the high tone are also produced with creaky voice. In

cases like this, it can be unclear whether it is the tone, the voicing type, or the pairing of the two that is being used for phonemic contrast. This combination of phonetic cues (i.e. phonation, tone, stress) is known as *register* or *register complex*.

Vowel - Tenseness/Checked Vowels vs. Free vowels:

- Tenseness is used to describe the opposition of *tense vowels* as in *leap*, *suit* vs. *lax vowels* as in *lip*, *soot*. This opposition has traditionally been thought to be a result of greater muscular tension, though phonetic experiments have repeatedly failed to show this.
- Unlike the other features of vowel quality, tenseness is only applicable to the few languages that have this opposition (mainly Germanic languages, e.g. English), whereas the vowels of the other languages (e.g. Spanish) cannot be described with respect to tenseness in any meaningful way.
- In most Germanic languages, lax vowels can only occur in closed syllables. Therefore, they're also known as checked vowels, whereas the tense vowels are called free vowels since they can occur in any kind of syllable.

Accent

- Stress or Accent is an important feature of English. It is generally defined as the degree of prominence a syllable has. In words having more than a single syllable, all the syllables are not and cannot be equally prominent. Therefore, those which are more prominent than others receive an 'accent' or 'stress'. All stressed syllables in words are little longer than the other syllables. Hence, the voice for a stressed syllable becomes the heaviest.
- Stress therefore can be defined as the degree of force that a speaker uses to pronounce a word or a syllable. During transcription, stress is shown by a vertical bar or stroke. Stress has a lot of significance in the spoken language as a change in stress can change the very meaning of a word or utterance.

Word Stress and Sentence Stress:

- There are two kinds of stress: **word stress** and **sentence stress**. The following examples show the nature of stress. The vertical stroke before each syllable shows that the stress is applied on that particular syllable.

  For example:

  1. Fable = /'feibl/
  2. Selection = /silekʃ'ən/
  3. Possibility = /pɔsi'biliti/

Primary Stress and Secondary Stress:

- There are two degrees of stress: **Primary** and **Secondary**. In a word of two or more syllables, the most prominent syllable receives Primary stress, while other syllables may be stressed or unstressed. When more than one syllable also needs to be stressed, but only slightly, it is denoted with a vertical stroke at the foot of the syllable.

  For example:

  1. Cultivation = /,k^lti'veiʃən/
  2. Examination = /iɡ,zæmi'neiʃən/
  3. Imagination = /i,mæd3i'neiʃən/
  4. Systematic = /,SISTə'mætik/

Grammatic Stress:

- Stress can shift or change according to the grammatic form of a word. Nouns and adjectives are stressed on their first syllables, while verbs on the second syllable. This change in stress helps the listener distinguish meanings. When the first syllable of a word is stressed, it is treated as a Noun. But if we stress the second syllable of the same word, it becomes verb. As these changes are due to grammar, this type of stress is known as Grammatic stress.

1. Object (N) = /'əbd3ikt/
   Object (V) = /əb'd3ekt/
2. Perfect (N) = /p3:fIkt/
   Perfect (V) = /pə'fekt/
3. Conduct (N) = /'kɔndəkt/
   Conduct (V) = /kən'd^kt/
4. Increase (N) = /'inkriz/
   Increase (V) = /in'kri:z/
5. Record (N) = /'rekɔd/
   Record (V) = /ri'kɔ:d/

Stress Change or Stress Shift:

- Stress in English is mostly fixed. This means that the primary stress is always on a particular syllable of a given word. That is why stress is denoted in dictionaries. There are certain rules for applying stress, thus it can be called 'fixed'. But it is also free as it does not always remain at the same place i.e. on the same syllables in all words. In a single sentence, it can be applied at any place, in accordance with the intention of the speaker. This is known as stress shift. Also, when there are many words with the same stem, primary stress is not applied on the same syllable.

  For example: 'Politics, Po'litical, Poli'tician.

  A'cademy, Aca'demic, A,cade'mician.
- Because of stress change or shift, in certain and even meaningful words lose stress and even less meaningful words get stress. This is because the speaker wants to give special emphasis on a particular word. In such circumstances, any word may be stressed heavily. Thus, stress can also shift according to the intention of the speaker.
- Stress shift can be found in sentences too. Such sentences are high in emotion and are usually called 'emphatic sentences'.

  For example: '<u>I</u> think he is nice (= others may not think so). I <u>'think</u> he is nice = (the speaker is not sure). I think '<u>he</u> is nice. (= the speaker thinks that the subject is nice and not any one else).

(B) Consonants

- Consonants include the sounds we represent as <p, b, t, d, m, n, f, v, s, z, l, r, h> in the ordinary alphabet. All **consonants** are produced by entirely or almost entirely stopping the airstream coming from the lungs. When we almost entirely stop the airstream we force it through such a narrow opening that the airflow at that point is turbulent and noisy.
- We classify consonants according to the following characteristics: (a)whether or not the vocal folds are vibrating (**voicing**); (b) whether the sound is made with a fully stopped or merely constricted airstream (its **manner of articulation**); (c) where in the mouth the stoppage or constriction is made (its **place of articulation**); (d) whether or not air is flowing through the nasal cavity (**nasality**); and (e) whether or not the lips are pursed (**lip rounding**).

Voicing

- As a warm-up exercise, make the sound fffff, and keep it going for a count of five. Now make the sound vvvvv, and keep it going for a count of five. Now alternate these two: fffffvvvvvfffffvvvvv. You probably noticed that vvvvv had a "buzz" that fffff did not have. That "buzz" is caused by the vibrating of your vocal folds—which you can check by putting your fingers on your throat or by covering your ears as you alternate fffff and vvvvv. Now try the same exercises with the first sounds of the following words: *thigh*, *thy*; *sip*, *zip*. You should be able to feel the vocal folds vibrate as you make the second sound of each pair.

- Sounds produced with vibrating vocal folds are said to be **voiced**; those produced without vocal cord vibration are **voiceless**.
- Lists the voiced and voiceless consonants of English. The letters in [ ] are the phonetic symbols for the sounds.

voiced	voiceless
by [b]	**p**ie [p]
my [m]	
wet [w]	
vie [v]	**f**ie [f]
thy [ ]	**th**igh [T]
die [d]	**t**ie [t]
nigh [n]	
zip [z]	**s**ip [s]
lie [l]	
rye [r]	
bei**ge** [Z]	ba**sh** [S]
jive [dZ]	**ch**ive [tS]
yet [j]	
guide [g]	**k**ite [k]
go**ng** [N]	
hive [h]	

Classification of Consonants:

- Most English consonants can be classified using three articulatory parameters:
  (a) Voicing: Vibration or lack of vibration of the vocal folds.
  (b) Place of Articulation: The point at which the air stream is most restricted.
  (c) Manner of Articulation: What happens to the moving column of air.

(a) Voicing: Consonants are classified as follows depending on the voicing:

- Voice or voicing is a term used in phonetics and phonology to characterise speech sounds, with sounds described as either voiceless (unvoiced) or voiced. Almost all consonant sounds come in pairs. For example, p and b are produced in the same place in the mouth with the tongue in the same position. The only difference is that P is a voiceless sound (no vibration of the vocal cords) while B is a voiced sound (vocal cords vibrate).
  (i) Voiced: Voiced consonants are produced when the vocal chords vibrate during its production. The following are the voiced consonants in English: b, d, th (as in then), v, l, r, z, j (as in Jane).
  (ii) Voiceless: Voiceless consonants are produced when the vocal chords do not vibrate during its production. The following are the voiceless consonants in English:
  p, t, k, s, sh, ch, th (as in thing).

(b) Consonants are classified as follows depending on the place of articulation:

(i) Bilabial: Made with the two lips (e.g. **/p/, /b/, /m/**).

(ii) Labiodental: The lower lip articulates with the upper teeth (e.g. **/f/, /v/**).

(iii) Dental: The tongue tip and rims articulate with the upper teeth (e.g. **/T/, /D/**).

(iv) Alveolar: The tongue tip and/or blade articulates with the alveolar ridge (e.g. the RP, English **/n/, /s/**).

(v) Palato-alveolar): The tongue blade articulates with the alveolar ridge, while the front of the tongue is raised towards the hard palate (e.g. the initial sounds in *ship* **/ʃIp/**, *genre* **/ʒa:nrə/**, *check* **/tʃek/**, and *jail* **/dʒeIl/**).

(vi) Retroflex: The tongue tip articulates with the back of the alveolar ridge. Many speakers of American English have a retroflex 'r' sound (**/saund/**) in, for instance, *car*.

(vii) Palatal: The front (or blade) of the tongue articulates with the hard palate (e.g. **/j/**).

(viii) Velar: The back of the tongue articulates with the soft palate (e.g. **/k/, /g/**).

(ix) Glottal: Sounds involving an obstruction or narrowing of the glottis (e.g. **/h/**).

(c) Manner of Articulation: Consonants are classified as follows depending on the manner of articulation:

- Depending on the **manner** in which sounds are produced, consonants are divided into different categories, depending on the **degree of stricture**. There are three such degrees: **closure** (i.e. at some point in the vocal tract, the airflow is completely stopped), **close approximation** (involving a constriction somewhere in the vocal tract, with the air being forced through the opening), and **open approximation** (sounds in which the airflow is smooth).
- Additional distinctions include whether the air flows through the nose - **nasal**, or mouth - **oral**, whether it runs along the centre – **central** or the sides of the tongue – **lateral.**
- The IPA recognises the following manners of articulation.

(i) Plosives (or occlusives): In plosives, a complete closure is made at some point in the vocal tract, behind which the air pressure builds up and can be released with a sudden burst (**plosion**): e.g. /**p, b, t, d, k, g**/.

(ii) Fricatives (or **spirants**): Fricatives involve a close approximation of two articulators so that the airstream is partially obstructed, resulting in **friction** (or **turbulence**) in the flow of air: e.g. **/f, v, θ, ∂ , s, z, ʃ, ʒ, h/**. The hissing sounds in this category are also called **sibilants**. English fricatives can be labiodental, dental, alveolar, or palat-alveolar.

(iii) Affricates: Affricates are a combination of a **plosive** and a **fricative** is called an **affricate**. In these sounds, the separation of the articulators in the release phase is slower than for ordinary plosives, and thus gives rise to friction. These sounds can be heard in the English words *church* (**/tʃɜ:tʃ/**) and *jam* (**/dʒxm/**).

(iv) Nasals: These are sounds in which the **velum** is **lowered**, while at the same time, there is an obstruction in the oral cavity: e.g. *man* (**/mxn/**). Some modern phoneticians place both **nasals** and **plosives** into one single **stops** category on the grounds that both involve a total obstruction of the airflow, which thus takes precedence over any secondary phenomena that may surround the stoppage.

(v) Trill (or roll): A trill is a series of rapid closures or taps (20 to 30 times per second), with, for instance, the tongue tip tapping against the alveolar ride (e.g. the Spanish **/r/** in *perro*).

(vi) Lateral: These sounds involve an obstruction of the airstream at some point in the mouth (usually in the denti-alveolar region and along the centre), with the air being allowed to escape on one or both sides of the tongue: e.g. **/l/** in the English word *lip*.

(vii) **Approximants:** In the production of an approximant, one articulator is close to another, but the vocal tract is not narrowed to such an extent that a turbulent airstream is produced. Among the approximants, we find the RP **/j/** and **/w/**, which are also referred to as **semi-vowels**. The voiceless companion to **/j/** is the palatal fricative **[ç]** – the sound which some English speakers use at the beginning of words like *huge* and *humour*.

(C) Dipthongs

- In phonetics, a vowel is a sound in spoken language that is characterized by an open configuration of the vocal tract so that there is no build-up of air pressure above the glottis. This contrasts with consonants, which are characterized by a constriction or closure at one or more points along the vocal tract. A vowel is also understood to be syllabic: an equivalent open but non-syllabic sound is called a semivowel.
- A vowel sound whose quality doesn't change over the duration of the vowel is called a monophthong. Monophthongs are sometimes called "pure" or "stable" vowels. A vowel sound that glides from one quality to another is called a diphthong, and a vowel sound that glides successively through three qualities is a triphthong.
- All languages have monophthongs and many languages have diphthongs, but triphthongs or vowel sounds with even more target qualities are relatively rare cross-linguistically. English has all three types: the vowel sound in hit is a monophthong /ɪ/, the vowel sound in boy is a diphthong /ɔɪ/, and the vowel sounds of flower, /aʊər/, form a triphthong.
- In phonology, diphthongs and triphthongs are distinguished from sequences of monophthongs by whether the vowel sound may be analysed into different phonemes or not.

Guidelines/Rules for Stress: Some of the rules for primary and secondary stress can be given as follows:

(a) Words with just one syllable take the stress on it and so it is not required to denote the stress with a stroke. For example, Book, Dog, Cat, etc.

(b) Words starting with a weak syllable (ə, I), take the stress on the second or the next syllables: For example, attempt=/ ə 'tempt/

(c) Words ending in -'-ion', '-ian', '-ious', '-ic', etc. take the stress on the second syllable from the end: For example, selection =/sI'lekəʃn/, fantastic =/fen'tæstik/, musician= /mju'zʃn/ etc.

(d) Words ending in '-ity' and '= ate' take the stress on the third syllable from the end. For example, Peculiarity = /pIkju'liærItI/, venerate =/'veneræt/ etc.

(e) Words ending in '-eer', '-aire', '-esque', etc. receive the primdry stress on the last syllable, For example, Engineer = /end3i'niə/, Millionaire =/milliɔ' neə/, Groteque = /grɔ' tesk/ etc.

(f) Suffixes like "ness, less, ment" do not affect the stress: For example, 'joyful = 'joyfullness.

Vowels and Consonants

- Every language is made of sounds which are called speech sounds. All speech sounds are broadly divided into two categories: **vowels** and **consonants**. English has twenty vowel and twenty-four consonants. For vowels, the air comes freely through the mouth. There is no closure nor narrowing for the articulation of them. While describing vowels, the position of the tongue and lips are to be considered. Vowels are classified as back, central and front vowels. They have two categories: **pure vowels** and **diphthongs**. They are also half open and half-close. They are all oral sounds, /a, e, U, ei/are some of the vowels which can be given as examples.
- The other sounds of speech are called consonants. While describing them, the following things are to be considered: place of articulation, manner of articulation and whether the vocal cords vibrate or not. Moreover, it is also to be indicated whether the air passes through the mouth or the nose. /p, b, k, t, d, g/ are all examples of consonants.

6.1.4 Stress and Intonation

Accent

- Every language has different accents; they are pronounced differently by people from different geographical places, social classes, and educational backgrounds. An accent may enable to identify the locality in which the speaker resides (a regional or geographical accent), the socio-economic status of the speaker, his ethnicity, caste or social class (a social accent), or influence from mother tongue. Accents typically differ in quality of the voice, pronunciation and distinction of vowels and consonants. Although grammar, semantics, vocabulary and other language characteristics often vary concurrently with accent, the word 'accent' may refer specially to the differences in pronunciation, whereas the world "dialect" encompasses the broader set of linguistic differences. The word dialect refers to a variety of a language which is different not only in pronunciation, but also in grammar, vocabulary, and word order. Hence, we can say that accent is a subset of dialect.

Rhythm

- Rhythm connote some noticeable event happening at regular intervals. One can detect rhythm in the working of a machine, heartbeat of a person, or music. Now, English is considered to be a very rhythmical language, so a speaker who can maintain rhythm of the language is more likely to sound both natural and fluent. Rhythm in English language is detected in the regular occurrence of stressed syllabus. The two components of the system which have the greatness influence on rhythm are sentence stress and the various features of connected speech i.e. what happens to words when we put them in utterance.

  (i) Sentence stress: In any sentence, some words carry a stress. These are the 'strong' or 'lexical' words (usually nouns, verbs, adjectives and adverbs). The remaining words are 'grammatical' words and are unstressed or 'weak' (conjunctions, pronouns, prepositions, auxiliaries, articles).

  'It's the worst thing that you could'

  The rhythm produced by this combination of stressed and unstressed syllables is a major characteristics of spoken English and makes English a stress-timed language. In stress-time languages, there is a roughly equal amount of time between each stress in a sentence. Sentence stress is an important factor in fluency, English spoken with only strong forms has the wrong rhythm, sounds unnatural and does not help the listener to distinguish emphasis or remaining.

  (ii) Connected speech: Speed is also a factor in fluency. When we speak quickly, we speak in groups of words which are continuous and may not have pauses between them. This causes changes to the 'shape' of words. Unstressed words always sound different when used in a sentence as opposed to being said to isolation. The most common features of connected speech are the weak forms of grammatical and some lexical words (and, to, of, have, was, were) and contractions, some of which are acceptable-in-written-English (can't, won't, didn't, I'll, he'd, they've, should've).

Intonation

- Intonation is the use of changing vocal pitch to convey grammatical information or personal attitude. Vocal pitch plays an important part in phonology of English. Intonation is the music of a language. It refers to the way the voice rises and fails as we speak. Normally we tend to vary our voice pitch as per the requirement of the information that we wish to convey.

Example: How will we tell someone it's raining"

It's raining, isn't it?

We're telling the person, so we give our speech a 'telling' melody. The pitch-level of our voice falls and we sound as it we know what we're talking about. We're making a statement. But now imagine we don't know if its raining or not. We think it might be, so we're asking someone to check. We can use the same words - but note the question-mark, this time.

It's raining, isn't it?

Now we're asking the person, so we give our speech an 'asking' melody. The pitch-level of our voice rises and we sound if we're asking a question".

Intonation can show which parts are regarded as being background, given, and which parts carry the information focus. Given material in a clause typically has some kind of rising intonation contour, indicating incompleteness - there is something still to come - while the new information that is added is more likely to carry a falling contour, indicating completion. This helps to make speech less dependent than writing on ordering.

The intonation system of English constitutes the most important and complex part of English. By combining different pitch levels (= unchanging pitch heights) and contours (= sequences of levels, changing pitch shapes) we express a range of intonational meanings: breaking the utterance into chunks, perhaps distinguishing between clause types 9such as statement vs. question), focusing on some parts of the utterance and not on others, indicating which part of our message is background information and which foregrounded, signalling our attitude to what we are saying.

Accent its meaning and significance and classification:

- Stress or Accent is an important feature of English. It is generally defined as the degree of prominence a syllable has. In words having more than a single syllable, all the syllables are not and cannot be equally prominent. Therefore, those which are more prominent than others receive an 'accent' or 'stress'. All stressed syllables in words are little longer than the other syllables. Hence, the voice for a stressed syllable becomes the heaviest.
- Stress therefore can be defined as the degree of force that a speaker uses to pronounce a word or a syllable. During transcription, stress is shown by a vertical bar or stroke. Stress has a lot of significance in the spoken language as a change in stress can change the very meaning of a word or utterance.

Word Stress and Sentence Stress:

- There are two kinds of stress: **word stress** and **sentence stress**. The following examples show the nature of stress. The vertical stroke before each syllable shows that the stress is applied on that particular syllable.

  For example:

  1. Fable = /'feibl/
  2. Selection = /silekʃ'ən/
  3. Possibility = /pɔsi'biliti/

Primary Stress and Secondary Stress:

- There are two degrees of stress: **Primary** and **Secondary**. In a word of two or more syllables, the most prominent syllable receives Primary stress, while other syllables may be stressed or unstressed. When more than one syllable also needs to be stressed, but only slightly, it is denoted with a vertical stroke at the foot of the syllable.

  For example:

  1. Cultivation = /,k^lti'veiʃən/
  2. Examination = /ig,zæmi'neiʃən/
  3. Imagination = /i,mæd3i'neiʃən/
  4. Systematic = /,SISTə'mætik/

Grammatic Stress:

- Stress can shift or change according to the grammatic form of a word. Nouns and adjectives are stressed on their first syllables, while verbs on the second syllable. This change in stress helps the listener distinguish meanings. When the first syllable of a word is stressed, it is treated as a Noun. But if we stress the second syllable of the same word, it becomes verb. As these changes are due to grammar, this type of stress is known as Grammatic stress.

1. Object (N) = /'əbd3ikt/
   Object (V) = /əb'd3ekt/
2. Perfect (N) = /p3:fIkt/
   Perfect (V) = /pə'fekt/
3. Conduct (N) = /'kɔndəkt/
   Conduct (V) = /kən'd^kt/
4. Increase (N) = /'inkriz/
   Increase (V) = /in'kri:z/
5. Record (N) = /'rekɔd/
   Record (V) = /ri'kɔ:d/

Stress Change or Stress Shift:

- Stress in English is mostly fixed. This means that the primary stress is always on a particular syllable of a given word. That is why stress is denoted in dictionaries. There are certain rules for applying stress, thus it can be called 'fixed'. But it is also free as it does not always remain at the same place i.e. on the same syllables in all words. In a single sentence, it can be applied at any place, in accordance with the intention of the speaker. This is known as stress shift. Also, when there are many words with the same stem, primary stress is not applied on the same syllable.

  For example: 'Politics, Po'litical, Poli'tician.

  A'cademy, Aca'demic, A,cade'mician.

- Because of stress change or shift, in certain and even meaningful words lose stress and even less meaningful words get stress. This is because the speaker wants to give special emphasis on a particular word. In such circumstances, any word may be stressed heavily. Thus, stress can also shift according to the intention of the speaker.
- Stress shift can be found in sentences too. Such sentences are high in emotion and are usually called 'emphatic sentences'.

For example: 'I think he is nice (= others may not think so). I 'think he is nice = (the speaker is not sure). I think 'he is nice. (= the speaker thinks that the subject is nice and not any one else).

Syllabic Structure and Placement of Nucleus:

- In a language, a "phoneme" is the "minimal contrastive unit". A phoneme, therefore, is the smallest linguistic unit except its variants called "allophones". The unit at the next higher level than the "phoneme" is called "syllable". It is "one full sound". It may consist one or more speech sounds. For example, the word "ah" has the syllable which has only one speech sound i.e. /a:/. On the contrary, in "cat" we have one syllable consisting of three speech sounds /k/, /æ/ and /t/. When the syllable has only one speech sound, the sound is usually a vowel. For example, /aI/ as in "I". However, it may be, though very rarely, a consonant. For example, "sh!" has only one consonant /ʃ/.
- In a syllable, one sound is more prominent than others. It is usually a vowel. It is called the "**nucleus**" of the syllable. The consonant is called a "marginal element". This is because; the consonant may occur either at the beginning or at the end or at both the places of the syllable.

The consonant at the beginning of the syllable is called releasing consonant and the consonant at its end is called the arresting consonant. For example, in the word "rat" there is one syllable i.e. /ræt/. In this syllable /æ/ is the *nucleus,* /k/ is the *releasing* consonant and /t/ is the *arresting* consonant.

- However, sometimes a consonant too can occupy the central or nuclear position in a syllable. For example, in the word "button", there are two syllables - /b^/ and /tn/. In the /tn/ both are consonants, hence /n/ is the nucleus here. Such consonants are called syllabic consonants because they can form a syllable without a vowel.
- A word having only one syllable is called monosyllabic, while the word having two syllables is called *disyllabic* and with three syllables is called *trisyllabic*. Words with more than three syllables are called *polysyllabic* words. Usually, the division of syllable is marked with a hypen. For example, the word "possibility" has five syllables which can be mentioned a /pɔ-sI-bI-lI-tI/.

Syllabic Structure:

- The structure of a syllable is represented by the "CVC" formula. Here, C stands for "consonant" and V stands for "vowel".

For example:

1. V=a = /ə/
2. VC=is = /Iz/
3. CV=tee = /ti:/
4. CVC=boss = /bɔs/
5. CCV=play = /plei/
6. CCCV=straw = /strɔ:/
7. CCCVC=street = /stri:t/
8. CCCVCC=strange = /streInd3/
9. CCCVCCC=strands = /strændz/
10. VCCCC=ankles = /æŋktz/
11. CVCCCC=texts = /teksts/
12. CCVCCCC=twelfths = /twelfθs/

Thus, the syllable is a small but important unit of a language. Its additional significance is it frames the smallest meaningful unit of a language, i.e. word.

Intonation and Intonation Patterns

- Intonation means the melody of speech. It is the changing pitch of the voice. It is the term given to the rise and fall in the pitch of the voice in speech. In English, intonation adds meaning to a spoken sentence. To some extent, the spoken voice resembles subtlety. Stress and intonation are linked concepts. They work together to give prominence. Intonation can be used to convey the emotions of the speaker. Intonation adds to the rhythm of language also. The rising and falling of the sound in tone is really melodious. Intonation has the following tone patterns.

  (a) The Falling Tone and its Uses: In this pattern, there is a high pitch level on the final syllable. It is slightly lower on the second syllable and lowers till the end of the sentence. The tone is the lowest on the last stressed syllable. This tone is shown with ` mark (symbol).

  The falling tone is used in the following types of sentences:

  (i) Statements: e.g. The 'cup is empty'.

  (ii) Wh-questions: e.g. 'What is the' time?

(iii) Commands: e.g. 'Bring me a 'cup of' tea.

(iv) Tag question: e.g. He is 'huge', isn't' he?

(v) Exclamations: e.g. 'How boring he' is!'

(b) The Rising Tone and its Uses: In the rising tone pattern, there is a series of descending steps on the stressed syllables. But there is the rise on the last stressed syllable, rising tone is marked as (I). It is used in the following types of sentences:

(i) Yes-No questions: e.g. 'Do you like, coffee?'

(ii) Requests: e.g. 'please, don't go there.

(iii) Apologies: e.g. 'I'm, sorry'.

(iv) Doubtful remark: e.g. The 'boy is, handsome.

(v) Incomplete utterances: 'When she will, say......

(c) The Falling Rising Tone and its Uses: Apart from the above patterns of tone in English, there are two more patterns: Fall-Rise and Rise-Fall. The Fall-Rise is used for special implications not verbally expressed. The term 'special implications' can include instinct ions, valid insult, apology etc. It is marked as (v).

For example:

(i) She is ᵛ beautiful. (She is beautiful but not cultured).

(ii) She is very ᵛ beautiful (but not cultured).

(iii) The 'house is ᵛ good (but the people living in it may not be good).

(iv) I am not an ᵛ Indian (but I am Spanish).

(v) His father will ᵛ come, (so we should not worry as he is also good).

(d) The Rising-Falling Tone and Its Uses: This tone is used in the following types of expressions:

(i) Enthusiasm showing statements:

e.g. A: It was ᵛwonderful.

B: ᵛYes. Of ᵛCourse.

(ii) Questions implying suspicions or mockery:

e.g. What was she doing in my ᵛhouse?

Will he speak ᵛclearly?

(iii) Sentences showing irony and sarcasm:

e.g. Good ᵛriddance! (Ironical)

Oh, ᵛwhatever! (Sarcastic).

Thus, the various patterns of intonation make speech live and rhythmic. That is why intonation is called the melody of speech.

6.2 DISCUSSION SKILLS

What is a Group Discussion?

- Group discussion is an important activity in academic, business and administrative spheres. It is a systematic and purposeful interactive oral process. Here the exchange of ideas, thoughts and feelings take place through oral communication. The exchange of ideas takes place in a systematic and structured way. The participants sit facing each other almost in a semi-circle and express their views on the given topic/issue/problem.

How does Group Discussion differ from a Debate?

- Debate is competitive in nature while group discussion is a co-operative group process. In a debate, a speaker can speak either 'for' the topic or 'against' the topic whereas in a GD, the speaker can express both. The final decision or result in a debate depends on voting while in a GD, the group reaches group consensus.

Why is a group discussion an important activity at college level?

- As a student, it helps you to train yourself to discuss and argue about the topic given, it helps you to express your views on serious subjects and in formal situations. It improves your thinking, listening and speaking skills. It also promotes your confidence level. It is an effective tool in problem solving, decision making and personality assessment. GD skills may ensure academic success, popularity and good admission or job offer. Thus it is important to be able to take part in a GD effectively and confidently. Participants should know how to speak with confidence, how to exhibit leadership skills and how to make the group achieve the goals. The panel which normally comprises of the technical and HR (Human Resource) experts of the company will observe and evaluate the members of the team. The rules of the GD – the time limit, panel's expectations etc., are explained after the initial introduction by the panel, soon after the topic or case to be discussed is given to the group members. The panel does not interfere during the discussion, it only observes. The panel at its discretion may provide some time to think over the topic or may ask them to start immediately. Each candidate is supposed to express their opinion either supporting or against the topic. The discussion carries on till the panel signals termination. It is left to the discretion of the panel to extend or cut short the given time.

The Objective of Selection in GD

- It is mainly to check your team playing skills. You have to understand the other persons' point of view, while making your point and ensure that your team as a whole reaches a solution or agreement that is both feasible and accepted by all team members.
- There are four major areas of evaluation in selection GDs: subject knowledge, oral communication skills, leadership skills and team management.

1. Subject Knowledge:

- Participants must possess a thorough understanding of the topic on which they are supposed to speak. You must prepare yourself to talk on a wide range of subjects. Be abreast of the current events, national and international affairs, burning social and economical topics, scientific and environmental issues, key newspapers' controversial topics and any experience that may be expected of an educated person. As a member of the group, you are expected to contribute substantially to the discussion. The originality of your ideas, your knowledge and initiative and your approach to the topic or case contribute to your success in the group discussion.
- The best way to equip yourself is to read daily newspapers, good magazines, national and international journals and also watch new bulletins and informative programmes on the television. Internet is the greatest boon which provides you with everything you are looking for. The World Wide Web is a vast database of current authentic materials that present information in multimedia form and reacts instantly to a user's input. The greater your knowledge of the subject, the more enthusiastic and confident you will be during the discussion. Once you have understood the topic or issue, you should be able to generate ideas as well as organize them so that you present it well. You will have the ability to analyze facts or information in a systematic way. A person putting forward new ideas that may work will be accepted as the natural leader of the group. The panel will observe the ideas put forward, their originality, the depth of analysis and their relevance to the topic. Problem solving skills are essential and do not hesitate to give solutions. Your approach to the case study will be observed keenly by the evaluators.

2. Oral Communication Skills:

- If subject knowledge is important, communication skills is more important as without expression, the knowledge is of no use. As the exchange of ideas in a group discussion takes place through speech, one of the pre-requisites of success in a GD is the ability to speak confidently and convincingly. Good communication skills include active listening, clarity of thought and expression, apt language and proper non verbal clues.

3. **Listening Skills:**

- One of the weaknesses of most human beings is that we love to listen to our own voice rather than listen to others. Listening is as important as speaking in a GD, unless you listen, you cannot contribute to the stated purpose of communication. It is extremely important to listen very carefully, only then you will be able to pick up the thread of discussion and continue. Only active participation as a listener in a group makes a person a good leader. A leader is identified by the panel.

4. **Clarity of Thought and Expression:**

- Clarity is the art of making yourself clear to the audience. Only when your expressions are clear, you can convince your team and the panel. More than words, it is the tone in which they are spoken that conveys the message. You should not be too loud or too soft. A lively and cheerful voice with appropriate modulations will attract the audience. Proper articulation of words, achieved through phonetic accuracy is very essential slang, and artificial accents are to be avoided.

5. **Apt Language:**

- The flow of language must be smooth. Use simple language and avoid long winding sentences. Appropriateness of language demands that there should be no errors of grammar. Do not use unfamiliar phrases and flowery language. Be precise. Be polite and courteous.

6. **Proper Non - verbal Clues:**

- Non verbal clues include eye contact, body movements, gestures and facial expressions. The panel very keenly watches the non verbal behavior of the team. They generally evaluate the body language cues of the team to determine personality factors such as nervousness, co-operation, frustration, weakness, insecurity, self confidence, defensiveness, etc. A candidate who appears professional is more likely to be noticed by the panel. A confident posture, appropriate facial expressions and meaningful eye contact with the team will create a good expression.

7. **Team Behaviour:**

- Your group behavior is reflected in your ability to interact with the other members of the group. You must be mature enough to not lose your temper even if you are proved wrong. You must be patient and balanced. Your success in a GD depends on how well you play the role of initiator, information seeker, information giver, procedure facilitator, opinion seeker, opinion giver, clarifier, summarizer, social-supporter, tension reliever, compromiser, attacker, humorist and dominator. The selection panel notes the differences in the amount of participation of the members. They observe the silent spectators, the ever dominating but not contributing much, member who participates actively exhibiting his knowledge and the moderate ones. Your ability lies in analyzing the problem well and making others to endorse your view. Finally while appreciating others point of view, you should effectively present yours without contradicting other's opinions. Your ability in convincing the team is your success

6.2.1 Importance of Group Discussion

- Group discussion (GD), by virtue of the term, means exchange of views by participants on a given subject. The whole idea is to bring together a set of people on a common platform to share their ideas, thereby giving an opportunity to exhibit not only their knowledge, but also to understand and enhance their learning by absorbing the thoughts of other people.

Importance of Group Discussions

Better understanding of the subject:

- GD offers a great opportunity to the discussants to take a deep dive into the subject, and understand it profoundly. It expands the horizon of knowledge and learning. In addition, it extends a chance to the participants to improve their abilities to think critically and develop their analytical skills.

Solve problems:

- Since GD brings together a group of people holding diverse opinions, it also perceived as one of the sought after ways to arrive at solutions to problems. Because of the fact that people bring in a wide range of insights, GDs help in changing the thought process from a narrow to a wider view, thereby paving the way to solutions.

Improve listening skills and confidence:

- Since GDs necessitate the contribution of ideas and thoughts, the participants are required to listen keenly to others' observations. In this sense, group discussions are a fantastic way to enhance listening skills, and also boost the confidence of speakers by engaging them and encouraging them to express their thoughts.

Enhance communication and interpersonal skills:

- While one aspect of GDs is that it encourages people to share their knowledge, the other important aspect is that it gives an opportunity to people to build their communication skills. The participants, in the process, learn how to express their ideas while also respecting viewpoints of others. This in turn contributes to the development of interpersonal skills among the peer group.

Team building exercise:

- GDs help in developing teamwork and leadership attributed in people, which are extremely important in the current context. Working in a group helps them increase tolerance and respect towards others, and also how to unanimously arrive at conclusions.
- After learning various benefits which people may derive from participation in group discussions, it is important to highlight that this activity holds a special importance in academic and professional set ups. On one hand, GDs reflect the values of coexistence, motivation, tolerance and acceptance for others, on the other hand, they are a key tool for recruiters. During the selection process, they help the hiring team to identify if the job applicants possess the values which mirror the principles and the vision of the organisations. The organisations today not only stress on the knowledge pool of the candidates, but also give lot of importance to enthusiasm and adaptability in the workforce. This helps us explain why academic institutions, and especially management schools emphasise GDs.

6.2.2 Leadership Skills

1. **Leadership Qualities:**

- **Taking initiative:** The qualities looked for is the leadership quality. The leadership qualities necessitates that the individuals have the capacity to take initiative during the course of the interaction. This could be achieved adopting strategies such as beginning the discussion, picking up the threads at a later stage, etc.
- **Ability to give direction:** If one is not familiar with the topic. It is not necessary to take the initiative. The quality of possessing the ability to give direction to the entire discussion. It follows naturally that the interactant should have the power and ability to sum up all, that is

being said in a manner which is conducive to the growth of the discussion. It happens that one is led astray by emotions or by an inability to comprehend the topic. This is coupled with the feeling that one must make a perceptible contribution right at the beginning of the discussion in order to get noticed. In situation such as these, if an individual is able to give some kind of direction, it becomes easy to follow the flow of the discussion. Therefore, to give direction an individual would be viewed as a leader with the capacity of chalking out a strategy, filtering and assimilating ideas while leading and controlling.

- **Taking the group along:** To sum up discussion not only at the end but also in between, is a major quality of the interactant. This helps from straying away from the topic. In the course of the discussion it is not important to be the first speaker, but is important to make regular contributions throughout the discussion. This can only be achieved when there is some one, who is willing to take up the rather tedious and monotonous task of conjoining all the ideas and presenting them in a nutshell to the participants at regular interviews.
- **Listening:** The individual should possess capabilities of listening to what the other interactants are saying. We must emphasize the difference between hearing and listening. Listening would only be evident when the listener shows signs of absorbing, assimilating and presenting the spoken material to the rest of the participants. On the contrary, if only hearing has taken, it indicates that the speaker has not been able to penetrate the screen of indifference.
- **Goal to fulfillment:** The leadership qualities observed in the group, indicate that the goal has been achieved or is in the process of being achieved. We can say that the topic has been thoroughly discussed by all the participants with appropriate input from the leader and all of them have been able to perform the important feat or task of bringing into focus the main/ancillary points related to the topic. Trying to contribute the efforts of all the participants in a fruitful manner, would reveal all of them to be part of a cohesive group. This is definitely a Herculean task which only a leader could perform.

2. Knowledge of the Subject:

- The individual should be well—read about the issues under discussion. His knowledge of the subject matter necessitates that the quantitative and the qualitative aspect of the topic. Where the former is concerned, figures and numbers should not be reeled off merely to prove one's point or knowledge about an issue, unless and until one is absolutely confident of the same. If not in the group, then definitely among the experts, there is bound to be someone who would be aware of the details. Keeping quiet, because of a lack of information or knowledge, is really not as bad as trying to impress the experts by spouting incorrect information. The quality of presentation would be an appropriate assessment of the topic and the issues discussed.

3. Clarity of thought:

- Clarity of thought is extremely important and can be brought about by a distillation of the essentials and abandonment of the peripherals. The move in the discussion could be either from the core to the periphery or from the periphery to the heart of the issue. The participants should not get hooked to peripheral issues. This could lead to unnecessary emphasis on trivial matters. There has to be concentration in the moves which would indicate awareness on the part of the participants regarding the subject matter and delineation of the topic. The shift from the general to the specific will indicate awareness along all lines and areas pertaining to the topic which are bound to creep in.

4. Analytical Ability:

- The next in the sequence of trying to present an acceptable picture of the self, is the capacity to use one's analytical ability to the optimum. 'While it is relatively simple to present data on an issue, it gives the appearance of a well thought out and analyzed presentation in the GD.

5. **Conviction and Flexibility:**

- **Conviction:** It often happens that the participants communicate their ideas in a group without really believing in them. This is more than evident at the face level and is easy for the experts to decipher and identify. It normally happens when the interactants harbours under the misconception that to be heard in the group is more important than positing of concrete ideas.
- **Flexibility:** Conviction while speaking goes hand-in-hand with flexibility in approach and ability to appreciate the viewpoint of the other person. Often a participant makes an error in interpreting the topic which he realizes, much to his dismay, somewhere in the middle of the interaction. The need then arises to rectify the mistake and proceed along correct lines. The transition, which must be made, needs to be extremely subtle. Without really admitting that a mistake has been made, the speaker has to accept the view point of the other interactant and change sides to be one with those who have a more positive understanding of the topic. In the first instance, there should be no errors made. Careful thought should be given to the topic floated, before any kind of discourse is begun on the same. This strategy has been presented as a rescue measure for extreme cases in which an error has been made, albeit unwittingly, and the situation demands a rectification. If unfortunately this be the case, the interactant would not be at a loss to make suitable amends.

6.2.3 Team Management in Group Discussion

- Team management is the ability of an individual or an organization to administer and coordinate a group of individuals to perform a task. Team management involves teamwork, communication, objective setting and performance appraisals. Moreover, team management is the capability to identify problems and resolve conflicts within a team. There are various methods and leadership styles a team manager can take to increase personnel productivity and build an effective team.[1]
- In any functional team, cohesion amongst team leaders and decision makers is vital. Cohesive leadership means that team leaders are acting together as a unit and making decisions as a leadership team instead of each branching off into their own work and operating individually. This will ensure that the team will be steered in one direction instead of multiple due to team leaders not being concise and consistent with their instructions. Cohesive leadership will require team leaders to have strong communication skills.

Effective Communication:

- There must be an effective channel of communication from the top to the bottom of the chain of command and vice versa. An effective channel of communication will allow messages to be transferred accurately without delay to the intended recipient, this will speed up decision making processes and the operations of the team. Furthermore, effective communication will increase the flexibility of an organisation and cause it to be less susceptible to changes in the external environment; as a faster decision making process will allow organisations a longer time period to adapt to the changes and execute contingency plans

Common Goal:

- When team members first come together they will all have different ideas, however the key to a successful team is the alignment of objectives within the team. It is essential that the team leader sets a common goal the entire team is willing to pursue. This way, all the team members will put in effort in order to attain the goal. If there is not a common goal, team members who disagree with the objective in hand will feel reluctant to utilise their full effort, leading to failure to achieve the goal. In other cases, team members might divert themselves to other tasks due to a lack of belief or interest in the goal

Defined Team Roles and Responsibilities:

- Poorly defined roles is often the biggest obstacle to a successful Team. If team members are unclear of what their role is, their contributions towards the team will be minimal, therefore it is the team leader's duty to outline the roles and responsibilities of each individual within the team and ensure that the team is working together as an integral unit.
- In a successful team, a team leader will first evaluate the mission of the team to understand what is needed to accomplish the task. Then, they will identify the strengths and weaknesses of their team members and assign roles accordingly. Lastly, they must ensure that all team members know what each other's responsibilities are to avoid confusion and to create an effective channel of communication.]

Methods of Team Management

"Command and Control":

- The method "Command and Control" as an approach to team management is based on the concept of military management. "Command and Control" was a commonly used system in the private sector during the 20th century.
- In this method, the team leader instructs their team members to complete a task and if they refuse, they will yell or punish them until they no longer refuse and comply with the instructions. The team leader has absolute authority and utilises an autocratic leadership style. There are considerable drawbacks to this team management method. Firstly, the team morale will be lowered due to team members being constantly belittled by the team leader at the slightest mistake; punishments will also lead to a lack of confidence in team members resulting in poor performance. Second, in modern organisations roles are often specialised, therefore managers will require the expertise of the employee, elevating the value of the employee. Implementing the "Command and Control" team management method will lead to a high rate of employee turnover. In addition, in large organisations managers don't have the time to provide instructions to all employees and continuously monitor them, this will impede an organisations performance as managers are not spending time on their core responsibilities.

"Engage and Create":

- Due to the limitations and ineffectiveness of "Command and Control", managers developed an alternative management strategy known as "Engage and Create". In this method team members are encouraged to participate in discussions and contribute. Furthermore, they are advised to engage with other team members to build a stronger sense of teamwork and unity. This will lead to increased productivity and accountability of each team member, driving the team towards success.

Problems in Team Management

Absence of Trust:

- In Patrick Lencioni's book The Five Dysfunctions of a Team, the absence of vulnerability-based trust – where team members are comfortable being vulnerable with each other, trust each other to help when asking for guidance, and are willing to admit their mistakes – within a team is detrimental to a team. Team leaders have to assist each other when they are vulnerable and also allow team members to see their vulnerable side, which is contradictory to the orthodox belief. If a team lacks vulnerability based trust, team members will not be willing to share ideas or acknowledge their faults due to the fear of being exposed as incompetent, leading to a lack of communication and the hindering of the team.

Fear of Conflict:

- Contrary to the general belief, conflict is a positive element in a team as it drives discussion. The fear of conflict is the fear of team members to argue with one another and the fear of disagreeing with the team leader. If team members hold back and are afraid of confronting the leader or their teammates, then the concept of a team is non-existent because there is only one person who contributes and no new ideas are generated from discussions.
- The fear of conflict in a team stems from an absence of trust, more specifically vulnerability based trust. If team members are afraid to be vulnerable in front of one another, disputes can be manipulative and a means to overthrow and shame the other team member. However, if team members trust each other and are comfortable being vulnerable in front of one another, then debates can be a pursuit of a better and more effective method to achieve a task.

Lack of Commitment:

- When team members don't provide input on a decision, it shows that they do not agree or approve of the decision, leading to a halt in team activity and progress. Furthermore when team members don't express their opinions, views and potential ideas are lost, hurting the project and the company.

Avoidance of Accountability:

- The avoidance of accountability in a team is the failure of team members to be accountable for the consequences of their actions. When team members do not commit to a decision, they will be unwilling to take responsibility for the outcomes of the decision.
- In addition, if a lack of trust exists within the team then there will be an absence of peer to peer accountability; team members will not feel accountable towards their team members and hence will not put effort into their tasks. The team must trust and hold each other responsible so that the intention will always be for the benefit of the team and for the team to succeed.
- Team leaders who are afraid of confrontation might avoid holding team members accountable when in fact they have made a mistake. Team leaders must develop the confidence to hold team members accountable so that they will feel the sense of responsibility and entitlement to the team, and learn from their mistakes. If not, then errors will not be corrected and might lead to worse problems, causing a defective team.

Inattention to Results:

- If team leaders and team members do not hold each other accountable then they will not be concerned about the outcome of the team and whether they have achieved their goal, as they do not have a drive to obtain great results. Inattention to results causes a loss of purpose and brings into question the existence of the team.

Resolving Problems through Team Management

Building Trust:

- An approach to resolving fundamental trust problems within teams is to build trust amongst team members. A team leader can build trust by persuading team members to ask questions and seek guidance from other team members, so that they are more familiar and comfortable in being vulnerable with one another, questions such as "May you teach me how to do this" or "You are better than me at this". However, in order to achieve vulnerability based trust within the team the team leader must be vulnerable first, if the team leader is unwilling to be vulnerable, nobody else in the team will be willing to follow.

Appraisals:

- Appraisals can be a way for team members to provide feedback to one another or for team members to provide advice to the leader. This will allow individual members of the team to reflect on their performance and aim to do better by amending their mistakes; furthermore appraisals create an environment where the chain of command is non-existent and team members can be honest towards one another. This is effective in a way that the team can provide progressive feedback towards other members and can advise the leader on how he or she can improve their leadership. After each member reads their appraisals, they will understand how they can strive to improve, benefitting the team in reaching its objectives. The commonly used forms of appraisals are performance appraisals, peer appraisals and 360 degree feedback

Team building Activities

- Team-building activities are a series of simple exercises involving teamwork and communication. The main objectives of team building activities are to increase the trust amongst team members and allow team members to better understand one another. When choosing or designing team building activities it is best to determine if your team needs an event or an experience. Generally an event is fun, quick and easily done by non-professionals. Team building experiences provide richer, more meaningful results. Experiences should be facilitated by a professional on an annual basis for teams that are growing, or changing.

Back to Back Drawing:

- Back to back drawing is a team building activity where two members of a team sit back to back. One member is given a picture while the other is given a blank piece of paper. The member with the picture has to describe to the other member what the picture includes and the other member has to sketch what is described. At the end of the game, both members compare the picture and the drawing to see how similar they are. This game aims to improve verbal communication between team members.

The Human Knot:

- The human knot is a team building activity where team members stand in a circle and grab hold of the hands of other participants that are not immediately next to them. The objective is to unravel the circle of entangled hands. This game, through the physical touch between team members, will allow them to feel more comfortable with each other. Furthermore, this game can improve the verbal and physical communication between team members and enable the team to identify which communication process is the most effective within the team.

6.3 PRESENTATION SKILLS

- It is hard to determine what a good presentation should be like because various situations require different styles. If you are training yourself to be good presenter, you can develop a certain way of speaking, using your body language and of course visual aids.
- One of the most useful natural skills we have is the ability to communicate using different visual elements like our body language. Facial expressions, hand movements and postures are all related to how we feel or what impression we want to give the people we are talking to.
- A formal presentation is one where the time, subject, participants and so on are structured. informal presentations range from a progress meeting with the project team or a supervisor to a two-minute talk with your boss after office hours, in the lift or in the parking lot.

6.3.1 Importance of Presentation

(a) Knowing the Audience:

- o Know your audience, their demographic characteristics such as age, gender, education, occupation, experience level, socio-cultural background, position in the organisation and their role in decision making etc.
- o For better effect, the audience could be prepared for the meetings or presentations by giving the basic details in advance.

(b) Level of Understanding:

- Learn about their level of understanding of your topic and how familiar they are with the technical terms being used.

(c) Clear Concept of the Purpose:

- The manager should be clear about the purpose of the presentation, whether it is just to pass on information or to seek some action at the end of it.
- That will determine his strategy in determining the content and appeal.

(d) Gathering of Information:

- Gathering all relevant information that is valuable for the audience takes time. It should be done meticulously.

(e) Format of Presentation:

- The format for the presentation should be dynamic and not static. This involves making it lively, forceful and interesting and interactive. Their doubts should be cleared and their questions answered. In short, it should not be tightly planned to the last detail.
- Visual aids if necessary should be prepared and used effectively. The start and the closing should be strong.

(f) Outline of the Message:

- The outline of the message should be simple and clear.
- The basic concept or the main idea should be explained clearly. If the situation requires, the manager could practice his presentation by writing down the message and reading it aloud. This will indicate to him if the message has been formulated clearly, it is easy to understand, sounds effective and interesting.
- This practice session gives the manager the opportunity to check the choice of words and language. It also helps him to manage the time properly.

(g) Brevity:

- The message should be brief but appropriate. It should neither be too short nor too long. The time of both the speaker and the audience should not be wasted. Lengthy sentences confuse the listeners and may lead to misunderstandings. Conversation in short sentences provides the listeners the time and opportunity to follow what the speaker endeavours to put across.
- Using precise words, simple and familiar words, and avoiding superfluous words are important factors in effective communication.

(h) Sequence:

- Another important principle of effective communication is the presentation of matter in a logical sequence.
- The speaker should not jump points or change the sequence.
- Consistency, continuity, and logical development of the subject-matter are important aspects to be kept in mind.

(i) Availability of Time:

- The total amount of time available should be kept in mind. If there are a number of points to be covered, the content should be organised in segments and put in a proper sequence, indicating the weightage given to each. This will help the speaker decide on the amount of time that should be given to each segment and the corresponding supporting data.
- He could prepare a written note on these segments to help him cover all points in a sequential manner. These notes should be written legibly, so that it can be used without any problem.

(j) Opening:

- To gain attention, the speaker should start by talking about something the audience is familiar with and is connected to the subject matter.

(k) Introduction:

- A good introduction is very important for the following reasons:
  - (i) To arrest the attention of the audience,
  - (ii) To awaken their interest in the subject,
  - (iii) To introduce the subject being presented,
  - (iv) To make a smooth transition to the presentation.

(l) Main Idea:

- The main idea should be stated and taken up right at the beginning. It should be emphasised by repeating.
- The manager should remember that the audience cannot see boldly typed or underlined words.
- The importance of an idea or an argument can be perceived only by the way in which it is stated.
- Each idea should be in a different segment and should be numbered by using key words. Each idea should be closed with a repetition to reemphasise the point and to indicate that the speaker is moving to a new segment.
- Transition from one segment to the next should be smooth. The adage is: *"Tell them what you are going to tell them, tell them, and tell them what you told them."*

(m) Choosing Precise Words:

- Precision is the most important principle in effective communication.
- Using precise words means giving exact details using the right words at the right place, pertinent to the context.
- The speaker should possess a good vocabulary and complete knowledge of the synonyms and antonyms of words and use them effectively.
- The vocabulary should be such that the audience can understand comfortably.

(n) Verbosity should be avoided:

- To convey meaning is more important than impressing the audience by using high sounding and superfluous words.
- Using more words does not provide greater clarity or improve effectiveness.

(o) Prepositions, Adjectives and Adverbs:

- Care should be taken to avoid use of unnecessary prepositions.
- Use of adjectives and adverbs should be made where necessary to emphasise the meaning.
- For example, 'the problem is under *active* consideration'; 'a *positive* decision will be taken'; '*definite* results can be seen in a few weeks', etc., convey emphasis.

(p) Avoid Jargon or idiom relating to specific professions:

- Terminology used in professions such as medicine, law, commerce, sports, defence etc. should be used carefully.
- They should not be used in a mixed group as not all of them will be able to understand them.

(q) Avoid Cliché or formula statements:

- A cliché is a stereotyped, commonplace phrase or expression, used without much thinking. It may have no specific meaning to the context.
- An effective communication avoids clichés as they may not stand to logic and may irritate the listeners.

- For instance, while discussing an accident caused by a woman driver, the remark may be, 'Oh these women!' indicating that the person knows all about women and according to him all women are careless or reckless and prone to all sorts of accidents.

(r) Seven Cs of communication:

- The 7 Cs of communication explained by Francis J. Begin are very much applicable to spoken communication. Give adequate importance to all of them. They are: (1) Clarity, (2) Correctness, (3) Completeness, (4) Concise, (5) Courteous, (6) Concrete, (7) Candid.

(s) The Closing segment:

- It should be planned carefully.
- The speaker may take some questions at this point. He should finally sum up the entire message, review the main points, emphasise the main idea and also list out the actions that need to be taken.
- It should be a positive end.
- The purpose is lost if the audience do not have any significant message to take away with them.

(t) Pleasant Demeanor:

- During the entire presentation, the speaker should have a pleasant demeanor. A smile is a mood elevator.
- He should use all elements of silent language effectively. He should practice effective delivery skills. That is the key to success.

6.3.2 Planning the Presentation

- Every time you stand in front of a group, you must achieve two basic goals. First, you need to communicate a message. And second, you need to communicate your personality who you are as a professional and an individual. If "the medium is the message," your personality is the window through which the message must travel to be received, understood and acted upon by the audience.
- You convey your message and your personality every day of your life in relaxed conversation. And relaxed conversation is, therefore, your best possible communication style.
- By understanding the following principles, you can sharpen the skills you've acquired one-on-one and transfer those skills to group presentations, thereby increasing your effectiveness in both.

Ten Principles of Effective Presentation:

(a) Your best style is relaxed conversation,
(b) Be yourself,
(c) Relaxed conversation is two way,
(d) Relaxed conversation is receiver-driven,
(e) Less is more,
(f) People can listen or think, but they can't do both.
(g) People won't remember what you say,
(h) Be conversational in your delivery,
(I) Think before you speak,
(j) Silence is essential.

Principle 1: Your Best Style is Relaxed Conversation:

- Every day of your life, you convey your messages and your personality while engaged in conversation. You should therefore emulate conversation in your presentations. You · re not there to "download" information. You are there to create understanding that is based on a two-way exchange, and to facilitate an environment in which people can apply what you tell them to their personal or professional life.

- If you see a puzzled expression, don't wait for them to ask a question. Handle it the same way you would in a conversation. Ask them if there is something you can explain more effectively. Allow questions throughout your presentations. But be brief with your answers. Questions are an opportunity to create milestones of mutual understanding. But remember, you pass milestones. You don't camp at them.

Principle 2: Be Yourself:

- You are unique. You have your own way of speaking and your own mannerisms how you talk, how you stand, how you hold your hands. To convey your personality to a group, you must express yourself in a manner similar to the ways in which you express yourself one-on-one. If you are expressive with your hands one-on-one, it's OK to be expressive with your hands when talking to a group.
- In fact, it's essential. You know it's important to be on your best behaviour. You know there are certain standards that you must meet. You must dress appropriately.
- You must be attentive when someone asks a question. You must answer the question. But worry less about how you "present" yourself, than how you communicate with the members of the group.
- Your body language must be natural. And what is natural for you are probably not natural for someone else, or vice-versa, which is why we hesitate in setting rules for gestures you should use or the body language you should attempt to convey.
- To understand this, think of the gestures you make when you are enthusiastically explaining a concept to a friend over the telephone. Who are these gestures for? The person on the other end? Understand that these gestures are part of who you are as an individual. Bring them to your presentations and let them happen naturally

Principle 3: Relaxed Conversation is Two Way:

- To be effective, relaxed conversation must be two-way. Indeed, by definition, all communication must be two-way. Even if one person does most of the talking in a conversation, he or she is looking for the nods, listening for the "uh-huhs," and stopping to answer questions. The sender quickly recognizes that a blank look means the receiver is not listening. He or she will respond by changing tactics pausing to let the listener catch up or asking if there is a question.
- Your presentations, like your conversations, must be two-way. If you treat people with respect, and create a two-way process in which their questions are answered clearly and concisely, you stand a better chance of having them use or act on the information you present.

Principle 4: Relaxed Conversation is Receiver - Driven:

- In a relaxed conversation, the speed at which information goes from sender to receiver is driven by the receiver's needs, not the sender's. During a conversation, if the person listening doesn't signal that he or she understands with a "nod" or by saying "uh-huh" the sender stops to create a mile stone of mutual understanding before moving on. If the sender doesn't do this, the receiver will stop listening.
- The same applies to your presentations. If you talk non-stop, you will quickly lose your audience. Instead, make sure the information you're sending is driven by the audiences needs, not yours. If you throw out an idea that creates puzzled expressions, it's probably a good time to stop and ask: "Are there any questions?".

Principle 5: Less is More:

- The less you say, the more your audience remembers. If you try to cram too much information into your presentations, you will not create a two-way exchange. And you certainly won· t be receiver-driven. How can you be? The speed at which information travels from you to the audience is not driven by their need for understanding, but your need to get through it all in time.

- If you have one hour for your presentation, bring 30 minutes of information. This leaves plenty of time for questions, enables you to finish on time or a bit early, and allows you time for networking at the end.

Principle 6: People can Listen or they can Think, but They can· t do Both:

- As human beings, we can listen or we can think. But virtually none of us can listen and think at the same time. By definition, this means that you must "pause" when delivering your presentations. And those pauses must be as full and as frequent in your presentations as they are in your conversations. You want your seminars to be thought provoking. You want people to think about what you're saying and apply it to their personal situation. But while they're thinking, if you're talking, they won't hear a word you say.
- If you talk non-stop, members of the audience will miss large portions of what you say. They'll rush to catch up once or twice. After that, they · ll give up. And if they give up your chances of doing further business with them decreases proportionately.

Principle 7: People won't Remember What you Say:

- Participants at your presentations will not remember your exact words. Instead, they will remember what they thought about what you said how they took your information and applied it to their frame of reference. But this process can only occur in silence, whether you give them that silence, or they take it for themselves. And remember, if they take that silence while you're talking, they won't hear a word you say.

Principle 8: Be Conversational in Your Delivery:

- If you've ever read the transcript of an interview or conversation, you have probably noticed that people rarely talk in complete sentences. And if you participated in the conversation from which the transcript was drawn, you were probably shocked at what you saw written down.
- There is a basic pattern in relaxed conversation. In the first step, which we refer to as the first pause, the sender thinks about what he or she is going to say. Once the idea is formed, the sender expresses it. If the sender is enthusiastic, the words come tumbling out at a rapid rate of word delivery.
- Once the idea is delivered, the sender stops talking and allows the listener to absorb the idea and relate it to a meaningful frame of reference. During this second pause, the sender watches and listens for the receiver's reaction. Once there is a nod or "uh-huh", the sender forms the next idea. And so on.

Principle 9: Think Before you Speak:

- In a conversation, you take time to form each thought before you say it. You should attempt to do the same thing at your seminars, although that can be much more difficult because of the impact of adrenaline. But stop talking. Think through the thought. Start talking again. Use good notes.
- Focus on delivering one idea at a time. This will get you off to a strong start. And, as for nervousness, by encouraging questions, you take the emphasis off a presentation and switch it to an extended conversation.
- Never forget that the greatest reducer of nervousness in public speaking is a question or two that you can answer clearly and concisely.

Principle 10: Silence is Essential:

- There are two types of silence. The first is for you to think. This is the first pause. In a conversation, after you express each idea, you look to see if the other person has "gotten it". Again, you do so while pausing. We call this the second pause. The first pause is for you to think.
- The second is for them to think. During your presentations, remember that you · re not there to prove that you can talk non-stop. You're there to provide information that people can think about and apply to their own personal circumstances. But remember, they can only think in silence whether you provide that silence or they take it for themselves.

- If you lose your place or your audience, pause. If you're lost, the pause allows you to think about where you are, where you're going, and what you need to say. If they • re lost, the pause will help them find their way back so they can listen to your ideas again, and relate those ideas to their personal frame of reference.

6.3.3 Handling Stage Fright

Ways to Overcome Stage Fright When Speaking in Public

- There are two types of people in the world: those who love speaking in public and those who are scared stiff at the thought of it.
- Performance anxiety and stage fright are perfectly normal phenomena that occur to many people. It is important for you to understand what stage fright is, so that you can fully overcome it.
- Stage fright or performance anxiety is a persistent phobia which is aroused in an individual when required to perform in front of an audience.
- So how do you overcome stage fright when speaking in public?
  1. **Know Your Stuff:** Nothing will stop stage fright in it's gripping tracks like being prepared. Know your content, your speech and more importantly your audience. If you know what you are talking about then you have no reason to be nervous.

     Understanding your topic will enable you to speak more naturally and hence more confidently. Also, should a technical hitch occur, this won't faze you as you are already confident on the subject.
  2. **Practice, Practice, Practice:** Knowing your stuff helps, but it doesn't necessarily eradicate the problem. You need to practice as much as you can before the performance or public speaking d-day.

     Really know your content inside out and practice (preferably in front of a live audience) as much as possible to build your confidence.
  3. **Talk Yourself Down:** You need to realize that even though stage fright is "all in the mind," the fear manifests itself in physical ways. The best offence is to change your negative talk. Stop worrying about, "What if I forget the content?"

     Change that into positive talk like, "What if I am great at this?" It may sound simplistic or too easy, but positive affirmation will go a long way in reducing stage fright when speaking in public.
  4. **Wallow in the Worst:** If you can't calm yourself down with positive talk, then maybe it is time you thought about the worst case scenario. Once you do this, you'll realize that the worst case scenario isn't really that bad. This might help calm your nerves.
  5. **Visualize the Outcome:** Call it what you will: reflection, visualization, meditation. Whatever you call it, just do it. Spend time visualizing yourself giving a perfect presentation and speaking in public – filled with humor, warmth, confidence and intelligence.

     The more you imagine being great, the more likely you will achieve it.

     Though you might feel like everyone
  6. **It is Not All About You:** It is not about to laugh, criticize or judge you, that is not the case. Get over the feeling that the world is going to hang on your every mistake.

     Focus on your speech, audience and what they deserve from you. This will ease the pressure that is already accumulating.
  7. **When Things go Wrong:** Sooner or later, something will go wrong. Your projector or microphone might stop working. If you already know your content, then chances are that

this won't faze you as much. If, for instance, your microphone stops working, don't stress over it, carry on with a louder voice. Chances are the technical people are already stressing and working to sort the problem out, so you getting worried over the same issue won't help.

8. **Keep Calm, Don't Rush It:** Don't rush your presentation. Start slow and allow yourself time to get into a comfortable pace. You need time to get used to the audience and the audience also needs time to get used to you.
9. **Focus on Getting through the First 5 Minutes:** Imagine your entire presentation is only five minutes long. This will make it less stressful. Focus on just getting through the first five minutes and by this time you will have already calmed down and the rest is downhill.
10. **Never Apologize for Being Nervous:** Three quarters of the time, no one will notice you are nervous. Why tell them? You may feel yourself shaking and shivering, but your audience might not be aware of it. Don't mention it. It will make your audience nervous too and they will be too worried about your performance to get much out of your presentation.
11. **Don't Share Your Mistakes:** You have prepared, practiced and feel good about your speech or presentation. Suddenly, on stage you realize you mixed the order of topics or you forgot an important point. But remember, you're the only one who knows about this. Your audience doesn't. So, don't make them aware of a mistake that they didn't even know existed. If you bring it up, some people might start looking for more holes, which ultimately distracts from the whole purpose of your presentation in the first place.
12. **Arrive Early:** Obviously, if you are late, this will only heighten your anxiety. Arrive early and acclimate to your surroundings. You can even check out the stage and the auditorium as get yourself used to the environment.
13. **Stretch:** If you are nervous, odds are your body will be stiff and your muscles tight. Fifteen minutes before speaking in public and going on stage, do a few stretches. This will loosen the tense muscles and relax your body.
14. **Breathe:** Nervousness is always accompanied by fast, short breaths and if this is not addressed, it will throw you off balance. Minutes before you go on stage, take some slow, deep breaths, so that by the time you get to the stage your breathing is relaxed.
15. **Double Check Everything:** Do you have a laptop or notes? Check that everything works. When you walk on stage and suddenly realize that you forgot your notes, it's too late. Of course, your nerves will take over. Know your speech or presentation so well that should this happen, you can continue without a hitch.
16. **Don't Fight Your Stage Fright ... Work With It:** You have to expect and accept the fact that you will feel anxious, especially the first few minutes of your presentation. The more you resist your anxiety, the more it will work against you. Again, focus on the presentation when speaking in public and the anxiety will slowly ease off.

6.4 MOCK INTERVIEW

- A mock interview is an emulation of a job interview used for training purposes. The conversational exercise usually resembles a real interview as closely as possible, for the purpose of providing experience for a candidate. It can help a job applicant to understand what is expected in a real job interview, and can help an applicant improve his or her self-presentation.
- Mock interviews can be videotaped; the candidate can view the tape afterward, and get feedback. There are coaches who can provide feedback on aspects of the interview process. Mock interviews are most common for job interviews, but may also be used to train public

figures to handle interrogations by journalists, as well as help candidates for office prepare for debates. Some organizations schedule mock interview events to help many students prepare for job interviews.

- For example, some schools have mock interview training days, often organized by career and guidance counselors While the usual sense of the term is an exercise done as a form of preparation prior to applying for jobs, there is another sense of the term which describes a playful or non-serious interview.
- Mock interviews can help a person gain confidence for real interviews, as well as provide the interviewee with information about how to handle an upcoming interview.
- A mock interview, also known as a practice interview, is a simulation of an actual job interview. It provides you with an opportunity to practice for an interview and receive feedback.

What is a Mock Interview?

- A typical mock interview is a practice job interview held with a professional career counselor. A mock interview helps you learn how to answer difficult questions, develop interview strategies, improve your communication skills and reduce your stress before an actual job interview.
- During a mock interview, the interviewer may use a semi-structured interview format rather than asking a formal list of questions.

In-Person Mock Interviews

- Many college career centers and career counselors offer in-person mock interviews. If you're a college student or graduate, check with your career office to see if they provide in-person or phone or video mock interviews. If you're not affiliated with a college or university, a career coach or counselor is another option for practice interviewing.
- You can make an appointment with a mock interviewer, and provide her with information on either a specific company with whom you are interviewing, or your general career field.
- The more specific information you can provide for your mock interviewer, the better. The mock interviewer will often record the interview with a video camera.
- Some career centers and career counselors also offer mock phone and online video interviews, and will record your interview on a tape recorder or with a webcam.
- After the mock interview (which usually lasts about 30 minutes), the interviewer will then review the interview with you, and provide feedback.

Examples of Mock Interview Questions

- The questions asked during a mock interview are typically the more general employer interview questions that are asked during a first round or screening interview.
- However, for candidates who know what type of position, career field or industry they are interested in, job specific interview questions may be asked.
- If you are using a mock interview to prepare for a job interview you have scheduled with a company, the interviewer may be able to ask you actual questions that are asked by hiring managers at the company. Glassdoor.com has company interview questions contributed by site visitors for many companies.
- When you schedule a mock interview with your college career center or a career coach, provide them with as much information as possible about your interests and goals. The more information the counselor has, the better able he or she will be to tailor the questions to fit the actual interviews you will have.
- For example, if you are searching for a tech job, the interviewer can ask a series of tech interview questions to familiarize you with the types of questions that will be asking and to help you frame good responses.

- If you're interviewing for a summer job, as another example, the interviewer can ask you the same questions you will be asked by the people who hire for summer positions.
- Do take the time to prepare responses to the questions you will be asked during a mock interview.
- This is an opportunity to practice interviewing and to make sure that you have the interview skills to make the best impression on your interview.

Interview Questions

Job Interview Questions and Answers

- Review examples of the best answers for the most frequently asked interview questions in several different categories, and advice on how to answer.

Most Frequently Asked Interview Questions

  - Top 10 Interview Questions and Answers

Interview Questions About You

- Interviewers will ask questions about you to gain insight into your personality, and to determine whether you're a fit for both the job and the company.
  - Tell me about yourself.
  - What is your greatest strength?
  - What is your greatest weakness?
  - Tell me about something that's not on your resume.
  - How will your greatest strength help you perform?
  - How do you handle failure?
  - How do you handle success?
  - Do you consider yourself successful? Why?
  - How do you handle stress and pressure?
  - How would you describe yourself?
  - Describe a typical work week.
  - Do you take work home with you?
  - How are you different from the competition?
  - How do you view yourself? Whom do you compare yourself to?
  - How does this job fit in with your career plan?
  - Are you a self motivator?
  - What do you find are the most difficult decisions to make?
  - What has been the greatest disappointment in your life?
  - What are you passionate about?
  - What are your hobbies?
  - What are your pet peeves?
  - What is your dream job?
  - What will you miss most about your last job?
  - What won't you miss about your last job?
  - Would you rather be liked or respected?

Questions About Leaving Your Job

- Employers almost always ask about why you left, or are leaving, your job. Be prepared with an explanation for why you're moving on.
  - Why are you leaving your job?
  - Why do you want to change jobs?
  - Why were you fired?

- Why were you laid-off?
- Why did you quit your job?
- Why did you resign?
- What have you been doing since your last job?

Interview Questions About Salary

- Some of the hardest questions to answer during a job interview are about compensation.
- Here's what you will be asked and examples of the best answers.
  - What were your starting and final levels of compensation?
  - What are your salary expectations?
  - What are your salary requirements?
  - Why would you take a job for less money?

Questions About Qualifications

- The most important thing for interviewers to determine is whether you're qualified for the job. Here's what they will ask to find out.
  - What applicable experience do you have?
  - Are you overqualified for this job?
  - How did you impact the bottom line?
  - Interview questions about your abilities.
  - Sell me this pen.
  - What can you do better for us than the other candidates for the job?
  - What part of the job will be the least challenging for you?
  - Which parts of this job are the most challenging for you?
  - What philosophy guides your work?

Questions About Job Performance

- How you performed in previous roles can indicate how you will perform in the job for which you're applying. Be prepared to answer questions about what you did well - and what you didn't.
  - What do people most often criticize about you?
  - What is the biggest criticism you received from your boss?
  - What is the worst thing that you have ever gotten away with?
  - What makes you angry?
  - What problems have you encountered at work?
  - What strategies would you use to motivate your team?
  - What would you be looking for in an applicant?
  - When was the last time you were angry? What happened?
  - Why weren't you promoted at your last job?
  - Tell me about something you would have done differently at work.
  - If the people who know you were asked why you should be hired, what would they say?

Questions About Why You Should Be Hired

- Why should you be hired over the other candidates? Here's when you'll have the opportunity to make the case for getting a job offer.
  - Why should we hire you?
  - Why shouldn't we hire you?
  - Why should we hire you instead of the other applicants for the job?
  - Why are you the best person for the job?
  - What can you contribute to this company?

Interview Questions About the New Job and the Company

- What do you know about the company, why do you want the job, and what would you do if you were to be hired, are just some of the questions you'll be asked about the position and employer.
  - How is our company better than your current employer?
  - What interests you about this job?
  - What do you know about this company?
  - Why do you want this job?
  - Why do you want to work here?
  - What challenges are you looking for in a position?
  - What do you see yourself doing within the first 30 days on the job?
  - What can we expect from you in the first 60 days on the job?
  - Are you willing to travel?
  - What is good customer service?
  - What would be your ideal company culture?
  - When could you start work?
  - Is there anything I haven't told you about the job or company that you would like to know?

The Final Question

- The last question you'll most likely be asked is whether you have any questions. Here's how to respond.
  - Do you have any questions for me?

Behavioral Interview Questions

- In addition to being ready to answer these standard questions, prepare for behavior-based interview questions. This is based on the premise that a candidate's past performance is the best predictor of future performance. You will need to be prepared to provide detailed responses including specific examples of your work experiences.

Interview Questions Employers Should Not Ask

- There are some interview questions, typically known as illegal interview questions, that employers should not ask during a job interview. Here are questions that shouldn't be asked during a job interview and how to best respond.

Phone Job Interview Questions

- Have a phone interview on the agenda? Here are common questions asked during a telephone interview, plus tips on how best to answer so you can move to the next stage of the interview process.

Interview Questions to Ask

- The last job interview question you may be asked is "What can I answer for you?" Have an interview question or two of your own ready to ask. You aren't simply trying to get this job - you are also interviewing the employer to assess whether this company and the position are a good fit for you.

6.4.1 The Interview Process

- *"An interview is a formal meeting in which a person or persons question, consult or evaluate another person or persons."* **– Murphy and Peak**
- An interview reveals the views, ideas and attitudes of the person being interviewed as well as the skills of the interviewer. It is a medium of two-way communication between people. There are various purposes of interviews.
- According to **E. C. Eyre**, *"an interview is a face-to-face verbal exchange which endeavours to discover as much information as possible in the least amount of time about some relevant matter"*.

- According to **Vivien Palmar**, *"an interview constitutes a social situation between the two persons, the psychological process involved requiring both the individuals mutually respond through social research. The purpose of the interview calls for a varied response from the two parties concerned"*.
- Depending on the company background, purpose, level etc., the depth, style and location for the interview can be decided. For instance, for a graduating class, campus interview would be the best method, while for senior levels, the interview will be held at the senior-most level and it can be held over a lunch or dinner.

Types of Interviews

Job Interview/Placement Interview

- In the context of a **job interview**, the candidate and the interviewer each have a need: the candidate wants the job and the interviewer wants to find the right person to fill a position. Companies need to recruit people at various levels. This need may arise due to internal promotions, normal retirement process, when the older employees make way for the younger generation or due to attrition, when the serving employees leave the organisation looking for greener pastures.
- **Job/Placement interviews** are conducted to recruit new employees in the organisation. It is a formal discussion between a hirer and an applicant or candidate, in which information is exchanged, with the intention of establishing the applicant's suitability for a position. They are aimed at assessing the strengths and weaknesses of prospective candidate. Though these are typically held in person, the current trend in many companies is to have a telephonic interview prior to a face-to-face interview, especially if the candidates are from far off places. Employing a new candidate for a company is a very challenging task for a manager. The job market is becoming extremely competitive and so, newer interview methods are being introduced to sift the candidates.
- There are many types of job interviews such as:
  - **(a) The Screening Interview:** Companies use screening tools to ensure that candidates meet minimum qualification required. Some organisations use computer programmes to weed out unqualified candidates. This is the reason why they require a digital resume that is screening-friendly. However, in most companies, professionals are assigned the role of gatekeepers who are skilled enough to determine whether there is anything that might disqualify a candidate for the position. They do not need to know whether the person is the best fit for the position, but only determine if the candidate fits in. For this reason, screeners tend to be very critical in looking for gaps in the employment history of the candidate and look for pieces of information that look inconsistent. They also will try to find out if the candidate is over-qualified and hence be too expensive for the company.
  - **(b) Face to Face Interview:** This is the traditional interview method in which job seekers meet the employers in person whether it is a fresh recruitment or the candidate is being considered for a higher post requiring some experience. The advantage of this is that the employer and the job seeker can get to know each other. The candidate also gets to assess the work environment. The questions generally pertain to the candidate's academic qualifications and experience, and are aimed at evaluating the strengths and weakness of the candidates. There may also be questions about the expected salary and other benefits. Before interview the candidates should gain as much knowledge about the company as possible, including their product or service range and financial positions.
  - **(c) Group interview or Group Discussions (GD):** These are part of the recruitment process for fresh candidates. These are conducted by most large companies for graduates who are all evaluated at one time. As they all come from similar educational background, it is easy to compare them and pick the best for further interviews. They may be given a topic for discussion or some exercises to solve in a group and the moderators observe how the

candidates react with each other. Group interviews are helpful in assessing the candidates on the following parameters:

- Knowledge level,
- Communication skill,
- Listening Capacity,
- Leadership qualities,
- Team work,
- Reaction under stress.

Normally a group consists of 8 to 10 students. The normal time duration for any GD is about 10 to 15 minutes. For a GD relating to any particular topic, 2-3 minutes thinking time may be given. For GDs relating to case studies, 15 minutes time should be given for studying the case. The valuation of the participants is normally done by experts (usually professors from business schools). All these experts possess vast experience and expertise in their respective fields.

Tips for Effective Participation in GDs:

1. Give proper respect to the views expressed by the other speakers.
2. Be courteous while speaking.
3. Do not use very insulting, harsh or aggressive language.
4. Speak clearly and pleasantly to all the members of the group.
5. Stick to the main point of the GD. Do not speak irrelevant matter in the discussion.
7. Do not use loud and angry tone while speaking.
8. Do not interrupt other members of the group while they are speaking.
9. Learn to disagree politely.
10. Use positive body language while speaking.

Note for Participants in GD

There are no fixed rules for a GD held for candidate selection. The participants are generally eager to be the first to speak. The first speaker should mention the topic and state the issues. He should not give any opinion at this point. Later, the person may offer clarifications. This should be in the form of a statement, and not a question. Care should be taken to address the entire group. The candidates should maintain calm even if there are opposing arguments. The candidate should acknowledge this viewpoint using a meta language.

Situations may arise, when one's point of view has already been expressed by someone else. The best approach would be acknowledge the contribution of the other person and then add to it and thus become the centre of attention.

If there is interruption from someone else, the approach should be to request the person to allow you to complete your statement and offer the field to that person.

The group should move towards a consensus but due to the all round anxiety to make one's point, this may not happen at all. The idea is to exhibit some leadership qualities in steering the group while making one's contribution. A person with leadership qualities will look at the clock and when it is almost nearing closing time, will start summing up. If the group is too noisy, the facilitator may allot one minute to each candidate to sum up the discussion. This is an opportunity to put on one's best effort. Without criticising the group, one can sum up and give one's own views. The candidates should not stray from the topic.

The candidates are evaluated on how they speak. Fluency, meaningful contribution, depth of knowledge on various topical issues, and leadership qualities are what the selectors would be looking for. Whatever personal views one may have, it is important to know both sides of the argument. One must be able to defend one's viewpoints convincingly and therefore the need for acquiring knowledge about a wide range of subjects.

As preparation for a GD, candidates should select some topics, prepare and talk in front of the mirror, family members and friends to understand where they need to strengthen their arguments or presentation skills. They can also record their speech to know where they falter, and this will indicate the areas for further improvement. The voice quality, body language etc., can also be assessed during this process. The practice sessions can be used for ensuring that the thought process is well organised. One way to practice this is to write down the points and keep it in front of you. By periodically looking at it, you can arrange your thoughts mentally.

(d) **Behavioural interview**: These are intended to gain insight into the personality traits and attitude of the candidates. The interviewer will attempt to predict their future behaviour based on their past experiences. The candidates are asked to explain their skills, experience, activities, hobbies, school projects, family life etc., as examples of their past behaviour. This interview shows the candidate's self-confidence, creativity, motivation, self-discipline, willingness to learn, train, work effectively in groups, and undertake travel etc.

(e) **Stress interview**: A stress interview is aimed at creating discomfort for the candidate with the objective of testing his ability to function under such situations. It may take the form of making the candidate wait for an hour or more, asking questions continuously without giving the candidate time to think and answer, openly challenging his beliefs or judgement etc. The interviewer is not negatively inclined, but is performing a task.

(f) **Panel interview**: In Panel interviews or Committee interviews candidates will meet several persons holding higher authority. This method is used when top positions are being filled. Questions may be asked by all the panel members. The candidate can expect any type of critical questions from them. He must establish eye contact with each member of the panel while answering the question.

The Panel members may be:

- The supervisor,
- The manager,
- The human resource officer,
- The union representative,
- Employees who are in the recruiting team.

(g) **Telephonic interview**: The main purpose of conducting a telephonic interview is to save the time and expenses for those who live very far from the location. A telephone interview is conducted in a professional manner like other interviews. The most important aspects for the candidate are his mastery over the subject, tone of voice and confidence in speech. The interviewer may ask all sorts of questions about the candidate's academic achievements, hobbies, interests, references etc. This method is adopted in the globalised scenario, where the company can hire suitable candidates from across the globe. Especially in the IT sector, many companies including Microsoft adopt this method to their advantage.

As a preparation, the candidate should do adequate research about the company. He should take the phone call in a quiet place. He should be comfortable and relaxed. He should have his resume in front of him to answer the questions arising from it. He should inform the referees ahead of the interview and keep their names and addresses ready, though the information is already given in the application. Right in the beginning, he should ask the interviewer's exact name with spelling and title as also postal address or email ID, so that a 'thank you' note could be sent. The advantage for the interviewer is the opportunity to judge the capability of the candidate to respond to an interview set up at a very short notice. Most companies use this method to eliminate those who fall short of the expectations and call only those who do well for the next level of interview. Even universities use this method for selecting students for professional courses.

(h)Campus Interviews: As a source for entry level recruitment, and depending on the nature of their requirement, the personnel department of various companies visit campuses towards the end of the academic year as an opportunity for screening the outgoing students. This screening process help the interviewer to form an impression regarding the candidates and their suitability for the organisation in terms of qualifications, personality, poise, attitude, ability to communicate etc. The candidate may be confident of his specific job-related skills and knowledge, whereas the interviewer may be more interested in the candidates' capability to understand the job requirements, whether his knowledge is based on realistic understanding of the job content and also his level of motivation. The interviewer may also want to know the other accomplishments of the candidates apart from the academics and how they value these accomplishments, and how they relate these experiences to their career interests. This campus selection gives the interviewer an opportunity to make a short list of candidates for a second round of more in-depth interview, prior to final selection of the candidate. However, if the interviewer is convinced of the suitability of a candidate at this stage, the job offer is made immediately.

Significance of Interviews as a Process of Selection:

- As can be seen, interviews involve high cost and time, and require that the interviewer is highly skilled in this technique. There is a lot of preparation involved. Secondly, sometimes, even with the best interview techniques, the result may not be error-free. For instance, the candidate at the job interview may impress the interviewer by a super display of flashy communication skills, supreme self-confidence, random knowledge about the subject and excellent references. Hidden behind this façade may be a loser, an inefficient and ineffective employee, who is not capable of taking even routine decisions. Though such outcomes cannot be ruled out, by and large, interviews provide the best possible opportunity to understand and evaluate the prospective employees. The advantages far outweigh the occasional setbacks.

Advantages of the interview method

1. As has been discussed, for hiring and other HR related issues, interviews provide an excellent opportunity for both the parties to evaluate the situation. In research, it can prove to be a very good technique for getting necessary information about complex, emotionally laden subjects. Apart from the core subject of the interview, the interviewer is able to observe the body language of the interviewee and thus, is in a better position to judge the overall situation better.
2. It can be easily adapted to the ability of the person being interviewed.
3. It yields a good percentage of returns.
4. In marketing or social research, It yields perfect representation of the general population if the sample is selected judiciously, and the tools are properly prepared and vetted. The data collected by this method is likely to be more correct compared to the other methods that are used for this purpose.

Role of the Interviewer while setting up the Interview

- Hiring a new employee is like asking an unknown individual to come to the hirer's home to spend some thirty years or more there. The new employee will be expected to join the team and show results. Hence, choosing an employee based on a person's or a panel's very brief interaction in a single interview has its risks. Thus, with the authority for hiring comes a lot of responsibility.
- The interviewer starts the entire exercise with a purpose and in order to make it a success, he will have to approach it with adequate planning, preparation and execution. The interviewer has to take care of the following:

1. The interviewer must be very clear in his mind about the purpose of the interview. He should come fully prepared for the interview and does not waste either his own time or that of the candidate.

2. The interviewer must ensure that the conditions are suitable for a fair assessment of the candidates called in, so that the purpose of the interview is fulfilled.
3. The interviewer should welcome the candidate with courtesy and make him feel at ease.
4. The interviewer should not abuse his position by asking personal and irrelevant questions.
5. The interviewer should go through the bio-data of the candidate carefully before framing the questions.
6. The interviewer must conduct the interview in a carefully planned manner.
7. The interviewer should adopt a positive attitude towards the candidate while interviewing him and not intimidate him/her.

General Tips for the Interviewer

1. Use a quiet and comfortable place chosen by the interviewer. (In contrast, in the case of marketing research, the meeting place can be decided mutually.)
2. Put the interviewee at ease. (In marketing research, give a brief background of the company, and explain the objectives of the interview. Sometimes, the respondents are offered some monetary incentive for answering all the questions. This has to be stated clearly. If the conversation is being recorded, it should be told to the respondent.)
3. Be interested in the person as well as the purpose of the meeting.
4. Outline clearly the requirements of the job or purpose of the interview.
5. Explain fully the conditions of employment.
6. Tell about the benefits, development and promotion opportunities and so on. Do not make false or flashy promises just to attract a particular candidate.
7. Avoid questions which are sensitive to the feelings of the respondent, issues like gender, religion etc. For instance, it is generally reported that women are asked how they would balance both personal and career related activities. This would put the candidate on the defensive, unless she has been called for the interview only as a formality, without any serious intentions.
8. Encourage the applicant to ask questions. This will tell the interviewer how casual or professional the interviewee is. It will also tell the interviewer about the candidate's communication skills, and level of self-confidence.
9. Guide the interviewee through the interview.
10. Listen and let him/her talk freely.
11. Be natural, and use a conventional, neutral tone. Body language should indicate a pleasant demeanour.
12. Know when and how to close the interview.
13. Announce your decision, or explain your next step. The interviewee should be sure of what to expect and when.
14. Part with goodwill.

The Interviewer should avoid the following:

1. Keeping the applicant waiting.
2. Building false hopes.
3. Overselling the job.
4. Interrupting the candidate while he is talking.
5. Rush through the interview.
6. Repeat questions already answered on the application form. However, some interviewers may deliberately repeat a question just to make sure that the written communication is not random.
7. Use a phony excuse for rejecting him/her.
8. Use bad words and send him away with a bad taste in his mouth.

Role of Interviewee in the Interview Process

- The interviewee at the job interview has a lot at stake. Hence it is necessary for him/her to prepare for the interview and go with a positive approach. The questions asked and the order in which they are asked may not match the method worked out by him/her. This should not disturb the candidate. He should realise that the interviewer is a representative of the organisation where he is seeking a placement and comes with a lot of experience and also the authority either to accept or reject the candidate. Along with that, the person (or interview panel) has a tremendous responsibility on his shoulders and so will be very observant throughout. The interviewer will constantly be on the lookout for warning signs or red flags, such as lies, lack of preparation, poor attitude, insincerity and so on.
- The candidate should realise that an interview is a major part of a well-thought-out, consistent, employee selection process. This realisation should make the candidate take utmost care while preparing for the occasion. Some indicative list of things on which care should be exercised are:
  1. **Exhibit appropriate Communication Behaviour:** Though the employers realise that the candidate would be anxious, they would not appreciate poor communication skills. Effective communication is critical for success in most jobs. In fact, it is one of the skills most frequently listed by employers in their job postings. If the candidates talk too little, the interviewers would not be able to learn enough about them to make a favourable hiring decision. Others may talk way too much, which would appear to be aggression. Eye contact should be maintained with the interviewer or the panel, whether the interviewer is a woman or a man. The candidates should remember that the person who conducts the interview could be a man, but the candidate may have to report to a lady. Hence, the candidates should develop good communication skills appropriate both for women and men.
  2. **Respond Effectively to Follow-up Questions After The Initial Answers:** Most candidates prepare answers to questions pertaining to the resume very meticulously. They are very articulate about it. However, it is the follow-up questions that provide an opportunity for the interviewer to assess the candidate adequately. The proof of experience, appropriateness, and knowledge is demonstrated in these answers. They provide the specific details the organisation would be looking for in order to assess his or her competence in the area for which he is being evaluated. If the candidate shows signs of panic or the answers are not up to expectations, it would give the impression that the candidate had got his resume prepared by someone else, even a professional agency and there is nothing beyond that.
  3. **Plan to Stay Very long at the Job:** It is obvious that the organisation would be looking for a stable work force, as hiring and training involve costs. Clues about how long the candidate may work for the organisation can be obtained in answer to questions about their family, the location of close family members and the reasons for the candidate choosing the particular location. In case a candidate is over-qualified for the job offer, it is likely that he would consider this as only a stop-gap arrangement. Candidates provide all sorts of clues about their plans in their answers.
  4. **Talk Appropriately About Former Employer:** It is likely that the candidate has had a bad experience at the previous job. It could be a stifling experience in a hierarchy driven organisation or the boss was irrational or biased. There could be many such reasons for job switching. The interviewer may ask leading questions, like - Why are you leaving your current job or employer? Why were you fired from your last job? Describe your relationship with your former boss. How did you handle your relationship with a difficult co-worker? The candidate should be diplomatic and not waste precious interview time dwelling on bad experiences.

Unfortunately, if the candidate falls into this trap and gives a negative picture, he is sending a powerful message. The interviewer may think that he would do the same with this organisation as well, perhaps even during his employment tenure. A reputed organisation would stay clear of such candidates.

- **Dress with Care and Dignity for The Interview:** Neat appearance, appropriate accessories, cleanliness and good taste matter a lot. Dirty, unkempt, wrinkled, and inappropriate clothes are loud signals to skip this prospective employee.

General Tips for the interviewee

(A) Preparation Stage for the Interview:

1. At the preparation stage, the interviewee should collect all relevant information about the organisation in which he seeks a job.
2. The interviewee should try to visualise the sort of questions that may be asked keeping in mind the nature of the post and the qualifications required. This will help the candidate prepare for the interview.
3. The candidate should make sure that the dress and style are appropriate for the occasion. He should carry his resume, certificates and testimonials in a neat binder or file.
4. The interviewee must arrive ahead of the scheduled time of interview, so that he can be calm and composed. There are many factors to be considered, such as traffic situation, weather conditions etc., while determining when to start and what mode of transport to take, so that he reaches the venue ahead of time. The interview venue may be many stories above and the lift could be out of order, requiring the candidate to use the staircase. This situation should also be factored in while setting the estimated time for departure for the interview. In short, all eventualities should be considered and nothing should be left to chance.

(B) At the Interview:

1. The candidate should not be nervous while entering the interview room. He should greet the interviewers with a polite smile and occupy the seat in a natural, composed manner. The body language, such as the way person sits (either on the edge of the seat or too relaxed), can convey a lot about the personality and attitude of the candidate. He should not do anything to indicate nervousness, such as playing with the papers in the hand, shaking the legs continuously, fiddling with the fingers, wiping the face unnecessarily etc.
2. The interviewee should pay minute attention to the surroundings during the conversation without interrupting the interviewer. This should be done so casually that the interviewer will not think that he is distracted.
3. The interviewee should answer all the questions in a clear and natural voice. The answers should be to the point. Unwanted or unsolicited information should not be given. It is not a friendly conversation but a professional meeting.
4. The interviewee should answer the questions with confidence. He should speak slowly and clearly, maintaining eye contact with the interviewer during the entire course of the interview. If the candidate is facing a panel, he should address all the members while answering, with special attention to the person who has asked the question. If the candidate does not have the answer for a question, it is better to acknowledge it instead of giving an irrelevant answer.
5. During the entire interview, the candidate should project a confident and positive image about himself.
6. At the end of the interview, when the interviewer thanks him, the candidate should get up, shake hands (if it is appropriate), thank the interviewer with a smile and leave.

Discipline Interview

- These types of interviews are conducted to maintain discipline in the organisation. These interviews can be either to explain the standards of discipline followed in the organisation and answer the questions that arise out of them, or to chastise in case of repeated breach of discipline by an employee. The former type of interview is on a more positive note and there is no unpleasantness in the interaction. On the other hand, an interview to point out breach of disciplinary norms of the organisation can be stressful for both the parties. The disciplinary interview can be conducted by an individual or by a committee. Such an interview should be conducted behind closed doors, away from prying eyes of other employees.
- The Interviewer will have prepared a dossier of all events of indiscipline by the employee, and the notices issued to him from time to time. He will have also studied similar cases in the organisation in the past. With this evidence, the interviewer confronts the erring employee and tries to find out the reasons for such indiscipline and why action should not be taken against him. The interviewer should limit his statement to the specific cases in front of him and not depend on any hearsay. He can read out from the file, but not repeat what others have said orally, as that can lead to perception of bias, favouritism etc. The objective here is to give a chance to the employee to explain his position and not to hurt him or antagonise him. Hence, even while asking for an explanation, the interviewer should be respectful. He can start by praising the employee for his overall contribution for the growth of the department or the company and then say that notwithstanding all that, it his duty to enquire into the complaints.
- The interviewee or the erring employee called to face the interview should have his objective clear – whether he is going to continue with the organisation, or is preparing to quit. Either way, he should aim ay maintaining goodwill in the company. If he wants to continue, he will stay calm, listen to the charges and answer them clearly. He may also have his own justifications. He might be facing some personal problems, or some shortcomings or obstacles within the organisation which is forcing him to adopt devious methods. His problems may be totally unrelated, such as attitude of some colleagues, parking space for his vehicle etc., but may be disturbing him emotionally. He will have to put forth his arguments convincingly, apologise if called for and seek help if necessary. He should limit his arguments to the situation on hand and not rake up unconnected issues. Any indication of aggression, raising of voice or temper can ruin his chances of continuing in the organisation.

Appraisal Interviews

- This type of interview is also called as assessment interview. It is conducted to evaluate the achievements of subordinates during a set period, to provide the basis for promotion, salary increase, production bonus etc. Such an interview is more in discussion form rather than question-answer form. It provides the organisation an opportunity to have an in-depth dialogue with the subordinates to not only assess their performance but also obtain a feedback about the overall work environment, working conditions, hiring policy etc. This information could be used to bring in policy changes aimed at improving the efficiency of the work place.
- The following matters are usually discussed at these interviews:
  - (a) Objectives and targets set for the review period and achievement,
  - (b) Review of the organisation policy, work environment and conditions etc., which have contributed positively or negatively to the performance levels observed,
  - (c) Any lacuna by way of training, staff strength etc.,
  - (d) Any innovative intervention by the employee and the way it was received by the organisation,
  - (e) Targets for the ensuing period and the support expected by the employee,

(f) Suggestions for improving the performance of the individual, the department and the company,

(g) If the employee is expecting better rewards, the basis for such expectations etc.

- These appraisal interviews can be successful only if there is an environment of mutual trust and both the interviewer and the interviewee adopt a positive and constructive attitude. The manager should not misuse the information, personal or about the work environment, provided by the subordinate, during the course of these discussions.

Exit Interviews

- An exit interview is a kind of grievance interview. It is held when the organisation is severing relation with an employee. The exit of an employee can be voluntary on the part of the employer and employee, or involuntary, i.e., when an employee retires after attaining a certain age, when he cannot contribute to the organisation any more.
- When the exit is involuntary, and the employee has been dismissed, the interview should be held in a pleasant atmosphere or not at all. If it is voluntary, i.e., the employee has decided to seek greener pastures, the organisation should aim at knowing the circumstances leading to the resignation. This interview can help the organisation obtain valuable feedback about its policies, based on which it can design improvements to gain employee loyalty. Hence, the manager in such interviews plays the role of a listener. Such interviews should be held after the farewell party is given by his colleagues. he chief advantage of this type of interview is that it enables the management to make a proper assessment of the real working of the organisation.

Promotion Interviews

- Promotion Interview is conducted when the organisation is trying to fill vacancies internally. If there are a number of candidates for promotion, the organisation can use this to gauge the suitability of each contestant and the time and energy required to train him for the higher post. Even if there is only one candidate, a promotion interview can be held, in which case the interview is likely to be more informal. During the interview, the organisation obtains vital clues about the candidate, such as the nature of work he does, targets achieved, any innovative ideas he has implemented for improving the work environment or efficiency, the group or team activities he has participated in, customer responses about his work and working style, personal issues he is facing, his expectations from the organisation etc.

Qualitative Research Interview

- In detailed marketing research or social research, apart from quantitative data collected through questionnaires, there is need for qualitative research to describe the situation and provide meanings of the central themes. The main objective here is to understand the meaning of what the interviewees say with reference to the prevailing context or circumstances.
- One essential element of all interviews is the verbal interaction between the interviewer/s and the interviewee/s. Hitchcock (1989:79) stresses that '*central to the interview is the issue of asking questions and this is often achieved in qualitative research through conversational encounters*.' Consequently, it is important for the researchers to familiarise themselves with the questioning techniques before conducting interviews.

Aspects of Qualitative Research Interviews

- The interviewer can use any of the following tools:

(a) Structured questionnaires which the interviewee answers in a very short time,

(b) Some structured questions and some open ended questions, so that the interviewee gets some scope to give elaborate answers,

(c) Checklists, for in-depth interviews, with the help of which, the investigator frames his questions and has the freedom to change the order in which the points are addressed and also add new ones in response to answers obtained during the interview. This requires a very highly knowledgeable and experienced interviewer to be able to go back and forth and yet cover all the points.

- In-depth interviews are a far more personal form of research than structured questionnaires administered by investigators or sent by post or email.
- In the personal interview, unlike in mail surveys, the interviewer has the opportunity to probe or ask follow up questions.
- Interviews are time consuming and are resource intensive.
- The interviewer has to be well trained in how to ask questions and respond to any contingency.
- Sometimes, an organisation may call a group of respondents together for the purpose of understanding their views. This may be prior to launching a new product or entering a new market or if the organisation is going through a critical period, to arrive at solutions. Questions may relate to their perceptions, opinions, beliefs, and attitudes towards a product, service, concept, ad message and format, or packaging. The role of the manager is more as a listener, though from time-to-time, he may interrupt to ask questions or obtain clarifications. As the views are expressed and responded to by the others, the manager gets a quick view of the situation, which can help him/her take decisions. Such interviews can be extremely stimulating and productive.
- As there is a time limit set, the moderator or the host will first explain the concept/ product/service features and answer questions about it. He will then throw the meeting open for discussion. He will have a list of questions to be asked. To be effective, focus group questions should be open-ended and move from the general to the specific. The members are allowed to discuss both the positive as well as negative aspects of the issue. Care will have to be taken to ensure that all the members participate and all the questions listed by him/her taken up for discussion. It is essential to ensure that sub-groups are not formed. Generally, a questionnaire is filled up by the participants, right in the beginning, giving information about their background. Analysis of this would yield a profile of the participants. The entire session is recorded for later analysis. For instance, today washing machines are an essential item in most urban middle and affluent class households in India. However, about a couple of decades ago, it was a rare acquisition for these households. Hence, prior to the launch of this product, a marketing consulting company was hired by a manufacturer of white goods to conduct a series of focus group meetings to find out the attitude of the consumers towards this product. Most of the participants, except for the very affluent, or those who had seen such machines on their travels abroad, had many doubts: Where is the assured water and power supply? It will add to our chores. It is too expensive, etc.
- After the meetings, each meeting is summarised and analysed as soon as possible, so as not to miss out on important observations made by the participants. These summaries are then brought together for further analysis and preparation of a report.

6.4.2 The Intervew Preparation

- Be sure to take your mock interview as seriously as you would an actual interview. Get ready for the interview just as you would for an interview with a hiring manager:
- Arrive 10 - 15 minutes early, and bring your resume and any other materials you would bring to a real interview.
- Bring a notebook to take notes on what your mock interviewer tells you.
- Dress in professional interview attire.
- You should also prepare answers to general interview questions before arriving. Here's more information on mock interview questions including sample questions and answers you can review to get ready for your interview.

- If you have a mock interview to prepare for a specific job or career field, also review these job-specific interview questions.
- Mock interviews are an ideal way to practice for real job interviews, because you are in a situation that mirrors an actual interview with a company. When you review your interview with the interviewer, you'll be able to modify your responses and interview behavior, if necessary.

Set Up Your Own In-Person Mock Interview

- If you're not in a situation where you can participate in a mock interview with a professional counselor, recruit a family member or friend to help you practice interviewing. The more you prepare, the more comfortable you will be with interviewing.

Online Practice Interviews

- Another option for mock interviewing, is utilizing an online program or application. Online practice interview programs provide job seekers with a pressure-free way to prepare and practice for upcoming job interviews. Some of these programs are very basic; users are given a series of random interview questions (either verbally or in writing) and type in answers. While these programs get users thinking about how to answer various questions, they do not allow users to practice verbalizing responses.
- More sophisticated interview practice programs allow users to select questions related to their particular career field or the type of interview for which they are preparing (i.e. behavioral interview, group interview, etc.). A pre-recorded video of a mock interviewer asks a series of questions; the user must then orally answer each question.
- Sometimes users are given a time limit, so they learn to answer questions concisely. The program will record these audio answers, or users may be able to record themselves via webcam. After the simulated interview, the user can review his recording, or email the file to a friend, family member, or career counselor for review.
- Some programs even allow users to conduct live online interviews via webcam with actual career counselors. The interviewer records the interview and then critiques the user's performance.

Benefits of Practice Interviews

- Online practice interviews familiarize users with the interview process and allow users to practice answering common interview questions with confidence. Webcam practice interviews are particularly helpful in that you can review not only your answers, but your body language, eye contact, and interview attire.

6.4.3 Answering Strategies

How to Answer Job Interview Questions

- During the interview, the employer will ask you a variety of questions to determine your interest in the job and your qualification for the job. Make sure your answer the questions clearly and concisely, supported by example, and that you maintain strong eye contact. Below you will find strategies for answering questions in a job interview and a link to sample questions that are often asked by employers.

Tell Me About Yourself: The 90-Second Response

- An employer may begin a job interview with an open-ended question such as, "tell me a little about yourself" or "why are you interested in my organization?" Respond as though the employer were asking, "why do you want to be in this interview room?" Responses to this question should show how your skills, interests, and experiences would contribute to the position and organization.

- Use a 90-second guideline when answering this question:
- Focus the first 15 seconds on relevant and appropriate personal information you wish to share (e.g., where you are from).
- Focus the next 30 seconds on your academic experience (e.g., what you are studying, relevant courses, study abroad experience, relevant research experience).
- Focus the next 30 seconds on your professional experience (e.g., leadership positions, relevant extra-curricular involvement, internships, part-time jobs).
- Use the last 15 seconds to discuss why you are interested in the position, given the background you just discussed.
- Remember, this is a short answer. Give highlights of your experiences and focus your answer. Prepare by writing down the experiences you wish to discuss and then practice how you want to answer the question.

The Job Interview: Strategies for Success

Preparation

- Savvy interviewees have laid the foundation for a successful interview before ever setting foot in their future employer's office. Listed below are several strategies that will help you navigate the interview process successfully and land the job that's right for you.

Start early

- Preparing for an interview takes time. The less familiar you are with a particular job market or prospective employer, the more time you should allow yourself to prepare.
- Know what you want.
- Identify the job characteristics and benefits that are most important to you (e.g., specific population or service, geographic location, healthcare or financial benefits, job autonomy, supervision) and what features are preferable, but not essential.

Know the market

- What changes are occurring, what new skills are important to have and what opportunities and challenges are likely to emerge in the near future?
- Understand the needs of the employer.
- Browse the organization's website to familiarize yourself with its mission , core values and leaders. Skim any relevant publications, including the organization's annual report and articles or books by key staff members. Additionally, talk to any colleagues who are familiar with the organization to get an insider's view of the company's culture and priorities.

Know yourself

- Assess your strengths and weaknesses. Think about the personal characteristics that help make you successful, as well as the professional skills you bring to the table. Be able to describe how these skills and attributes relate to the job you are interested in.
- Be prepared to discuss your weaknesses.
- Avoid using the typical, "I'm a perfectionist" response. Choose a relatively benign area in need of development, acknowledge your awareness of the weakness and be able to describe what you are doing to actively address the shortcoming.

Tune up your CV

- Include the most updated information and tailor it to the position you are seeking. Consider which clinical or research experiences are most relevant to the job and elaborate on them. If the position has a supervisory or administrative component, take care to include professional activities that helped you grow your skills in these areas.
- Prepare several pertinent questions.

- Be ready to ask questions about both the position and the organization during the interview. Show that you've done your homework by integrating information you have read about the organizational culture, the services offered, and the current job opening into your questions.

Practice

- Based upon your understanding of the culture and needs of the employer, anticipate the kinds of questions they may ask you. Prepare answers to these questions, but don't worry about memorizing them. It will be easier to build rapport with the interviewer if you allow yourself to speak naturally during the real interview. Finally, grab a friend or colleague and do a mock interview.

Call ahead

- Telephone the administrative assistant or office manager a few days before the interview to confirm the appointment time and ask if there are any materials the interviewer wants you to bring to the interview.

Have a tested travel plan

- Take the time to map out directions to the interview site and do a test run to determine your travel time and parking location before the interview day. On the day of the interview, remember your directions and your contact person's name, address and phone number.

Be centered

- Whether it's a morning workout or a good breakfast, do whatever will set you at ease and put you in a positive frame of mind going into the interview. Arriving early and giving yourself enough time for a restroom break will also help to put you at ease.

First Impression

- The first few minutes of an interview are typically the most important. Below are a few tips for starting out on the right foot:
  1. **Dress appropriately:** The rule in terms of interview attire is to dress at least one level higher that the position you are interviewing for. For many positions in the field of psychology, a suit is most appropriate on interview day. Project a professional image and be conservative in terms of your hair style and accessories. You want them to remember your personality rather than your blouse or tie.
  2. **Relax:** The interviewer will be more at ease with you if you are at ease yourself. Trust that you are well-prepared, and project confidence without coming across as arrogant.
  3. **Cultural consideration:** In American culture, good eye contact, a firm handshake, and a smile are typically viewed as signs of honesty and professionalism in the workplace. When working in multicultural contexts, be aware of cultural factors and social rules that govern appropriate social interaction.
  4. **Use your skills as a psychologist:** Quickly assess your interviewer's mood, body language, tone and rate of speech, and personality style, and adjust your interactions in a way that increases the comfort level in the room.
  5. **Pay attention:** Remember, even though you are being interviewed, you are also interviewing the organization to see if it is a good fit for you. Begin to look for clues about the work environment and organizational culture. How are you greeted when you arrive? What does the reception area look like? How do coworkers relate to one another as they pass in the hallway?

The Interview

1. **Tailor your communication:** Is your interviewer a psychologist? Would this person be your direct supervisor or is this an interview from an outside human resources company? Based on the answers to these types of questions, think about the language you use and the kinds of questions you will want to ask the interviewer.

2. **Seek input. Remember the questions you prepared:** Early in the interview, ask about the position and the type of person who would be most successful in the job. This key information will help you to best sell your strengths later on in the interview.
3. **Be positive, but be honest:** Promote your strengths when the opportunities arise, but don't dance around your weaknesses. It's best to be honest and professional when fielding tough questions that may come your way.
4. **Close the interview:** Remember three important points: be ready to ask a few questions when given the opportunity; thank the interviewer; and inquire as to what the next step in the process will be.

Following Up

- The interview process doesn't end when you walk out the door. Following up appropriately can help you seal the deal.
  1. **The thank you letter:** Within a day or two of interviewing, send a letter to personally thank your interviewer. In the age of e-mail and overloaded voice-mails, letters send a message that you took the time to think about the interview and the position. Your letter should generally be no longer than one-typed page and provides an opportunity to not only thank the interviewer, but also to restate how your expertise would benefit the organization.
  2. **Be visible but not a nuisance:** If you still have not heard from the organization after about two weeks, consider sending an e-mail or making a follow-up call to show continued interest in the position.

     A successful job interview requires a considerable amount of thought and preparation. However, devoting the necessary time and energy to each of these stages in the interview process will help you put your best foot forward when meeting prospective employers.

Answering Strategies

How to Answer Job Interview Questions:

- During the interview, the employer will ask you a variety of questions to determine your interest in the job and your qualification for the job. Make sure your answer the questions clearly and concisely, supported by example, and that you maintain strong eye contact. Below you will find strategies for answering questions in a job interview and a link to sample questions that are often asked by employers.

Tell Me About Yourself: The 90-Second Response

- An employer may begin a job interview with an open-ended question such as, "tell me a little about yourself" or "why are you interested in my organization?" Respond as though the employer were asking, "why do you want to be in this interview room?" Responses to this question should show how your skills, interests, and experiences would contribute to the position and organization.

Use a 90-second guideline when answering this question:

- Focus the first 15 seconds on relevant and appropriate personal information you wish to share (e.g., where you are from).
- Focus the next 30 seconds on your academic experience (e.g., what you are studying, relevant courses, study abroad experience, relevant research experience).
- Focus the next 30 seconds on your professional experience (e.g., leadership positions, relevant extra-curricular involvement, internships, part-time jobs).
- Use the last 15 seconds to discuss why you are interested in the position, given the background you just discussed.

- Remember, this is a short answer. Give highlights of your experiences and focus your answer. Prepare by writing down the experiences you wish to discuss and then practice how you want to answer the question.

Common interview questions and how to answer them

- Interviews 19 September 2013
- Common interview questions , Sample interview questions , inteview questions , Interviewing like a boss
- Gulp. The interview is looming. How can you make sure you get the job? One way is to be ready with winning answers to common interview questions.
- You'll do better if you understand the psychology behind these questions. They're designed to tease information out of you about:
  - Who you really are
  - What you can do for the organisation
  - Why you're switching jobs
  - How much you expect to be paid
- In this article we outline the top 10 most typical interview questions and how to answer them:

1. **Tell me about yourself?**

   Inside story: Don't you just hate this one? The employer doesn't want to know what level you have reached on Candy Crush or your boyfriend. What the employer really wants to know is your experience, your skills and, if you're new in the workforce, your education. Keep it short and sweet.

2. **What interests you about this job?**

   Inside story: This question weeds out people who just want any job. This is your chance to shine. Try: "there is a greater scope for me to use my skills" – which you then outline.

3. **What did you like and/or dislike about your current/previous job?**

   Inside story: The interviewer wants to find out whether your cup is half full or half empty. It's best to focus on what you liked. Talk about the culture and the organisation and post it in a good light.

4. **Why are you leaving?**

   Inside story: This question can give a lot away about you. Focus on the opportunities available to you if you take the job.

5. **What do you know about our company?**

   Inside story: This question uncovers candidates who haven't taken the time to find out about the organisation. If you've done your homework you should be able to talk about the products or services, opportunities or difficulties in the market, and some recent news.

6. **Why do you want to work for us?**

   Inside story: Employers want someone who'll be committed, not someone who just needs a job. If you can link your skills to what the company or organisation does you're on the right track.

7. **What are your strong points?**

   Inside story: We know you're perfect. But this could work against you if you brag. Think of three or four good points and give examples of how they have benefited previous employers.

8. **What are your weak points?**

   Inside story: If you're counting on your fingers, stop now. Only ever give one weak point – which you can shed in a good light. Have another up your sleeve in case you're pushed.

9. What have you learned from your mistakes?

Inside story: Everyone has made mistakes. If you're honest about one or two mistakes you'll come across as more believable and trustworthy than someone who denies all. Focus on the take outs from these mistakes.

10. How much do you think this position pays?

Inside story: The interviewers are trying to find out if you are affordable and how little they can get away with paying.

Whatever you do: practice, practice, practice when it comes to job interview questions. Have an answer to every common question ready that focuses on the positive.

Important Points

(A) Vowels

- The letters *a, e, i, o, u and their sounds are known as vowels.* Consonants are produced by constricting the airstream to various degrees as it flows through the oral tract.
- **Vowels** are produced with a smooth, unobstructed airflow through the oral tract.
- Characteristic vowel qualities are determined by
  - (a) the height of the tongue in the mouth;
  - (b) the part of the tongue raised (front, middle, or back);
  - (c) the configuration of the lips; and
  - (d) the tension of the muscles of the oral tract.

(B) Consonants

- Consonants include the sounds we represent as <p, b, t, d, m, n, f, v, s, z, l, r, h> in the ordinary alphabet. All **consonants** are produced by entirely or almost entirely stopping the airstream coming from the lungs. When we almost entirely stop the airstream we force it through such a narrow opening that the airflow at that point is turbulent and noisy.
- We classify consonants according to the following characteristics: (a)whether or not the vocal folds are vibrating (**voicing**); (b) whether the sound is made with a fully stopped or merely constricted airstream (its **manner of articulation**); (c) where in the mouth the stoppage or constriction is made (its **place of articulation**); (d) whether or not air is flowing through the nasal cavity (**nasality**); and (e) whether or not the lips are pursed (**lip rounding**).

Classification of Consonants:

- Most English consonants can be classified using three articulatory parameters:
  - **(a) Voicing:** Vibration or lack of vibration of the vocal folds.
  - **(b) Place of Articulation:** The point at which the air stream is most restricted.
  - **(c) Manner of Articulation:** What happens to the moving column of air.

Guidelines/Rules for Stress: Some of the rules for primary and secondary stress can be given as follows:

(a) Words with just one syllable take the stress on it and so it is not required to denote the stress with a stroke. For example, Book, Dog, Cat, etc.

(b) Words starting with a weak syllable (ə, I), take the stress on the second or the next syllables: For example, attempt–/ ə 'təmpt/

(c) Words ending in -'-ion', '-ian', '-ious', '-ic', etc. take the stress on the second syllable from the end: For example, selection =/sI'lekəʃn/, fantastic =/fen'tæstik/, musician= /mju'zʃn/ etc.

(d) Words ending in '-ity' and '= ate' take the stress on the third syllable from the end. For example, Peculiarity = /pIkju'liærItI/, venerate =/'veneræt/ etc.

(e) Words ending in '-eer', '-aire', '-esque', etc. receive the primdry stress on the last syllable, For example, Engineer = /end3i'niə/, Millionaire =/milliə 'neə/, Groteque = /grə 'tesk/ etc.

(f) Suffixes like "ness, less, ment" do not affect the stress: For example, 'joyful = 'joyfullness.

(a) The Falling Tone and its Uses: In this pattern, there is a high pitch level on the final syllable. It is slightly lower on the second syllable and lowers till the end of the sentence. The tone is the lowest on the last stressed syllable. This tone is shown with ` mark (symbol).

(b) The Rising Tone and its Uses: In the rising tone pattern, there is a series of descending steps on the stressed syllables. But there is the rise on the last stressed syllable, rising tone is marked as (I). It is used in the following types of sentences:

The Objective of Selection in GD

- It is mainly to check your team playing skills. You have to understand the other persons' point of view, while making your point and ensure that your team as a whole reaches a solution or agreement that is both feasible and accepted by all team members.
- There are four major areas of evaluation in selection GDs: subject knowledge, oral communication skills, leadership skills and team management.

Importance of Group Discussions

- Better understanding of the subject:
- Solve problems:
- Improve listening skills and confidence:
- Enhance communication and interpersonal skills:
- Team building exercise:

Importance of Presentation

(a) Knowing the Audience
(b) Level of Understanding
(c) Clear Concept of the Purpose
(d) Gathering of Information
(e) Format of Presentation
(f) Outline of the Message
(g) Brevity
(h) Sequence
(i) Availability of Time
(j) Opening
(k) Introduction
(l) Main Idea
(m) Choosing Precise Words
(n) Verbosity should be avoided
(o) Prepositions, Adjectives and Adverbs
(p) Avoid Jargon or idiom relating to specific professions
(q) Avoid Cliché or formula statements
(r) Seven Cs of communication
(s) The Closing segment
(t) Pleasant Demeanor

Ten Principles of Effective Presentation:

(a) Your best style is relaxed conversation
(b) Be yourself
(c) Relaxed conversation is two way
(d) Relaxed conversation is receiver-driven
(e) Less is more

(f) People can listen or think but they can't do both.
(g) People won't remember what you say
(h) Be conversational in your delivery
(i) Think before you speak
(j) Silence is essential

Ways to Overcome Stage Fright When Speaking in Public

1. Know Your Stuff:
2. Practice, Practice, Practice:
3. Talk Yourself Down:
4. Wallow in the Worst:
5. Visualize the Outcome:
6. It is Not All About You:
7. When Things go Wrong:
8. Keep Calm, Don't Rush It:
9. Focus on Getting through the First 5 Minutes:
10. Never Apologize for Being Nervous:
11. Don't Share Your Mistakes:
12. Arrive Early:
13. Stretch:
14. Breathe:
15. Double Check Everything:
16. Don't Fight Your Stage Fright … Work With It:

Job Interview/Placement Interview

(a) The Screening Interview:
(b) Face to Face Interview:
(c) Group interview or Group Discussions (GD):
(d) Behavioural interview:
(e) Stress interview:
(f) Panel interview:
(g) Telephonic interview:
(h) Campus Interviews:

Common interview questions and how to answer them

1. Tell me about yourself?
2. What interests you about this job?
3. What did you like and/or dislike about your current/previous job?
4. Why are you leaving?
5. What do you know about our company?
6. Why do you want to work for us?
7. What are your strong points?
8. What are your weak points?
9. What have you learned from your mistakes?
10. How much do you think this position pays?

Practice Questions

1. What are vowels and Dipthongs?
2. What are Consonants? What are its features.
3. How are Consonants classified.
4. Give the classification of vowels.

5. What is Accent. Give the meaning and significance and classification.
6. What is Intonation.
7. What are objectives of selection in GD?
8. State the importance of GD.
9. What are Leadership skills?
10. What are Presentation skills? State its importance.
11. What are the principles of effective presentation?
12. State the ways to overcome stage fright.
13. Write a detail note on Mock Interview.
14. State the Interview preparation and answering strategies.

■■■

Appendix

Vocabulary

The meaning of the word vocabulary as stated in the Oxford Advanced Learner's Dictionary, in so far as it relates to an individual, is 'all the words that a person knows or uses'. Depending upon the number of words a person knows or uses, his vocabulary is categorised as 'wide vocabulary' or 'limited vocabulary'. These two terms are quite self-explanatory and need no explanation. Vocabulary of an individual is classified as 'Active Vocabulary' and 'Passive Vocabulary'. Active vocabulary is the stock of words an individual knows and uses, while passive vocabulary relates to the number of words a person understands but does not use. In other words, the more the number of words a person knows and uses, the richer is his vocabulary. Generally speaking, most people do not have a rich vocabulary to understand or comprehend simple ideas. For enriching one's vocabulary, it is absolutely essential to have a grasp over factors like:

(a) Relative importance of the structure of language and vocabulary;

(b) Kinds of vocabulary a person uses;

(c) The exact nature of words and their meaning;

(d) The implications of learning and using new words; and

(e) The various methods of learning vocabulary.

(a) Relative importance of the structure of language and vocabulary: It is generally held that the learning of the structure of language is more important than vocabulary. No doubt the structure of language is to be studied first for understanding words. That does not mean that vocabulary is less important. For effective communication both structure and vocabulary must be mastered. In day-to-day experience when the context is clear to the speaker and the listener, it is quite possible to communicate with vocabulary alone rather than indulging much in the structure of language, the learning of which is more difficult than the mastering of vocabulary.

(b) Kinds of vocabulary a person uses: There is a very useful classification of words that has been made by scholars. Words are classified into three broad groups which are: (i) the small number of words a person knows intimately and uses them effectively (active vocabulary), (ii) relatively large number of words he understands but does not normally use (passive vocabulary) and (iii) the vast number of words he does not know. Active vocabulary or the words in group (i) are almost like friends. Therefore, active vocabulary is termed as friendly and comprises the most useful words. A majority of people fall in the group where vocabulary (ii) is like a commander with too many servants who are not under one's full control.

(c) The exact nature of words and their meanings: Semantics deals with the study of different aspects of meaning of words. To fully understand the meaning of a word, it is not enough to just know what it stands for but also its relationship with other words in the language. Thus, understanding words in their proper context is more important than merely increasing word power.

(d) The implications of learning and using new words: For making a word a part of one's active vocabulary, one must master not only its meaning but also its spelling, pronunciation and usage. Usages are concerned with conventions or customs in language. They are different in different languages. Hence, for appropriate use of any word it is essential to understand what is and what is not acceptable in the language. Many teachers adopt the practice of teaching meanings of words in the student's mother tongue. But this method is not quite satisfactory since there are very few words in the mother tongue which reflect and have exact translation equivalents.

(e) The various methods of improving vocabulary: Although there are several ways to increase vocabulary, the best and perhaps the only method is extensive reading. Extensive reading helps students understand and learn families of words from basic words by analogy. For example, from the basic word 'care', other words like careful, carefully, careless, carelessly, carelessness etc. can be derived. It is also possible to improve vocabulary by learning word patterns, word-building games etc. These are the methods in addition to the traditional method of learning, remembering and properly using words and keeping them at our command. Once a person gets accustomed to this practice, one can gradually shift the emphasis on the other aspects of increasing word power.

In addition to the sincere efforts to develop the habit of extensive reading, the following aspects of language studies would also help to enrich vocabulary:

Word Formation

In the English language, word formation is concerned with derivatives, formed from simple words (roots) by adding either a prefix or suffix or both. Simple words which are also called **Primary Words** belong to the original words in the language. They are not derived or compounded or developed from other words. Derivatives are either **Primary Derivatives** or **Secondary Derivatives**. Primary derivatives are formed by making changes in the body of the simple words. It should be noted that the Past Tenses of primary verbs formed by internal change do not fall in the category of primary derivatives. Some of the primary derivatives are listed here:

(I) Primary Derivatives

(a) Nouns from Verbs and Adjectives

Verb	Noun	Verb	Noun
Advise	Advice	Gape	Gap
Bind	Bond	Grieve	Grief
Bless	Bliss	Live	Life
Break	Breach	Lose	Loss
Choose	Choice	Prove	Proof
Float	Fleet	Sing	Song
Wake	Watch	Speak	Speech
		Strike	Stroke

Adjectives	Nouns
Dull	Dolt
Hot	Heat
Proud	Pride

(b) Adjectives from Verbs and Nouns

Verb	Adjectives	Verb	Adjectives
Float	Fleet	Milk	Milch
Lie	Low	Wit	Wise

(c) Verbs from Nouns

Nouns	Verbs	Nouns	Verbs
Bath	Bathe	Glass	Glaze
Blood	Bleed	Grass	Graze
Brood	Breed	Knot	Knit
Cloth	Clothe	Price	Prize
Drop	Drip	Tale	Tell

(II) Secondary Derivatives

As stated earlier, the secondary derivatives are obtained by using affixes. Affixes may be prefixes or suffixes. An addition to a word at the beginning is a prefix, while at the end is a suffix. Some of the prefixes and suffixes used for getting secondary derivatives indicating their purport and words obtained by adding them are given on the next page.

(a) Prefixes (Inclusive of Latin and Greek Prefixes) in common use

Prefix	Purport	Words obtained by adding Prefixes
A	on, in	aboard, ashore, asleep
A	out, from	arise, awake, alight
A (an)	without, not	atheist, apathy, anarchy
Ab (a, abs)	from away	avert, abstract
Ac (ad, af, ag, al, an, ap, ar, as, at, a)	to	accord, adapt, affect, aggrieve, allege, announce, appoint, arrest, attach, avail
Ambi (amb, am)	on both sides	ambiguous, ambivalent, ampute
Amphi	around, on both sides	amphibious, amphitheater
Ana	up, back	anachronism, analysis
Ante (anti, an)	before	antedate, anticipate
Anti (ant)	against	antipathy, antagonist
Arch (archi)	chief	archduke, archbishop, architect
Auto	self	automobile, autonomy, autograph
Be	by	beside, betimes
Bene	well	benefactor, benefit, benevolent
Bis (bi, bin)	twice, two	bisector, binocular, bigamy
Circum (circu)	around, round	circumference, circumlocutory, circular
Con (col com, cor)	together, with	consent, colleague, combine, correlate
Contra (counter)	against	contraception, counterfeit, counteract
De	down	devaluation, degenerate, deplete
Demi	half	demigod
Di	two, double	dioxide, dilemma
Dia	through	diameter, diagonal
Dis (diff, di)	apart, reverse, opposite, negative	dishonest, disfigure, differ, divide, disease
En (em)	in	enclosure, emblem, encyclopaedia
Epi	upon	epitaph, epilogue
Eu	well	eulogy, euphony, eugenics
Ex (ec, ef, e)	out of	extract, exodus, eccentric, effect, educe, extract
For	thoroughly	forgive, forget, forbear
Fore	before	foregone, forefather, forecast
Hemi	half	hemisphere
Homo (hom)	like	homogeneous, homosexual, homophone
Hyper	over, beyond	hyperbole, hypertension, hypercritical
Hypo	under	hypocrite, hypothesis
In (im, il. ir, en, em)	in into	income, inland, illustrate, immerse, irrigate, enact, embrace
In (il, im, ir)	not	indecent, illegitimate, immature, irreligious

Male (mal)	ill, badly	malpractices, malnutrition, malevolent
Meta	of change	metaphysics, metabolism, metamorphosis
Mis	wrong, wrongly	misguide, misuse, misfortune, misdeed
Mono	single, one	monologue, monogamy, monorail
Non	not	nonentity, non-fiction, nonsense
Over	above, beyond	overact, overdose, overcharge
Para	by the side, beside	parasite, paramilitary, parameter
Peri	around	periphery, perimeter, periscope
Philo (phil)	love, liking, fond of	philology, philosophy, philanthropy
Pre	before	prefix, pre-historic, prenatal
Pro (por, pur)	for, in favour of, supporting	prophesy, pro-German, programme, pro-chancellor
Quasi	to a certain extent, not really	quasi-government, quasi-rent
Re	back, again	regain, rewind, reclaim, refund, rewrite, regenerate
Retro	backward	retrospective, retrograde
Semi	half	semi-skilled, semi-precious
Sub (suc, suf, sugt sum, sup, sur, sus)	under	subsidy, succeed, suffer, suggest, summon, support, surmount, sustain
Super	above	supernatural, superfluous, superintendent, superfast
Syn (sym, syl, sy)	with, together	synonym, sympathy, syllable
Trans (tra, tres)	across	transworld, traverse, trespass
Un	not	unkind, unbelievable, unruly
Un	to reverse an action	unfold, undo, unveil, untie
Under	beneath, below	underground, undergo
Vice	in place of	viceroy, vice-captain, vice-president

(b) Suffixes (inclusive of Latin and Greek) in commone use

Suffix	Purpose	Words with these suffixes
Of Nouns	**Denoting agent or doer of a thing**	
- ain (-an, -en, on)		chieftain, artisan, citizen, surgeon
- ar (-er, -eer, ier, - ary)		scholar, preacher, engineer, financier, missionary
- ate (-ee, -ey, -y)		advocate, trustee, attorney, deputy
- er (-ar, -or, -yer)		painter, driver, baker, sailor, lawyer
- or (-our, -eur, -er)		emperor, saviour, amateur, interpreter
-ster		spinster, songster, punster
- ter (- ther)		daughter, father
- age		bondage, marriage, leakage
- ance (ence)		assistance, brilliance, excellence, intelligence
- cy		intimacy, lunacy, fancy
-dom		stardom, kingdom, wisdom
- hood(- head)		brotherhood, manhood, godhead
- ice vise)		service, cowardice, exercise
- ion		opinion, selection, union
- lock (- ledge)		knowledge, wedlock
- ment		treatment, judgement, punishment
- mony		testimony, matrimony, parsimony
- ness		fitness, business, brightness

- red		kindred, hatred
- ship		kinship, guardianship, lordship
- th		health, wealth, growth
- tude		multitude, magnitude, fortitude
- ty		frailty, cruelty, credulity
- ure		pleasure, forfeiture, verdure
- y		treasury, victory, misery
	Forming diminutives	
- cule (-ule, - sel, - cel, - el, - le.)		molecule, globule, damsel, parcel, chapel, circle
- el (- le)		kernel, satchel, girdle, handle
- en		chicken, maiden, kitten
- et		lancet, trumpet, owlet
- ett		cigarette, coquette
- ie		dearie, birdie, lassie
- kin		lambkin, napkin, bumpkin
- let		booklet, bracelet, leaflet
- ling		darling, duckling, weakling
- ock		hillock, bullock
- ary (-ery, -ry)		library, treasury, nunnery, dairy, dispensary
- ter (-tre)		theatre, cloister
Of Adjectives		
- al	belonging to	formal, legal, international, global
- an (- ane)		human, mundane, humane
- ar		similar, regular, familiar
- ary		honorary, contrary, temporary, necessary, customary
- ate		temperate, fortunate, obstinate
- ble	having the quality	capable, culpable, sensible, laughable
- ed	having	gifted, talented, learned
- en	made of	wooden, golden, earthen
- esque	in the style of, way of	picturesque, grotesque
- id	containing	humid, arid, lucid
- ile	like, belonging to	fragile, servile, juvenile
- ine	of, concerning	masculine, divine, canine
- ish	somewhat, like	reddish, childish, girlish, boorish
- ive	with the quality of	active, attentive, constructive
- lent	with the quality of	indolent, turbulent, virulent
- less	free from without	fearless, senseless, hopeless
- ly	like, having quality of	godly, humanly, manly
- ose(ous)	full of, possessing	virtuous, dangerous, copious, verbose
- some	with the quality of inclining to	gladsome, troublesome, wholesome, quarrelsome
- ward		forward, backward, wayward
- y	with the quality of, possessing	healthy, wealthy, greedy, thirsty, dirty
Of Verbs		
- ate		assassinate, captivate, terminate
- en	causative, forming transitive verbs	weaken, sweeten, strengthen
- er	intensive or frequentative	glitter, chatter, flutter
- fy	to make	simplify, purify, qualify, terrify
- ish	to make	publish, furnish, punish, banish
Of Adverbs		
- ly	like	wisely, shrewdly, boldly
- long		headlong, sidelong
- ward (– wards)	turning	homeward, upward, downward
- way (- ways)	to by towards	straightway, anyway, sideways
- wise	manner, mode	likewise, otherwise

Greek Suffixes		
- ic (- ique)	possessing the quality of	unique, angelic, cynic
- ist	concerned with, related with, dealing with	artist, chemist
- isk	of	asterisk, obelisk
- ism (- asm)	possessing the quality of	patriotism, despotism,
- ize	to make, to show	enthusiasm, civilize, criticize, sympathize
- sis (- sy)	belonging to	crisis, analysis, poesy, heresy
- e (- y)		catastrophe, monarchy, philosophy

(III) Compound Words

Compound words are those words which are formed by combining different parts of speech viz. nouns, adjectives, adverbs, gerunds, prepositions. The compound words got by these different combinations are mostly nouns, adjectives and verbs and are recognised as **Compound Nouns, Compound Adjectives** and **Compound Verbs**.

(a) Compound Nouns

Compound nouns are obtained by the following combinations:

(i) Noun + Noun

e.g.: armband, artwork, batsman, chairman, cowboy, doorbell, dockyard, dashboard, filmstar, godfather, gold-mine, hammerman, headgear, ink-pot, juke-box, keyboard, landlord, milkmaid, newspaper, nightdress, postcard, quartermaster, roadroller, railway, screwdriver, silkworm, sound-box, sunglasses, teapot, tablespoon.

(ii) Adjective + Noun

e.g.: big-bank, broadsheet, black-board, deadlock, nobleman, sweetheart, fineart, halftruth, stronghold, highclass, lowland, quickstep, slowcoach, tenderfoot.

(iii) Verb + Noun

e.g.: breakfast, cut-throat, cutpiece, daredevil, pickpocket, hangman, spendthrift, telltale.

(iv) Gerund + Noun

e.g.: dancing-club, bathing-beauty, hiding-place, singing-chair, burning-train, looking-glass, walking-stick, booking-office, cooking-class, writing-table, scribbling-pad, dining-hall, stepping-stone, working-class, swimming-pool, smiling-face.

(v) Adverb (or Preposition) + Noun

e.g.: afterthought, outswinger, inswinger, foresight, outlaw, bypass, off-beat, outside, overdose, underestimation, onlooker, inside.

(vi) Adverb + verb

e.g.: backcomb, backwash, outcry, income, outcome, forebear, forecast, outfit,

(vii) Verb + Adverb

e.g.: spin-off, die-hard, drawback, lock-up, send-off, go-between.

(b) Compound Adjectives

Compound adjectives are obtained by the combination of –

(i) Noun + Adjective (or participle)

e.g.: heart-warming, note-worthy, headstrong, homesick, lifelong, worldwide, snow-white, blood-red, pitch-dark, skin-deep, purse-proud.

(ii) Adjective + Adjective

e.g.: red-hot ,blue-black, dull-grey, white-hot.

(iii) Adverb + Participle

e.g.: everlasting, never-ending, well-dressed, down-hearted, long-suffering, inborn, outspoken.

(c) Compound Verbs

Compound verbs can be obtained by the combination of -

(i) Noun + Verb

e.g..: backbite, earmark, browbeat, waylay, typewrite.

(ii) Adjective + Verb

e.g.: Safeguard, fulfill, whitewash.

(iii) Adverb + Verb

e.g.: outcry, outwit, undertake, overthrow, upset, ill-use, ill-breed, in-breed, overtake, over-joy, over-act.

In most compound words, the first word tends to modify the second word. The accent is given upon the modifying word when the combination is complete. It may be noted that, when the two elements of the compound word are partially blended, a hyphen is put between them and the stress falls equally on both the words or elements.

Homonyms

In the English language, there are a number of words spelt and pronounced in the same way but carry different meanings. Such words are called *homonyms*. Similarly, there are words pronounced alike but with a different meaning or spelling. Those words are called *homophones*. These peculiarities of English words which have a great resemblance in spelling or pronunciation but differences in their meaning, create, confusion in the minds of students leading to mistakes in their usage. Some of the words belonging to these categories are given below indicating their different meanings in appropriate sentences bringing home those different shades of meanings.

(i) Words with same spelling and pronunciation but carrying different meaning. (Homonyms).

1. Act

(a) Decree on law made by a legislative body.

e.g.: The Government has passed an Act to prohibit the sale of drugs.

(b) Any of the main divisions of a play or an opera.

e.g.: The tempo of the play culminates in the Act III.

2. Agent

(a) Person who acts for or manages the affairs of other people in business, politics etc.

e.g.: He works as an insurance agent.

(b) Force or Substance that produces an effect or change.

e.g.: Yeast is used these days in many recipes as a raising agent.

3. Bar

(a) A long shaped piece of hard stiff material.

e.g.: A large number of metal bars were piled up in the godown.

(b) A counter at which drinks are served.

e.g.: They were found in an intoxicated condition at the bar.

4. Bat

(a) Small mouse-like mammal that flies at night and feeds on fruits and insects.

e.g.: He found bats flying in the house that had remained closed for years together.

(b) A wooden implement of a specified size and shape and with a handle used for hitting the ball in games like cricket, baseball and table tennis.

e.g.: He used a little heavier bat during the tournament.

5. Beat

(a) Stroke or regular sequence of strokes.

e.g.: We heard the drum beat early in the morning.

(b) Emphasis repeated regularly marking rhythm in music or poetry.

e.g.: That song has a very good and melodious beat.

6. Capital

(a) Wealth or property that may be used for producing more wealth.

e.g.: He started his business with a meagre amount of capital.

(b) Town or City that is the centre of the Government of a country.
e.g.: Mumbai is the capital of the State of Maharashtra.

7. Card

(a) Thick stiff paper on thin pasteboard.
e.g.: An identity card is compulsory for all college students.

(b) Wire brush or toothed instrument for cleaning or combing wool.
e.g.: He used his card very carefully and efficiently in the department.

8. Date

(a) Specific numbered day of the month or specific year usually given to show when something happened or is to happen.
e.g.: The date of the conference conflicted with the date of his marriage anniversary.

(b) Brown sweet edible fruit of a palm-tree common in N. Africa and S. W. Asia.
e.g.: Date is a very healthy fruit. Dates are good for health.

9. Drill

(a) Training in military exercises.
e.g.: N. C. C. cadets have three hours of drill everyday.

(b) Machine for making furrows, sowing seed in them and covering the seeds by soil.
e.g.: He purchased a new drill for his farm.

10. Even

(a) Level, smooth, flat.
e.g.: Highways should be perfectly even for smooth traffic.

(b) Divisible by two with no remainder.
e.g.: The teacher today taught the sums of even numbers.

11. Fan

(a) Device with rotating blades operated mechanically to create a current of cool air.
e.g.: We have installed a new ceiling fan in our drawing room today.

(b) Enthusiastic admirer or supporter.
e.g.: A lot of people are cinema fans.

12. Fence

(a) Structure of wire etc. put around a field, or garden to make a boundary or keep animals from straying.
e.g.: It was necessary to put a fence around our garden.

(b) Person who knowingly buys and resells stolen goods.
e.g.: He was arrested on the charge of being a fence.

13. Gin

(a) Trap or snare for catching animals.
e.g.: The hunter was aware about where the gin was placed.

(b) Colourless alcoholic drink often used in cocktails.
e.g.: He preferred gin with water as a tonic.

14. Gloss

(a) Brightness or shine on a smooth surface.
e.g.: Heavy polishing was necessary to bring gloss to the old wooden furniture.

(b) Explanatory comment; explanation, interpretation.
e.g.: The election candidate put a different gloss on the issue of corruption in the political field.

15. Hip

(a) Part on either side of the body below the waist where the bone of a person's leg is joined to the trunk.

e.g.: He was standing there with his hands on his hips.

(b) Berry-like fruit of the wild rose.

e.g.: Did you see hips on the fence of our garden ?

16. Jade

(a) Hard, usually green, stone from which ornaments are carved.

e.g.: He owned a beautiful jade vessel.

(b) Tired or worn-out horse.

e.g.: He came across a jade on his way back home.

17. Labour

(a) Physical or mental work.

e.g.: Workers naturally expect appropriate wages for their labour.

(b) Workers as a group or class.

e.g.: The factory engaged a large number of unskilled labour last month.

18. Law

(a) Rule established by authority.

e.g.: The old law has already become obsolete.

(b) Subject of study.

e.g.: He took up law after his graduation in commerce.

19. Magazine

(a) Paper-covered periodical, usually weekly or monthly.

e.g.: He purchased a magazine for reading during his long journey by train.

(b) Store of arms, ammunition, explosives etc. Receptacle for number of rounds loaded at one time in automatic guns.

e.g.: He enquired in the gun shop about how many bullets one magazine contained.

20. Margin

(a) Blank space round the written or printed matter.

e.g.: The teacher directed the students to leave sufficient margin before beginning to write on the paper.

(b) Difference between cost-price and selling-price.

e.g.: He discarded the deal because there was hardly any profit margin in it.

21. Mess

(a) Dirty or untidy state.

e.g.: Spectators made a mess in the auditorium on failure of electricity.

(b) Building in which people take meals together.

e.g.: It is the best mess in our locality.

22. Nature

(a) The whole universe and every created, not man-made, thing.

e.g.: The wonders of nature are beyond human imagination.

(b) Typical qualities and characteristics of a person or animal.

e.g.: Corruption is not in his nature.

23. Negative

(a) Word or statement that expresses or means denial or refusal.

e.g.: The words like *no, not, neither* are negatives.

(b) Developed photographic film from which photoprints are taken.

e.g.: All the negatives of our tour to Simla have been spoiled by the photographer.

24. Order

(a) Way in which people or things are arranged in relation to one another.

e.g.: The names were arranged in an alphabetical order.

(b) Command or instructions given by somebody in authority.

e.g.: The cadets obeyed the orders given by their commander.

25. Part

(a) Distinct portion of a human or animal body or of a plant.

e.g.: Which part of your body is hurt ?

(b) Role played by an actor in a play, film etc.

e.g.: His part in the film was appreciated by the public.

26. Play

(a) Activity done for amusement especially by children.

e.g.: The park was full of the happy sounds of children at play.

(b) Drama

e.g.: The new play staged yesterday for the first time was quite excellent.

27. Pole

(a) Either of the two ends of a magnet or the terminal points of an electric battery; either of the two points at the exact-top or bottom of the earth.

e.g.: The Arctic is the region around the North Pole.

(b) Long thin rounded piece of wood or metal used especially as a support for something or for pushing a boat etc. along.

e.g.: He is an expert in climbing the telegraph pole.

28. Raw

(a) Uncooked

e.g.: Eat vegetables like carrots, radishes etc. raw as they are good for health.

(b) Inexperienced, unskilled.

e.g.: Raw hands would not help to improve the conditions of the company.

29. Right

(a) Best in view of the circumstances, most suitable.

e.g.: We are on the right track of action.

(b) Contrasted with left.

e.g.: She was standing on my right.

30. Saw

(a) A long cutting tool with a sharp toothed edge.

e.g.: He installed a new saw in his factory.

(b) Saying, Proverb, old maxim.

e.g.: The old saw 'A stitch in time saves nine' is true even today.

31. Seal

(a) Animal with flippers that lives near and in the sea and eats fish.

e.g.: A number of seals were seen on the beach.

(b) Soft material such as wax etc. stamped with a design and fixed to a document to show that it is genuine and to prevent it to be opened by a wrong person.

e.g.: The envelope was affixed with the seal of the king.

32. Second

(a) Person or thing that comes next after the first.

e.g.: He stood second in the elocution competition.

(b) 60th part of a minute of time.

e.g.: The state of the game changed within five seconds.

33. Share

(a) Part or portion of a larger amount divided among many people or to which many people contribute.

e.g.: Everyone present will get a fair share of the food.

(b) Any of the equal parts into which the capital of a business company is divided.

e.g.: He held 500 shares of that co-operative credit bank.

34. Sight

(a) Ability to see, vision.

e.g.: He lost his sight in the recent accident.

(b) Range within which somebody can see or something can be seen.

e.g.: In a few seconds, his car was out of our sight.

35. Slip

(a) Act of slipping; false step.

e.g.: There is many a slip between the cup and the lip.

(b) A thing or small piece of paper.

e.g.: I noted down the phone number on a slip of paper.

36. Spirit

(a) Soul thought of as separate from the body; soul without a body; ghost.

e.g.: People say that the house on the other side of the road is haunted by an evil spirit.

(b) Strong, distilled alcoholic drink.

e.g.: My friend is habituated to drink spirits like rum, whisky, brandy etc.

37. Tank

(a) A large container generally for liquid or gas.

e.g.: His business is to give tanks on hire for transport of milk.

(b) Armoured fighting vehicle with guns which moves on caterpillar tracks.

e.g.: A large number of tanks of the enemy were destroyed in the battle.

38. View

(a) What can be seen from a particular place; fine natural scenery.

e.g.: You will get a better view of the valley from the top of the hill.

(b) Personal opinion or attitude.

e.g.: His views about today's political situation are very bitter.

39. Ward

(a) Separated part or room for a particular group of patients in a dispensary or hospital.

e.g.: You can meet him in the surgical ward.

(b) Division of a city that elects and is represented by a councillor in local government.

e.g.: The councillor from our ward has been chosen as a mayor of the city.

40. Wave

(a) Move regularly and loosely to and fro or up and down.

e.g.: Our tricolour was waving proudly in the air.

(b) Move one's hand to and fro or up and down in order to attract attention, or to make a signal or give a greeting.

e.g.: All the family members waved as the train moved from the station.

Homophones

(ii) Words with same pronunciation but different-spelling or meaning are called *homophones*. Some examples of such words are given below.

1. Advice - Advise.

(a) Advice - Opinion given about what to do, or how to behave.

e.g.: His advice to his friend proved to be beneficial.

(b) Advise - Recommend, give advice to somebody.

e.g.: The doctor advised him to take complete rest.

2. Aid-Aide.

(a) Aid - Help, food, money etc. sent to a country to help it.

e.g.: How much foreign aid did we receive during the recent floods ?

(b) Aide - Assistant.

e.g.: He is one of the chief aides to the President.

3. Beat - Beet

(a) Beat - Hit something repeatedly especially with a stick

e.g.: Somebody was beating at the door.

(b) Beet - Type of plant with a fleshy root which is used as a vegetable or for making sugar.

e.g.: Eating beets is very good for the health.

4. Canvas - Canvass.

(a) Canvas - Strong coarse cloth used for making tents etc. and by artists for painting on.

e.g.: He purchased a canvas for painting his mother's portrait.

(b) Canvass - Go around an area for political support.

e.g.: He actively participated in the canvassing campaign.

5. Cell - Sell

(a) Cell - A very small room; microscopic unit of living matter.

e.g.: The prisoners were kept in a single cell in the basement.

(b) Sell - To give goods etc. to somebody who pays money.

e.g.: He goes to the market every morning to sell books.

6. Dear - Deer

(a) Dear - Greatly valued, loved; expensive.

e.g.: His daughter was very dear to him.

(b) Deer - Any of the several types of graceful, quick-running animals.

e.g.: There were many beautiful deer in the zoo.

7. Die - Dye

(a) Die - Stop living, come to the end of one's life.

e.g.: His relative died of heart attack.

(b) Dye - Substance used for colouring.

e.g.: He uses hair dye every month to look young.

8. Desert - Dessert

(a) Desert - Barren land with very little water and vegetation, often covered by sand.

e.g.: The Sahara Desert is the largest one in the world.

(b) Dessert - Any sweet dish eaten at the end of a meal (e.g. ice-cream, fruit-salad etc.).

e.g.: If your meal is over, shall we now move on to the dessert.

9. Fair - Fare

(a) Fair - Large scale exhibition of commercial and industrial goods.

e.g.: A trade fair was recently held in the capital.

(b) Fare - Money charged for a journey by bus, taxi, ship etc.

e.g.: The bus fare has been increased exorbitantly.

10. Gait - Gate

(a) Gait - Manner of walking or running.

e.g.: He was walking with an unsteady gait.

(b) Gate - Means of entrance or exit; movable barrier which closes on opening in a wall, fence etc.

e.g.: The garden gate is completely broken and needs repairs.

11. Hair - Hare

(a) Hair - Mass of fine thread-like strands that grow from the skin of people and animals, esp. on the human head.

e.g.: She has beautiful long, black hair which attracts people.

(b) Hare - Fast running rabbit-like but larger mammal with long ears and divided upper lips, that lives in fields.

e.g.: Hares are very rarely seen even in forests.

12. Hall - Haul

(a) Hall - Building or large room for meeting, meals etc.

e.g.: The meeting was convened in the Town Hall.

(b) Haul - Pull or drag something with effort.

e.g.: Sailors hauled the boat up the beach.

13. Idle - Idol

(a) Idle - Doing or having no work; not employed.

e.g.: Idle people are a nuisance to the whole society.

(b) Idol - Person or thing that is greatly loved or admired.

e.g.: Actors or actresses are always the matinee idols for people.

14. Knight - Night

(a) Knight - Man to whom the sovereign has given a rank of honour having the title of 'Sir'.

e.g.: Rabindranath Tagore had surrendered the knighthood conferred on him by the British Sovereign.

(b) Night - Time of darkness between sunset and sunrise.

e.g.: People generally sleep during night.

15. Main-Mane

(a) Main - Most important, chief; principal.

e.g.: An increase in population is the main cause of concern for our country.

(b) Mane - Long hair on the neck of a horse, lion etc.

e.g.: Lions look graceful because of their manes.

16. Manner-Manor

(a) Manner - Way in which a thing is done or happens.

e.g.: The manner in which he behaved was not graceful.

(b) Manor - Unit of land under the feudal system, part of which was used by the lord and the rest farmed by his tenants.

e.g.: Tenants in manors were not always treated well by their lords.

17. Not - Nought

(a) Not - Used with auxiliary verbs and modals to form the negative.

e.g.: It is unsafe not to follow traffic rules.

(b) Nought - The figure zero (0); nothing.

e.g.: Put three noughts after the figure 5.

18. One - Won

(a) One - One less than two, single.

e.g.: He was the only one present in the hall.

(b) Won - Past tense and past participle form of the verb win.

e.g.: India won the Hero trophy match by a narrow margin.

19. Peer - Pier

(a) Peer - Person who is equal to another in rank, status or merit.

e.g.: He likes to mix only with his peers.

(b) Pier - Structure built out into the sea, lake etc. so that boats can stop and take on or put down goods or passengers.

e.g.: He was waiting at the pier for a boat to arrive.

20. Quote - Cote

(a) Quote - Repeat in speech or writing words previously said or written by another person.

e.g.: He always quotes lines from Sanskrit verses.

(b) Cote - Shed, shelter or enclosure for domestic animals or birds.

e.g.: All the chicken were enclosed in a cote during night.

21. Right - Rite

(a) Right - Morally good conduct, action; required by law or duty.

e.g.: His reaction to the issue was quite right.

(b) Rite - Religious or some other solemn ceremony.

e.g.: He took leave for performing certain religious rites at home.

22. Sew - Sow

(a) Sew - Make stitches in cloth etc. with a needle and thread.

e.g.: She has been sewing her gown since morning.

(b) Sow - Put or scatter seed in or on the ground.

e.g.: The farmer sowed cabbage seeds in the farm.

23. Sight - Site

(a) Sight - Ability to see, vision.

e.g.: Due to old age his sight has become a little weak.

(b) Site - Place where a building, town etc. was, is or will be situated;

e.g.: The site on the hill top was selected for putting up windmills.

24. Tail - Tale

(a) Tail - Movable part at the end of the body of a bird an animal, a fish or reptile.

e.g.: Dogs wag their tails when they are excited and happy.

(b) Tale - Narrative or story.

e.g.: Children love fairy tales.

25. Urn - Earn

(a) Urn - A tall vase usually with a stem and a base; Large metal container with a tap for making or serving tea, coffee etc. in hotels.

e.g.: The urn contained a large quantity of coffee.

(b) Earn - Get money etc. by working.

e.g.: He earns his wages once in a fortnight.

Same word used as different parts of speech

Sometimes a derivative is got from a root word or headword without adding a prefix or suffix. However, such derivatives carry different meanings and belong to a part of speech other than the one which the original word belongs to. Examples of some such zero derivatives are given below.

1. Advance (n): Forward movement.

e.g.: The advance of Iraqi forces into Kuwait was halted by the US allies.

Advance (adj): Going before others, done or provided in advance.

e.g.: He was given an advance notice about his dismissal.

2. **Advocate (v):** Speak publicly in favour of something, recommend.
   e.g.: He always advocated the Presidential type of government.
   Advocate (n): Person who supports or speaks in favour of a cause or policy.
   e.g.: Nehru was a lifelong advocate of 'Panchsheel'.
3. **Bale (n):** Large bundle of paper, straw, goods etc. pressed together and tied with rope or wire for transport etc.
   e.g.: The bales of hay in the farm caught fire.
   Bale (v): Make something into or pack in bales.
   e.g.: The attendant was busy baling cut-pieces of cloth.
4. **Balloon (n):** Brightly coloured rubber bag that is filled with air used as a child's toy or a decoration.
   e.g.: A number of different coloured balloons were brought home for celebrating Ram's birthday.
   Balloon (v): Swell out like a balloon.
   e.g.: His loose attire ballooned in the heavy wind.
5. **Cable (n):** Set of insulated wires for carrying messages by telegraph, or electricity overhead, message sent abroad.
   e.g.: The company used copper cables on the new lines.
   Cable (v): Send a cable to somebody abroad.
   e.g.: The news of his success was cabled to him.
6. **Camp (n):** Place where people live temporarily in tents or huts.
   e.g.: The party pitched its camp by the river bank.
   Camp (v): Put up a tent or tents.
   e.g.: Our plan was to camp in the village for a night.
7. **Ditch (n):** A narrow channel, dug at the edge of a field, road etc. esp. to hold or carry water.
   e.g.: The ditch proved to be dangerous for pedestrians.
   Ditch (v): Land an aircraft in the sea in an emergency.
   e.g.: The pilot was forced to ditch the aircraft in the Arabian sea due to engine failure.
8. **Documentary (adj):** Consisting of documents (documentary evidence, proof, sources).
   e.g.: The file had documentary importance in the court.
   Documentary (n): A documentary film, or radio or T. V. programme.
   e.g.: His documentary about AIDS received the critics' award.
9. **Drizzle (v):** To rain in many fine drops.
   e.g.: It has been drizzling all day.
   Drizzle (n): Fine misty rain.
   e.g.: There was a cold drizzle right since morning.
10. **Earth (n):** This planet, the planet on which we live; soil.
    e.g.: The Earth rotates around the Sun.
    Earth (v): Connect an electrical appliance etc. with the ground.
    e.g.: The connection of the electric iron was earthed.
11. **Eccentric (adj):** Unusual, peculiar, not normal.
    e.g.: The eccentric person next door is a nuisance to all.
    Eccentric (n): Abnormal person.
    e.g.: That group is overflowing with eccentrics.
12. **Flight (n):** Process or action of flying in the air.
    e.g.: The aeroplane was shot down in flight.
    Flight (v): Give a certain pitch to a ball through the air while bowling so as to deceive the batsman.
    e.g.: He made a well flighted delivery of his last ball of the over.

13. Flirt (v): To behave in a romantic way but without serious intentions.

e.g.: He always tries to flirt with girls in his office.

Flirt (n): A person who flirts with many people.

e.g.: He is known as a terrible flirt in the whole office.

14. General (adj): Affecting all or most people, places or things.

e.g.: The proposal received a general approval of the people.

General (n): An army officer of a very high rank.

e.g.: He retired as a general from the Indian Army.

15. Gesture (n): Expressive movement particularly of a hand or head.

e.g.: He tried to communicate his desire through gestures.

Gesture (v): Make expressive movements; convey something by gestures.

e.g.: She gestured her happiness on the proposal of her friend.

16. Handicap (n): A thing that makes progress difficult; physical or mental disability.

e.g.: Blindness is beyond doubt a serious handicap.

Handicap (v): To give or be a disadvantage.

e.g.: He felt handicapped due to his scanty income.

17. Handle (n): A part of a tool cup, bucket, drawer etc. by which it can be held, carried or controlled.

e.g.: His vintage car cannot start without using a handle.

Handle (v): Deal with, manage or control (people, a situation, a machine etc.)

e.g.: The foreman knew how to handle the workers under him.

18. Independent (adj): Not dependent (on other people or thing); not controlled (by other people or things).

e.g.: India was once a British colony but now it is an independent country.

Independent (n): MP, candidate etc. who does not belong to any political party.

e.g.: He filed his nomination to contest the election as an independent.

19. Individual (adj): Single, separate.

e.g.: Each individual debt must be repaid.

Individual (n): A single human being.

e.g.: The rights of an individual are controlled in a dictatorship.

20. Jumble (v): Mix things in a confused way.

e.g.: Everything in the hall was jumbled up.

Jumble (n): Untidy group of things; muddle.

e.g.: There was a jumble of books on the table.

21. Key (n): Metal instrument for locking or unlocking.

e.g.: He lost the key of his house in the park.

Key (adj): Very important or essential.

e.g.: He is the key person in the company.

22. Lament (v): Feel or express great sorrow or regret.

e.g.: They lamented the loss of their popular leader.

Lament (n): Strong expression of grief.

e.g.: People were greatly moved by her laments on the loss of her only child.

23. Large (adj): Of considerable size, extent or capacity.

e.g.: A large country needs large sources of production.

Large (n): At full length, thoroughly and in great details.

e.g.: The matter was discussed at large in the report.

24. Minor (adj): Smaller, less serious, less important.

e.g.: The house requires, minor repairs.

Minor (n): A person under the age of full legal responsibility.

e.g.: He is not eligible to vote as he is a minor.

25. Misuse (v): Use in the wrong way or for the wrong purpose.

e.g.: Nobody should misuse public funds.

Misuse (n): You should always avoid misuse of power in public life.

26. Nail (n): Layer of thorny substance over the outer tip of a finger or toe; a small thin piece of metal.

e.g.: She defended herself with her strong pointed nails when the stranger tried to molest her.

Nail (v): Catch or arrest.

e.g.: The police nailed the culprit within 24 hours of his committing the offence.

27. National (adj): Of a nation; characteristic of a whole nation.

e.g.: Local and national newspapers always differ.

National (n): A citizen of a particular nation.

e.g.: Only Indian nationals are allowed to contest elections in India.

28. Objective (adj): Unbiased; fair; not influenced by personal feelings or opinions.

e.g.: The magistrate's decision in the case was perfectly objective.

Objective (n): Thing aimed at or wished for; purpose; aim.

e.g.: To become a lawyer was his sole objective.

29. Patent (n): Official document giving the holder the sole right to make, use or sell an invention and preventing others from imitating it.

e.g.: He has already applied to the government for a patent of the gadget he invented very recently.

Patent (v): Obtain a patent for an invention or process.

e.g.: The government has already patented his new gadget. He patented for his invention.

30. Poison (n): Substance causing death or harm if absorbed by a living thing (animal or plant).

e.g.: The actor committed suicide by consuming poison.

Poison (v): To give poison to a living thing; kill or harm something/somebody with poison.

e.g.: She poisoned her husband for obtaining his wealth.

31. Quarter (n): Each of four equal or corresponding parts of something.

e.g.: Only one quarter of the theatre was empty.

Quarter (v): Provide somebody with lodging.

e.g.: We were quartered in the government rest houses on payment.

32. Refill (v): To fill again.

e.g.: He had to refill the petrol tank before starting the journey.

Refill (n): New material used to refill a container.

e.g.: Buy two refills for my ballpen while coming home.

33. Reform (v): Become or make better by removing faults, errors etc.

e.g.: Many social organisations desire to reform the structure of our Constitution. **Reform (n):** Reforming or being reformed.

e.g.: They plan to bring about social reforms quickly.

34. Screen (n): Blank surface on which pictures or films are projected.

e.g.: He writes for both the big and small screen.

Screen (v): Show a film, scene on a screen.

e.g.: The film has been screened in the cinema and on TV.

35. Sheer (adj): Complete, thorough, utter.

e.g.: His efforts were a sheer waste of time.

Sheer (adv): Straight up or down.

e.g.: The sheer drop at the edge of the cliff scared me.

36. **Tape (n):** Magnetic tape on which recording is made.
e.g.: The police could not find the tape they needed.
Tape (v): Record on magnetic tape.
e.g.: The concert was taped from the programme shown on TV.

37. **Thrust (v):** Push violently or suddenly.
e.g.: He thrust the dagger into the body of the thief.
Thrust (n): Act or movement of thrusting.
e.g.: The soldier was killed by a bayonet thrust.

38. **Unlike (adj):** Different, dissimilar.
e.g.: They are quite unlike each other.
Unlike (prep): Not like; different from.
e.g.: His performance was quite unlike the earlier one.

39. **Venture (n):** Project or undertaking involving risk of failure.
e.g.: His friend embarked on a doubtful venture.
Venture (v): Dare to go to something dangerous.
e.g.: The mouse never ventured far from its hole.

40. **Yelp (n):** A short, sharp cry of pain, anger, excitement etc.
e.g.: The dog gave a yelp when the boy trodded on its paw.
Yelp (v): To give a short, sharp cry of pain, anger etc.
e.g.: The dog yelped when the boy trodded on its paw.

Synonyms and Antonyms

As per the 'Oxford Advanced Learner's Dictionary', a synonym is 'a word (or phrase) with the same meaning as another in the same language, though perhaps with a different style, grammar or technical use'; and an antonym is a word that is opposite in meaning in the same language. Some of the commonly used words with their synonyms and antonyms are given below - Students are advised to carefully go through them and practise to use them frequently in their own sentences. This will positively help them to increase their word power.

Word	Synonym	Antonym
Abnormal	Uncommon, Rare	Common, Normal
Absurd	Ridiculous, Foolish	Rational, Sensible
Abundant	Ample, Bountiful	Inadequate, Scarce
Accept	Receive, Approve	Refuse, Reject
Accuse	Blame, Censure	Admire, Appreciate
Admire	Respect, Appreciate	Despise, Condemn
Advantage	Benefit, Boon	Disadvantage, Harm
Agree	Accept, Consent	Disagree, Decline, Deny
Appreciate	Admire, Esteem	Despise, Depreciate
Attack	Assault, Charge, Encounter	Defend, Guard, Fortify
Backward	Retarded, Sluggish	Forward, Progressive
Bad	Evil, Corrupt, Wicked	Good, Chaste, Decent
Beautiful	Charming, Attractive	Ugly, Repulsive, Graceless
Benefit	Advantage, Boon	Loss, Damage
Best	Excellent, Highest	Worst, Meanest
Blunt	Dull, Flat, Rude	Sharp, Pointed, Polite
Borrow	Obtain, Receive, Take	Lend, Give, Deliver
Brave	Courageous, Gallant	Weak, Coward
Bright	Shining, Intelligent	Dull, Ignorant
Buy	Purchase, Procure	Sell, Dispose

Cheap	Worthless, Inferior, Low	Worthy, Superior, Costly
Cheerful	Delightful, Gay, Happy	Gloomy, Dull, Unhappy
Clever	Sharp, Talented, Skilful	Dull, Stupid, Fool
Competent	Able, Efficient, Skilful	Incapable, Incompetent
Condemn	Accuse, Blame, Censure	Appreciate, Praise
Constant	Firm, Perpetual, Stable	Flexible, Unsteady, Variable
Correct	Right, Accurate, True	Wrong, Incorrect, False
Courage	Boldness, Valour, Bravery	Cowardice, Fear, Timidity
Cruel	Brute, Callous, Unkind	Merciful, Kind, Humane
Curse	Bane, Malediction	Boon, Benediction
Damn	Condemn, Denounce, Curse	Admire, Praise, Bless
Dark	Black, Gloomy, Ebony	White, Bright, Clear
Decrease	Reduce, Lessen, Diminish	Increase, Augment, Enhance
Defeat	Failure, Retreat, Collapse	Victory, Triumph, Success
Danger	Peril, Risk, Hazard	Safety, Security
Deny	Decline, Disown, Refute	Accept, Admit, Confirm
Dirty	Filthy, Impure, Stained	Clean, Pure, Spotless
Dismal	Cheerless, Gloomy, Sad	Cheerful, Jolly, Happy
Divine	Heavenly, Holy, Godly	Earthly, Unholy, Satanic
Dull	Dim, Dry, Stupid, Sad	Bright, Clever, Jovial
Easy	Simply, Convenient	Difficult, Tough
Effective	Active, Operative, Powerful	Ineffective, Futile, Incapable
Efficient	Competent, Capable	Deficient, Incapable, Incompetent
Eminent	Famous, Distinguished	Anonymous, Obscure
Encourage	Inspire, Persuade, Urge	Discourage, Depress, Dissuade
Enemy	Foe, Opponent, Antagonist	Friend, Fellow, Associate
Entire	Full, Complete, Whole	Incomplete, Partial
Essential	Basic, Necessary, Primary	Subsidiary, Secondary, Auxiliary
Eternal	Everlasting, Infinite, Permanent	Transitory, Perishable, Temporary
Evil	Bad, Corrupt, Wicked	Good, Honest, Noble, Moral
Fail	Drop, Miss, Disappoint	Pass, Gain, Attain, Succeed
Fair	Bright, Clean, Proper	Brunette, Dirty, Unfair
False	Deceptive, Illusory, Incorrect	Correct, Real, True
Familiar	Intimate, Accustomed	Strange, Unaccustomed
Famous	Distinguished, Eminent	Notorious, Anonymous
Fast	Agile, Quick, Rapid, Speedy	Dull, Inert, Slack, Slow
Fat	Fleshy, Bulky, Plump	Slim, Thin, Slender, Lean
Fierce	Savage, Wild, Ferocious	Docile, Gentle, Tame
Forward	Progressive, Advanced	Backward, Primitive
Friend	Associate, Ally, Companion	Foe, Enemy, Rival
Happy	Blessed, Cheerful,	Unhappy, Sad, Disappointed
Hard	Stiff, Rigid, Solid, Tough	Soft, Flexible, Easy, Mellow
Harmful	Injurious, Mischievous	Harmless, Innocent
Hasten	Accelerate, Quicken	Delay, Retard, Hinder
Hate	Abhor, Despise, Dislike	Adore, Love, Like
Heavy	Massive, Weighty	Light, Petty, Trivial

High	Great, Lofty, Tall	Low, Small, Short
Holy	Divine, Godly, Pious	Unholy, Impious, Satanic
Honest	Truthful, Sincere, Frank	Deceitful, Dishonest, Tricky
Honour	Repute, Glory, Respect	Disrespect, Despise, Infamy
Hope	Desire, Expectation, Trust	Despair, Disappointment
Ignorance	Stupidity, Folly, Absurdity	Wisdom, Knowledge, Cleverness
Illegal	Unlawful, Illicit, Forbidden	Legal, Lawful, Legitimate
Important	Significant, Prominent	Unimportant, Trivial
Inactive	Idle, Indolent, Lazy	Alert, Active, Industrious
Incomplete	Unfinished, Deficient, Imperfect	Complete, Finished, Perfect
Indolent	Idle, Lethargic, Lazy	Active, Dynamic, Energetic

Word	Synonym	Antonym
Inferior	Lower, Subordinate, Mean	Superior, Higher, Greater
Injustice	Unfairness, Partiality	Justice, Fairness, Impartiality
Innocent	Honest, Simple, Artless	Cunning, Clever, Tricky
Insult	Abuse, Humiliate, Offend	Admire, Honour, Respect
Intelligent	Bright, Brilliant, Clever	Dull, Foolish, Unintelligent
Irregular	Uneven, Eccentric, Inconsistent	Regular, Systematic, Orderly
Jealousy	Envy, Suspicion, Doubt	Love, Appreciation, Belief
Joy	Ecstasy, Exultation, Pleasure	Grief, Pain, Sorrow, Anguish
Judicious	Wise, Prudent, Careful	Indiscreet, Irrational, Careless
Justice	Fairness, Honesty	Unfairness, Dishonesty
Keen	Eager, Acute, Enthusiastic	Dull, Flat, Lazy
Keep	Detain, Hold, Retain	Abandon, Reject, Discard
Kill	Slay, Murder, Assassinate	Protect, Produce, Restore
Kind	Compassionate, Generous	Cruel, Unkind, Inhuman
Knowledge	Wisdom, Learning	Ignorance, Illiteracy
Known	Familiar, Famous, Popular	Unknown, Ordinary, Simple
Lament	Weep, Mourn, Grieve	Rejoice, Enjoy, Celebrate
Last	Final, Ultimate, Extreme	First, Initial, Foremost
Lavish	Prodigal, Profuse, Excessive	Miserly, Economical, Thrifty
Lawful	Legal, Legitimate, Regular	Unlawful, Illegal, Irregular

Word	Synonym	Antonym
Lazy	Idle, Indolent, Slack	Active, Alter, Agile, Quick
Long	Lengthy, Prolonged, Extended	Short, Brief, Compact
Lose	Waste, Fail, Miss	Gain, Obtain, Recover
Loss	Damage, Detriment	Profit, Gain, Earning
Loud	Noisy, Resonant, Blaring	Low, Quiet, Soft
Love	Affection, Fondness, Liking	Hatred, Dislike, Antipathy
Lovely	Attractive, Beautiful, Cute	Ugly, Repulsive, Hideous
Loyal	Faithful, Obedient, Sincere	Disloyal, Treacherous, Unfaithful
Lucky	Fortunate, Successful	Unlucky, Unfortunate
Luxury	Abundance, Plentiful	Scarcity, Pausity, Wanting
Mad	Crazy, Insane, Lunatic	Wise, Sane, Rational
Malice	Hatred, Enmity, Resentment	Love, Passion, Sympathy
Master	Ruler, Chief, Leader	Servant, Slave, Subordinate
Mature	Ripe, Seasoned, Perfect	Raw, Unripe, Immature

Meagre	Inadequate, Scarce, Deficient	Abounding, Ample, Plentiful
Melancholy	Cheerless, Gloomy, Dismal	Cheerful, Gay, Joyful
Merciful	Kind, Gentle, Tender	Cruel, Harsh, Tyrannical
Miserable	Unhappy, Dismal, Poor	Happy, Cheerful, Prosperous
Modern	New, Recent, Progressive	Old, Ancient, Primitive
Natural	Normal, Usual, Regular	Abnormal, Unnatural, Superficial
Necessary	Binding, Essential, Compulsory	Optional, Voluntary, Unnecessary
Neglect	Carelessness, Failure, Omission	Attention, Vigilance, Carefulness
Nice	Fine, Lovely, Pleasing	Repulsive, Ugly, Deformed
Noble	Dignified, Honourable	Ignoble, Mean, Uncultured
Notorious	Scandalous, Ignoble	Famous, Reputable, Virtuous
Numerous	Many, Various, Manifold	Few, Scarce, Scanty
Obnoxious	Odious, Offensive, Perverse	Pleasant, Enjoyable, Nice
Obscene	Dirty, Lewd, Filthy	Pure, Virtuous, Clean
Obstinate	Stubborn, Adamant, Resolute	Amenable, Yielding, Accommodative
		Frequent, Regular, Repeated
Occasional	Sporadic, Casual, Incidental	Normal, Customary, Even
Odd	Abnormal, Anomalous, Singular	Defend, Please, Flatter
Offend	Hurt, Displease, Provoke	Rarely, Hardly, Seldom
Often	Usually, Repeatedly, Generally	Young, Modern, Immature
Old	Aged, Primitive, Elderly	Auspicious, Propitious
Ominous	Inauspicious, Menacing	Include, Insert, Enroll
Omit	Delete, Drop, Eliminate	
Optimistic	Hopeful, Promising, Inspiring	Hopeless, Disheartening, Pessimistic
		Abnormal, Exceptional, Unusual
Ordinary	Normal, Regular, Usual	Unimportant, Ordinary, Normal
Outstanding	Eminent, Important, Excellent	Joy, Pleasure, Happiness
Pain	Agony, Anguish, Pang	Dispassionate, Cool, Balanced
Passionate	Excited, Hasty, Impulsive	Amusing, Comic, Farcical
Pathetic	Pitiable, Sad, Touching	War, Turmoil, Upheaval
Peace	Tranquility, Amity, Silence	Impious, Irreligious, Sinful
Pious	Religious, Devout, Godly	

Word	**Synonym**	**Antonym**
Pleasant	Attractive, Nice, Amiable	Repulsive, Unpleasant, Harsh
Poison	Venom, Toxin, Virus	Nectar, Honey, Antidote, Sweet
Polite	Courteous, Mannerly, Cordial	Impolite, Offensive, Rude
Positive	Affirmative, Categorical, Sure	Negative, Vague, Dubious
Powerful	Energetic, Strong, Mighty	Powerless, Weak, Dull, Frail
Precious	Valuable, Expensive, Esteemed	Worthless, Cheap, Contemptible
Profit	Advantage, Privilege, Gain	Loss, Damage, Deprivation
Prohibit	Debar, Forbid, Stop, Hinder	Permit, Authorise, Sanction
Protect	Defend, Guard, Shield	Expose, Disclose, Reveal
Quantity	Bulk, Measure, Amount	Quality, Deficit, Shortage
Quarrel	Dispute, Fight, Feud	Peace, Harmony, Amity
Question	Inquiry, Interrogation, Query	Answer, Reply, Solution
Questionable	Doubtful, Debatable, Uncertain	Assured, Certain, Unquestionable
Quick	Alert, Swift, Fast, Rapid	Slow, Slack, Lethargic, Sluggish

Rapid	Fast, Quick, Speedy, Swift	Slow, Sluggish, Indolent
Rare	Unique, Sparse, Exclusive	Common, Usual, Ordinary
Rational	Judicious, Sensible, Normal	Irrational, Erratic, Insane
Reasonable	Fair, Judicious, Sensible	Illogical, Absurd, Ridiculous
Religious	Holy, Devout, Pious, Saintly	Unholy, Sinful, Impious, Irreligious
		Forget, Ignore, Overlook
Remember	Recollect, Recall, Retain	Contemptible, Disreputable,
Respectable	Honourable, Reputable, Worthy	Unworthy
		Poor, Scarce, Inadequate
Rich	Wealthy, Abundant, Profuse	Wrong, False, Unfair, Bad
Right	Correct, Exact, Proper, True	Soft, Smooth, Decent
Rough	Hard, Coarse, Crude	Courteous, Refined, Mannerly
Rude	Impolite, Rough, Mannerless	Urban, Civic, Metropolitan
Rural	Rustic, Agrarian, Pastoral	Kind, Compassionate, Pitiful
Ruthless	Merciless, Cruel, Pitiless	Blasphemous, Unholy, Sinful
Sacred	Holy, Religious, Divine, Pious	Exposed, Unsafe
Safe	Protected, Secure	Prosperity, Abundance, Enormity
Scarcity	Dearth, Paucity, Shortage	Generous, Liberal, Charitable
Selfish	Greedy, Mean	Insensitive, Blunt, Robust
Sensitive	Tender, Delicate, Susceptible	Deep, Profound, Wise
Shallow	Superficial, Simple, Trivial	Blunt, Dull, Flat
Sharp	Edged, Alert, Keen	Mild, Flexible, Lenient
Strict	Austere, Firm, Stern	Inferior, Unimportant, Mean
Superior	Excellent, Higher, Finer	Simple, Negligent, Stupid
Tactful	Diplomatic, Skilful, Prudent	

Word	**Synonym**	**Antonym**
Temporary	Interim, Ephemeral	Permanent, Imperishable
Tender	Delicate, Soft, Fragile	Hard, Firm, Rigid
Tentative	Provisional, Experimental	Final, Definite, Firm
Thick	Dense, Solid, Condense	Thin, Slim, Slender
Thin	Delicate, Slim, Slender	Thick, Solid, Dense
Tough	Sturdy, Tenacious, Firm	Flexible, Soft, Tender, Weak
Transparent	Clear, Distinct, Glassy	Opaque, Hazy, Unclear
Treacherous	Disloyal, Deceitful, Unfaithful	Loyal, Faithful, Devoted
Ugly	Repulsive, Hideous, Vicious	Pretty, Beautiful, Charming
Unanimity	Agreement, Harmony, Unison	Conflict, Discord, Differences
Uncertain	Indefinite, Doubtful, Dubious	Definite, Certain, Sure
Uneasy	Disturbed, Restless, Worried	Steady, Quiet, Restful
Unfair	Unjust, Wrongful, Biased	Just, Fair, Impartial
Unfaithful	Disloyal, Deceitful, Treacherous	Faithful, Reliable, Loyal
Unfit	Improper, Incompetent, Unsuitable	Fit, Capable, Competent, Suitable
Unfortunate	Unlucky, Ruined, Unfavoured	Blessed, Fortunate, Lucky
Unholy	Impious, Sinful, Irreligious	Holy, Pious, Religious
Unique	Uncommon, Matchless, Rare	Normal, Ordinary, Common
Unity	Oneness, Harmony, Union	Diversity, Difference, Variance
Vacant	Empty, Unfilled, Void	Occupied, Packed, Inhabited
Valiant	Brave, Gallant, Heroic	Timid, Cowardly, Frightened
Vanity	Pride, Arrogance, Egotism	Modesty, Humbleness, Humility
Vast	Huge, Enormous, Boundless	Small, Limited, Scanty
Vertical	Erect, Perpendicular, Upright	Horizontal, Flat, Prone

Vice	Sin, Immorality, Corruption	Virtue, Morality, Honesty
Victory	Conquest, Success, Triumph	Defeat, Failure, Downfall
Vigilant	Alert, Cautious, Watchful	Careless, Negligent, Slack
Violent	Fierce, Furious, Passionate	Unexcited, Quite, Cool
Virtue	Morality, Purity, Holiness	Vice, Wickedness, Sin
War	Battle, Combat, Conflict	Peace, Harmony, Amity
Warmth	Cordiality, Earnestness	Apathy, Indifference, Coolness
Wary	Alert, Cautious, Prudent	Careless, Negligent, Rash
Watchful	Attentive, Alert, Careful	Careless, Negligent, Uncautious
Weak	Feeble, Fragile, Poor	Strong, Sturdy, Powerful
Whole	Complete, Entire, Total	Part, Fractional, Incomplete
Wicked	Evil, Sinister, Villainous	Noble, Virtuous, Chaste
Wisdom	Intelligence, Cleverness	Absurdity, Stupidity, Folly
Yearn	Crave, Desire, Languish	Contented, Satisfied, Cheerful
Young	Youthful, Fresh, New	Old, Mature, Ripe
Zeal	Enthusiasm, Passion, Fervour	Apathy, Reluctance, Idleness
Zenith	Climax, Apex, Crown, Culmination	Base, Bottom, Foundation, Minimum

One Word Substitution

This topic is a double-edged weapon in the hands of students since it helps to elaborate ideas, concepts, thoughts for writing as well as to bring precision when required to write or express things in short. Although it is impossible to list out all one-words, effort has been made to compile some important and commonly used groups of words and one-word substitutes for them in this section. Students are advised to study them and try to remember them by using them in practice whenever and wherever possible. This will also be one of the measures for enriching one's vocabulary.

1. People inhabiting a land from a very early period before the colonies were established. **Aboriginals**
2. Something that exists in thought or idea but does not have a physical or practical existence. **Abstract**
3. A person whose profession is to maintain or inspect financial accounts. **Accountant**
4. A word formed from the initial letters of a group of words. **Acronym**
5. Concerned with beauty and the appreciation of beauty. **Aesthetic**
6. Instance of unprovoked attacking or hostility by one country against another. **Aggression**
7. Medical conditions that produce an unfavourable reaction to certain foods, pollens, insect bites, etc. **Allergy**
8. A diplomat sent from one country to another, either as a permanent representative or on a special mission. **Ambassador**
9. Supply of bullets, bombs, grenades etc. fired from weapons or thrown. **Ammunition**
10. An animal able to live both on land and in water. **Amphibian**
11. Condition of lawlessness i.e. the complete absence of government law or control of society. **Anarchy**
12. Scientific study of the structure of animal bodies. **Anatomy**
13. With a name that is not known or not made public. **Anonymous**
14. Study of mankind esp. of its origin, development, customs and beliefs. **Anthropology**
15. A word that is opposite in meaning to another. **Antonym**
16. An artificial pond or glass tank where live fish and other water creatures and plants are kept. **Aquarium**

17. Study of ancient civilizations by scientific analysis of physical remains found in the ground. **Archaeology**
18. Art and science of designing and constructing buildings. **Architecture**
19. Part of a country's military forces that is organised and equipped for fighting on land. **Army**
20. Study of the position of stars and movements of planets in the belief that they influence human affairs. **Astrology**
21. Person who travels in a space-craft. **Astronaut**
22. Scientific study of the sun, moon, stars planets etc. **Astronomy**
23. Story of a person's life written by that person himself. **Autobiography**
24. Microscopic organisms existing in air, water, soil, living and dead creatures, plants, causing diseases. **Bacteria**
25. Style of dancing used to tell a story in a dramatic performance with music but without speech or singing. **Ballet**
26. A strip of material used for binding around a wound or an injury. **Bandage**
27. An instrument for measuring atmospheric pressure used especially for forecasting weather. **Barometer**
28. A lawyer who has acquired the degree of Bar-at-Law and has right to speak and argue as an advocate in higher law courts. **Barrister**
29. Stretch of sand or pebbles along the edge of the sea or a lake; shore between high and low water mark. **Beach**
30. Person who gives money or other help to a school, hospital, charitable institution etc. **Benefactor**
31. Being or wishing to be kind, friendly and helpful to or towards somebody. **Benevolent**
32. List of books or articles about a particular subject or by a particular author. **Bibliography**
33. System or crime of marrying a person when still legally married to someone else. **Bigamy**
34. Having two sides; affecting or involving two parties, countries etc. **Bilateral**
35. Able to speak two languages equally well. **Bilingual**
36. Scientific study of plants and their structure. **Botany**
37. A structure of wood, iron, concrete etc. providing a way across a river, road, railway etc. **Bridge**
38. Meal at which guests serve themselves from a number of dishes. **Buffet**
39. Powerful tractor with a broad steel blade in front used for moving earth or clearing ground. **Bulldozer**
40. System of government through departments managed by state officials and not by elected representatives. **Bureaucracy**
41. Piece of furniture with drawers or shelves for storing or displaying things; Group of the most important government ministers responsible for government administration and policies. **Cabinet**
42. Person whose job is to receive and pay out money at a bank, shop, hotel counter etc. **Cashier**
43. Small sealed case containing a reel of film or magnetic tape. **Cassette**
44. Book or a booklet containing a complete list of items usually in a special order and with a description of each item. **Catalogue**
45. Place serving food and drink in a factory, an office, a school, etc. **Canteen**

46. Economic system in which a country's trade and industry are controlled by private owners and not by state for profit. **Capitalism**
47. Branch of medicine concerned with the heart and its diseases. **Cardiology**
48. Branch of mathematics that deals with the problems involving rates of variation. **Calculus**
49. Strong smelling white substance used in medicines and mothballs and making plastics. **Camphor**
50. Official counting of a country's population or of other classes of things for statistical purposes. **Census**
51. Official printed or written statement that may be used as a proof or evidence of certain facts. **Certificate**
52. Ornamental hanging light with branches for several bulbs or candles. **Chandelier**
53. Fatty substance found in animal fluids and tissue thought to cause hardening of arteries. **Cholesterol**
54. Regular pattern of weather conditions such as temperature, rainfall, winds etc. of a particular region. **Climate**
55. Social and economic system in which there is no private ownership and the means of production belong to all members of society. **Communism**
56. Fellow member of a trade union or of a socialist or communist political party etc. **Comrade**
57. Person who has done something wrong. **Culprit**
58. Person who believes that people do not do things for good, sincere or noble reasons but only for their own advantage. **Cynic**
59. Facts or information used in deciding or discussing something. **Data**
60. Academic title given by a university or college to one who has passed an examination, written a thesis etc. **Degree**
61. System of government by the people through the representatives they elect. **Democracy**
62. Ruler with unlimited powers especially a cruel and oppressive one. **Despot**
63. Person in the diplomatic service e.g. ambassador etc.

    A person clever at dealing with people. **Diplomat**
64. Illness of the body, of the mind or of plants caused by infection or internal disorder. **Disease**
65. Legal ending of a marriage, separation. **Divorce**
66. Weekly payment made by the state to unemployed people. **Dole**
67. Model of the human figure used for displaying or fitting clothes. **Dummy**
68. Device for converting steam-power, water-power etc. into electricity. **Dynamo**
69. Blocking of the light of the sun or moon due to the specific position of the sun, the moon and the earth. **Eclipse**
70. Scientific study of the relation of plants and living creatures to each other and to their surroundings. **Ecology**
71. Science or principles of the production, distribution and consumption of goods especially with reference to cost. **Economics**
72. Person showing or holding a belief in equal rights, benefits and opportunities for everybody. **Eglitarian**
73. Minute particle of matter with negative electric charge found in all atoms. **Electron**
74. Poem or song expressing sorrow for the dead. **Elegy**
75. Art or style of speaking clearly and effectively especially in public. **Elocution**

76. Quick spreading of a disease among many people in the same place for a time. **Epidemic**
77. Information that proves something or gives a reason for believing something. **Evidence**
78. Collection of things such as works of art, industrial or commercial goods etc. shown publicly for advertisement. **Exhibition**
79. Departure of many people at one time. **Exodus**
80. Introduced from another country; not native. **Exotic**
81. Person with special knowledge, skill, or training in a particular field. **Expert**
82. Spoken or done without previous thought or preparation. **Extempore**
83. Person more interested in what is happening around him than in his own thoughts and emotions. **Extrovert**
84. Department or group of departments in a university etc. **Faculty**
85. Extreme right-wing dictatorial political system or views (as originally seen in Italy between 1922 and 1943) **Fascism**
86. Related to government money or public money. **Fiscal**
87. All the plants of a particular area or period of time. **Flora**
88. Remains of a pre-historic animal or plant preserved by being buried in earth and now hardened like rock. **Fossil**
89. Border between two countries. **Frontier**
90. Type of sugar found in fruit-juice, honey etc. **Fructose**
91. Ceremony of burning or burying a dead body. **Funeral**
92. Substance that kills fungus. **Fungicide**
93. Any of various types of plants without leaves, flowers or croon colouring matter. **Fungus**
94. Tube or pipe that is wide at the top and narrow at the bottom used for pouring liquids, powders etc. into a small opening. **Funnel**
95. Enclosed space or chamber for heating metal glass etc. to a very high temperature. **Furnace**
96. Scientific study of the ways in which characteristics are passed from parents to their off springs. **Genetics**
97. Person who has exceptionally great mental or creative ability. **Genius**
98. Substance used for killing germs. **Germicide**
99. Person who holds esp. the first or bachelor's degree from a university. **Graduate**
100. Soft black substance used in making lead pencils, in lubrication and for slowing down neutrons in atomic reactors. **Graphite**
101. Any thick semi-solid oily substance used for lubrication. **Grease**
102. Room in a theatre, TV studio etc. where the performers can relax or do their make-up. **Green-room**
103. Protective screen of metal bars or wires. **Grill**
104. Shopkeeper who sells food in packets, fins, or bottles and general small household goods. **Grocer**
105. Person who shows others the way esp. a person employed to point-out interesting sights on a journey or places of tourist. **Guide**
106. Exercises performed to develop the muscles or fitness or to demonstrate agility. **Gymnastics**
107. Scientific study and treatment of diseases and disorders of the female reproductive system. **Gynaecology**
108. Member of a wandering group of people who live in caravans. **Gypsy**
109. Substance carrying oxygen in the red blood-cells of vertebrates. **Haemoglobin**
110. Person with the legal rights to receive property when the owner dies. **Heir**
111. Animal that feeds on plants. **Herbivore**
112. Person who has withdrawn from society and lives completely alone. **Hermit**
113. Geometric figure with six sides and angles. **Hexagon**

114. Large scale destruction of life and property especially due to fire. **Holocaust**
115. Things said or done to show great respect to a person or his qualities. **Homage**
116. The line at which the earth and sky appear to meet. **Horizon**
117. The scientific study of art and science of growing flowers, fruit and vegetables. **Horticulture**
118. Friendly and generous reception and entertainment of guests or strangers in one's own home. **Hospitality**
119. Person/s held as captive and threatened to be killed or harm to be done unless certain demands are met. **Hostage**
120. Animal or plant that has parents of different species or varieties. **Hybrid**
121. Study and practice of cleanliness as way of maintaining good health and preventing disease. **Hygiene**
122. Huge mass of ice floating in the sea. **Iceberg**
123. Phrase or sentence whose meaning is not clear from the meaning of its individual words in isolation but must be learnt as a whole unit. **Idiom**
124. Small dome-shaped house built by Eskimos from blocks of hard snow as a temporary shelter. **Igloo**
125. Not following accepted standards of morality. **Immoral**
126. That cannot be harmed by a disease or illness. **Immune**
127. List of names or topics referred to in a book etc. usually arranged at the end in alphabetical order. **Index**
128. Type of small animal having six legs, no backbone and a body divided into three parts viz. head, thorax and abdomen. **Insect**
129. Guarantee of compensation for loss, damage, sickness, death etc. in return of regular payment of premium. **Insurance**
130. An organisation through which national police forces of different countries can cooperate with one other. **Interpol**
131. Meeting at which the applicant is asked questions to find out if he is suitable for what he has applied. **Interview**
132. Enter a country or territory with armed forces in order to attack damage or occupy it.**Invade**
133. Supply water to land or crops by means of streams, reservoirs, channels, pipes etc. **Irrigate**
134. Piece of land surrounded by water on all sides. **Island**
135. Creamy-white, bone-like substance forming the tusks of elephants, walruses etc. **Ivory**
136. Device for raising heavy weights off the ground esp. vehicles, motor cars etc. **Jack**
137. Hard, green stone from which ornaments are carved. **Jade**
138. Sweet substance made by boiling fruit with sugar until it is thick, preserved in jars. **Jam**
139. Technical or specialised words used by a particular group of people and difficult for others to understand. **Jargon**
140. Short, close-fitting jacket without sleeves worn by men or women. **Jerkin**
141. Aircraft powered by a jet engine. **Jet**
142. Newspaper or periodical, especially one that is serious and deals with a specialised subject. **Journal**
143. Celebration of a special anniversary of an event. **Jubilee**
144. Japanese system of unarmed combat in which arms, feet etc. are used as weapons. **Karate**
145. Metal instrument shaped to move the bolt of a lock. **Key**
146. Either of a pair of organs in the body that remove waste products from the blood and produce urine. **Kidney**
147. Of or produced by movement. **Kinetic**
148. Clothing and personal equipments of a soldier or of a traveller. **Kit**
149. Room or building in which food/meals are cooked or prepared. **Kitchen**
150. Room or building used especially for scientific research, experiments, testing etc.**Laboratory**
151. Existing but not active, developed or visible. **Latent**
152. Machine that shapes pieces of wood, metal etc. by holding and turning them against a fixed cutting tool. **Lathe**
153. Person who is trained and qualified in legal matters esp. solicitor. **Lawyer**

154. Story handed down from the past especially one that may not be true. **Legend**
155. Type of plant like peas, beans etc. which has its seeds in pods. **Legume**
156. Communication and co-operation between units of an organisation. **Liaison**
157. False, written or printed statement that damages somebody's reputation or status. **Libel**
158. Freedom from captivity, slavery or oppressive control. **Liberty**
159. Room or building where the books for reading or borrowing are kept. **Library**
160. Liquid medicine or cosmetic for use on skin. **Lotion**
161. Oily or greasy substance that is put on or in machinery so that it moves easily. **Lubricant**
162. Meal taken at the middle of the day. **Lunch**
163. Person who writes the words of esp. popular songs. **Lyricist**
164. Apparatus with several moving parts for performing a particular task and driven by electricity, steam, gas etc. or human power. **Machine**
165. Paper covered periodical published every week or month with articles, stories etc. by various authors. **Magazine**
166. Done with or controlled by the hands. **Manual**
167. Animal dung or other material spread over or mixed with soil to make it fertile. **Manure**
168. Head of the council of a city or borough, usually elected yearly, who enjoys the status of the ' first citizen ' of the town. **Mayor**
169. A monument, ceremony etc. that reminds people of an event or person. **Memorial**
170. A heavy silver-coloured metal (chemical element) found in liquid form used in scientific instruments like thermometer, Barometer etc. **Mercury**
171. Science or the properties of metals, their uses, methods of obtaining them etc. **Metallurgy**
172. Scientific study of the earth's atmosphere and its changes used especially for forecasting weather. **Meteorology**
173. A scientific instrument useful for making a very small object appear larger. **Microscope**
174. Person who moves from one place to another to live or work. **Migrant**
175. Person in trade union or politics using force or strong pressure or supporting their use to achieve desired aims. **Militant**
176. Place where coins are made usually under state authority. **Mint**
177. Sole right to supply or trade in some commodity or service. **Monopoly**
178. Scientific study of the form and structure of animals and plants. **Morphology**
179. Person born in a place, in a country etc. and associated with that place by birth. **Native**
180. Feeling of sickness or disgust for something. **Nausea**
181. Ornament of pearls, beads etc. worn round the neck. **Necklace**
182. Not supporting or helping either side in a dispute, contest, war etc. **Neutral**
183. Poisonous oily substance found in tobacco. **Nicotine**
184. Member of a tribe that wanders from place to place looking for pasture for its animals and having no fixed home. **Nomad**
185. Person with official authority to witness the signing of legal documents and perform certain other legal functions. **Notary**
186. Book-length story in prose about either imaginary or historical characters. **Novel**
187. Central part of a living cell. **Nucleus**
188. Place where young plants and trees are grown for transplanting later and use for sale. **Nursery**
189. Substance serving as or providing nourishment especially for plants or animals. **Nutrient**
190. Fertile place with water and trees in a desert. **Oasis**
191. Thing in the way that either stops progress or makes it difficult. **Obstacle**
192. A long poem expressing noble feelings and often written to a person or thing. **Ode**
193. Smooth greasy paste for rubbing on the skin to heal injuries or roughness or as a cosmetic. **Ointment**
194. Suggesting that something bad is about to happen. **Ominous**
195. Path followed by a planet, star, moon etc. round another body. **Orbit**
196. A large group of people playing various musical instruments together. **Orchestra**
197. Of or from the countries of the East. **Oriental**

198. Scientific study of birds. **Ornithology**
199. Person whose parents are dead. **Orphan**
200. Branch of surgery that deals with the correction of bone deformities and diseases. **Orthopaedics**
201. Instrument used to draw an exact copy of a plan, map etc. on any scale. **Pantograph**
202. Loss of control of a part of the body caused due to a disease of or injury to the nerves. **Paralysis**
203. Official document issued by the government of a country to its citizen to travel abroad under its protection. **Passport**
204. Person who gives money or other support to a person, cause, activity etc. **Patron**
205. Chemical substance used to kill insects. **Pesticide**
206. Deadly infectious disease that spreads quickly through large numbers of people. **Pestilence**
207. Implement with a curved blade used for digging furrows in the soil before sowing seeds. **Plough**
208. Artificial fabric used for making clothes etc. **Polyester**
209. Custom of having more than one wife at a time. **Polygamy**
210. Place where ships load and unload cargo. **Port**
211. Liked, admired or enjoyed by many people. **Popular**
212. Hens, ducks, geese, turkeys etc. kept for eating or for their eggs. **Poultry**
213. Amount or installment regularly paid for an insurance policy. **Premium**
214. Study of the mind and how it functions. **Psychology**
215. Place where stones, slate etc. is extracted from the ground. **Quarry**
216. Bitter liquid made from the bark of a tree and used in drinks or as a medicine against malaria, fever etc. **Quinine**
217. Competition especially on T.V. or radio in which people try to answer questions to test their knowledge. **Quiz**
218. Repeat in speech or writing words previously said or written by another person. **Quote**
219. Number obtained when one number is divided by another. **Quotient**
220. Contest of speed between runners, horses, cars, vehicles etc. to see which reaches the specific place first. **Race**
221. Large gathering of people with a common purpose. **Rally**
222. Fixed quantity of food-grains etc. especially in times of shortage. **Ration**
223. Person who fights against or refuses to serve the established government. **Rebel**
224. Best performance or highest or lowest level ever reached especially in sports. **Record**
225. Exact copy made by an artist of one of his own pictures. **Replica**
226. Class of cold-blooded, egg laying animals like lizards, crocodiles, snakes etc. **Reptiles**
227. Public place where snacks, meals, beverages etc. can be bought and eaten. **Restaurant**
228. Deliberate punishment or injury inflicted in return for what one has suffered. **Revenge**
229. Rise in rebellion against authority. **Revolt**
230. Overthrow of a system of government especially by force. **Revolution**
231. Wild or violent disturbance by a crowd of people. **Riot**
232. Person or thing competing with another. **Rival**
233. Part of a plant that keeps it firmly in the soil and absorbs water and food from it. **Root**
234. Information spread by being talked about but not certainly true. **Rumour**
235. Connected with or dedicated to God or religion. **Sacred**
236. Strong lockable box, cabinet etc. for storing valuables. **Safe**
237. Fixed, regular payment to employees doing other than manual or mechanical work. **Salary**
238. Liquid produced in the mouth that helps one chew and digest food. **Saliva**
239. An instrument with a long, sharp-toothed edge operated manually or mechanically for cutting wood, stone, metal etc. **Saw**
240. A frame of metal tubes or bamboo and wooden planks put up next to a building so that builders, painters etc. can work on it. **Scaffold**
241. Act, behaviour etc. that causes public feelings of outrage or indignation. **Scandal**

242. Either of the two divisions of the academic year in a college or a university. **Semester**
243. Caused by or causing infection with harmful bacteria. **Septic**
244. Outer covering of eggs, of nut-kernels of some seeds or fruits and of animals like oysters, crabs, snails, tortoise etc. **Shell**
245. Method of writing rapidly using special quickly-written symbols. **Shorthand**
246. Short-handled tool with a curved blade used for cutting grass, corn etc. **Sickle**
247. Scientific study of the nature and development of society and social behaviour. **Sociology**
248. Tool for gripping and turning nuts on screws, bolts etc. **Spanner**
249. Figure of a person, an animal etc. in wood, stone, bronze, usually life-size or larger. **Statue**
250. Work-room of a painter, sculptor, photographer or a place where cinema films are made acted and photographed. **Studio**
251. Doctor who performs surgical operations. **Surgeon**
252. Combining of separate parts, elements etc. to form a complex whole. **Synthesis**
253. Armoured fighting vehicle with guns, which moves on Caterpillar tracks. **Tank**
254. Sum of money to be paid by people, businessmen, workers etc. to a government for public purposes. **Tax**
255. Optical instrument shaped like a tube, with lenses to make distant objects appear larger and nearer. **Telescope**
256. Formal offer to supply goods or carry out work at a stated price. **Tender**
257. Instrument for measuring temperature. **Thermometer**
258. Medicine that gives strength or energy, taken after illness or when tired. **Tonic**
259. Powerful motor vehicle used for pulling farm machinery or other heavy equipment. **Tractor**
260. Passing of beliefs or customs from one generation to the next esp. without the written form. **Tradition**
261. Branch of mathematics dealing with the relationship between the sides and angles of a triangle. **Trigonometry**
262. Underground passage for a road, or railway through a hill or under a river or the sea. **Tunnel**
263. Person appointed to see that the rules are followed during play etc. and to settle disputes. **Umpire**
264. All existing things including the earth and its creatures and all the stars, planets etc. in space. **Universe**
265. Institution that teaches and examines students, awards degrees and provides facilities for academic research. **University**
266. Awkward to move or control because of its shape, size or weight etc. **Unwieldy**
267. White soluble crystalline compound contained especially in the urine of mammals. **Urea**
268. Imaginary place or state of things in which everything is perfect. **Utopia**
269. Substance that is injected into the blood-stream and protects the body from disease. **Vaccine**
270. Space that is completely empty of all matter or gas/gases. **Vacuum**
271. Wanderer or vagrant, especially an idle or dishonest person. **Vagabond**
272. Conveyance such as car, lorry or cart etc. used for transporting goods or passengers on land. **Vehicle**
273. Place where people agree to meet especially for a sports contest or match. **Venue**
274. At a right angle to another line or plane or to the earth's surface. **Vertical**
275. Metal tool with a pair of jaws that hold a thing securely tight while work is done on it. **Vice**
276. One who is alert and watchful against any possible danger, trouble in the offing. **Vigilant**
277. Person guilty or capable of great wickedness. **Villain**
278. Using, showing or caused by strong physical force. **Violent**
279. Mountain or hill with an opening or openings through which lava, cinders, gases come up from below the surface of the earth, or may come up after intervals or cease to come up. **Volcano**
280. Document showing that money has been paid for goods, services etc. received. **Voucher**
281. Regular payment made or received for work or services. **Wages**
282. Four wheeled vehicle driven by horses or oxens, or open railway truck or trolley used for carrying heavy loads. **Wagon**

283. Fighting between nations or groups within a nation using military force. **War**
284. A small instrument showing the time, worn on wrist. **Watch**
285. Thing designed or used for causing physical harm. **Weapon**
286. Condition of the atmosphere at a certain place and time with reference to temperature, rain, sunshine, wind etc. **Weather**
287. Disc or circular frame that turns on an axle as on vehicles or as a part of a machine. **Wheel**
288. Place in a river or the sea where there are whirling currents. **Whirlpool**
289. Woman whose husband has died and who has not married again. **Widow**
290. Mill worked by the action of wind on long projecting arms that turn on a central shaft. **Windmill**
291. Process of producing photocopies without the use of wet materials. **Xerox**
292. Type of short-wave electromagnetic radiation that can penetrate solids. **X-ray**
293. Time taken by the earth to make one orbit round the sun $\left(\text{about } 365\frac{1}{4} \text{ days}\right)$. **Year**
294. Fungous substance used in making of beer and wine or to make bread. **Yeast**
295. Round yellow part in the middle of the white of an egg. **Yolk**
296. Point in the heavens directly above an observer. **Zenith**
297. Turning right and left alternatively at sharp angles. **Zigzag**
298. Imaginary band of the sky containing the positions of the sun, the moon and the main planets, divided into 12 equal parts. **Zodiac**
299. Area, band or stripe that is different from its surroundings or area or region with a particular feature or use. **Zone**
300. Place like garden, park etc. where living, especially wild animals are kept for exhibition, study and breeding. **Zoo**
301. Scientific study of the structure, form and distribution of animals. **Zoology**
302. A person who initiates any task, business or organisation. **Entrepreneur**
303. One who takes pleasure in the pain of others. **Sadistic**
304. One who rides a horse in races. **Jockey**
305. Government by the king or queen. **Monarchy**
306. One who does not believe in the existence of God. **Atheist**
307. One who believes in God. **Theist**
308. A medicine which prevents infection by germs. **Antiseptic**
309. One who does something for pleasure and not professionally. **Amateur**
310. A statement open to more than one interpretation. **Ambiguous**
311. One who is a great lover of books. **Bibliophile**
312. The life history of a man written by someone else. **Biography**
313. People working together in the same office or department. **Colleagues**
314. A round about way of speaking. **Circumlocution**
315. To give one's authority to another. **Delegate**
316. The action of looking into one's own mind. **Introspection**
317. The action of looking back on past time. **Retrospection**
318. A work for which no salary is paid. **Honorary**
319. An officer with high salary but no work. **Sinecure**
320. One who helps a stranger in difficulties. **Samaritan**
321. One who looks at the bright side of things. **Optimistic**
322. One who looks at the dark side of things. **Pessimistic**
323. One who is unable to pay one's debts. **Insolvent**
324. Lover of mankind. **Philantrophist**
325. The first speech delivered by a person. **Maiden**
326. One who thinks of himself and talks of one self always. **Egoist**
327. One who changes one's principles. **Turncoat**
328. A peculiarity of temperament or constitution. **Idiosyncrasy**
329. One who abstains from intoxicants. **Teetotaller**
330. A person with excessive appetite/excessive eater. **Glutton**
331. A person able to do many different things. **Versatile**
332. From the text. **Textually**

Important Points

Vocabulary

All those words a person uses or remembers are known as vocabulary.

Classified in two parts:

1. Active vocabulary and
2. Passive vocabulary
3. If a person uses more number of words his vocabulary will increase.
4. Language and Vocabulary:
   (a) One has to learn the structure of language to increase vocabulary.
   (b) In communication, structure and vocabulary are more important and they are to be mastered.
5. Kinds of vocabulary: There are three categories in vocabulary
   (a) Active vocabulary
   (b) Passive vocabulary
   (c) Words not known
6. Words which are understood in context with other words.
7. One has to master the words, their meanings, spellings, pronunciation and usage. Usage of words in a particular language has different aspects.
8. Methods for improving/building vocabulary: There are various methods to increase one's vocabulary which are as follows:
   (a) Extensive reading
   (b) Learning word patterns
   (c) Word building
   (d) Remembering the usage of a word.
9. Additional ways to increase vocabulary:
   (a) Primary derivatives
   (b) Secondary derivatives
   (c) Prefixes and Suffixes
   (d) Compound words
   (e) Homonyms
   (f) Synonyms and Antonyms
   (g) One word substitution
   (h) Same word used in different parts of speech.

■■■

Reference

1. **SBTE, Mumbai, Text Book of Communication Skills, SBTE, Mumbai.**
2. **Website: www.khake.com/page66htm/-72k**
3. **Website: www.BMConsultantIndia.Com**
4. **Business Communication: M.K. Sehgal and Khetrapal, Exel Books.**

■■■